D1236836

Adaptive Herbivore Ecology

From Resources to Populations in Variable Environments

The adaptation of herbivore behaviour to seasonal and locational variations in vegetation quantity and quality is inadequately modelled by conventional methods. Norman Owen-Smith innovatively links the principles of adaptive behaviour to their consequences for population dynamics and community ecology, through the application of a metaphysiological modelling approach. The main focus is on large mammalian herbivores occupying seasonally variable environments such as those characterised by African savannas, but applications to temperate zone ungulates are also included. Issues of habitat suitability, species coexistence, and population stability or instability are similarly investigated. The modelling approach accommodates various sources of environmental variability, in space and time, in a simple conceptual way and has the potential to be applied to other consumer–resource systems. This text highlights the crucial importance of adaptive consumer responses to environmental variability and is aimed particularly at academic researchers and graduate students in the field of ecology.

NORMAN OWEN-SMITH is Research Professor in African Ecology and heads the Centre for African Ecology at the University of the Witwatersrand. He was awarded the Honorary Overseas Membership Award by the Ecological Society of America for his exceptional contribution to ecology. His previous book is *Megaherbivores: The Influence of Very Large Body Size on Ecology* (1988, 0 521 36020 X hardback, 0 521 42637 5 paperback).

CAMBRIDGE STUDIES IN ECOLOGY

Cambridge Studies in Ecology presents balanced, comprehensive, up-to-date, and critical reviews of selected topics within ecology, both botanical and zoological. The series is aimed at advanced final-year undergraduates, graduate students, researchers, and university teachers, as well as ecologists in industry and government research.

It encompasses a wide range of approaches and spatial, temporal, and taxonomic scales in ecology, experimental, behavioural and evolutionary studies. The emphasis throughout is on ecology related to the real world of plants and animals in the field rather than on purely theoretical abstractions and mathematical models. Some books in the series attempt to challenge existing ecological paradigms and present new concepts, empirical or theoretical models, and testable hypotheses. Others attempt to explore new approaches and present syntheses on topics of considerable importance ecologically which cut across the conventional but artificial boundaries within the science of ecology.

A selection of recent titles published in the series

J.S. Findley	*Bats: A Community Perspective*	0 521 38054 5
G.P. Malanson	*Riparian Landscapes*	0 521 38431 1
S.R. Carpenter & J.F. Kitchell	*The Trophic Cascade in Lakes*	0 521 43145 X
R.C. Mac Nally	*Ecological Versatility*	0 521 40553 X
R.J. Whelan	*The Ecology of Fire*	0 521 32872 1
K.D. Bennett	*Evolution and Ecology: The Pace of Life*	0 521 39028 1
H.H. Shugart	*Terrestrial Ecosystems in Changing Environments*	0 521 56342 9
M.R. T. Dale	*Spatial Pattern Analysis in Plant Ecology*	0 521 79437 4
D.W. Larson, U. Matthes & P.E. Kelly	*Cliff Ecology*	0 521 554896
P.A. Keddy	*Wetland Ecology*	0 521 78367 4
L.E. Frelich	*Forest Dynamics and Disturbance Regimes*	0 521 65082 8

Adaptive Herbivore Ecology

From Resources to Populations in Variable Environments

R. NORMAN OWEN-SMITH

PUBLISHED BY THE PRESS SYNDICATE OF THE UNIVERSITY OF CAMBRIDGE
The Pitt Building, Trumpington Street, Cambridge, United Kingdom

CAMBRIDGE UNIVERSITY PRESS
The Edinburgh Building, Cambridge CB2 2RU, UK
40 West 20th Street, New York, NY 10011-4211, USA
477 Williamstown Road, Port Melbourne, VIC 3207, Australia
Ruiz de Alarcón 13, 28014 Madrid, Spain
Dock House, The Waterfront, Cape Town 8001, South Africa

http://www.cambridge.org

First published 2002

Printed in the United Kingdom at the University Press, Cambridge

Typeface Bembo 11/13 pt. *System* LaTeX 2_ε [TB]

A catalogue record for this book is available from the British Library

Library of Congress Cataloguing in Publication data

Owen-Smith, R. Norman
Adaptive herbivore ecology : from resources to populations
in variable environments/R. Norman Owen-Smith.
p. cm. – (Cambridge studies in ecology)
Includes bibliographical references (p.).
ISBN 0 521 81061 2 (hb)
1. Ungulates – Ecology. 2. Herbivores – Ecology. I. Title. II. Series.
QL737.U4 O94 2002
591.5′4–dc21 2001043442

ISBN 0 521 81061 2 hardback

Contents

Acknowledgements

I am indebted firstly to Wayne Getz for introducing me to the metaphysiological modelling approach during his visit to South Africa in 1994. This opened the door to the synthesis between behavioural and population ecology that I had been seeking.

This book was begun while I was on sabbatical leave with the Department of Mathematics and Applied Mathematics of the University of Natal in 1995. I thank John Hearne and his colleagues for providing the serene environment needed to formulate the models needed to illuminate the analysis.

Numerous colleagues gave of their time to read draft chapters for this book, and made helpful suggestions as to how these could be improved. They include Peter Baxter, Johan du Toit, Wayne Getz, Iain Gordon, John Hearne, Andrew Illius, Emilio Laca, Martyn Murray, Tim O'Connor, Katherine Parker, Charlie Robbins, Tony Starfield and John Wilmshurst.

My wife Margaret and daughters Trishya and Lynne coped with the strain of the extended hours spent in my office during the period when finishing this book took priority over many other things.

Help with typing references and assembling figures was provided by Jean Mengel, Mamosa Nkoko, Hazel Khanyi and Herman du Preez.

Finally, I acknowledge my indebtedness to Tony Starfield for inducting me into modelling, and for guiding me to keep models simple, so that the biological detail does not obscure the allegory.

Acronym and symbol conventions

The list includes only those used widely in different chapters.

Acronyms

BM	Body mass
C–R	Consumer–resource
DBM	Diet breadth model
DDF	Density-dependent function
DM	Dry mass
DS	Dormant season (for plant growth, i.e. winter or dry season)
GS	Growing season (for plants, i.e. summer or wet season)
GMM	Growth, metabolism and mortality (model)
IBP	International Biological Programme
RGP	Relative growth potential
RGF	Resource gain function
RMF	Resource-dependent mortality function
SDP	Stochastic dynamic programming
ZGI	Zero (population) growth isocline

Symbols

In general, upper case indicates a state variable or rate function, and lower case a constant parameter (but what is constant over one scale may be variable over another).

Symbol	*Definition*	*Units*
a	acceptance coefficient	fraction
B	biomass for any consumer population	mass/area
b	half-saturation level (of vegetation) for intake rate	mass/area

b_m	bite mass	mass
C	carnivore (predator) biomass	mass/area
c	conversion coefficient from vegetation consumed into herbivore biomass	fraction
d_d	decay rate of vegetation necromass	proportion/time
d_g	death rate of green vegetation biomass	proportion/time
e	eating rate (short-term, identical to I_p)	mass/time
F	food availability, i.e. edible and accessible component of vegetation biomass	mass/area
f	any function	
$f_{1/2}$	half-saturation level for rate of resource intake	mass/area
G	resource (nutritional) gain rate, per unit of herbivore biomass	proportion/time
G_D	resource gain over daily cycle, per unit of herbivore biomass	proportion/time
G_F	resource gain rate during foraging spells, per unit of herbivore biomass	proportion/time
g_V	maximum rate of vegetation growth, per unit of vegetation biomass	proportion/time
$g_{1/2}$	half-saturation level for rate of nutritional gain	mass/area
H	herbivore biomass density	mass/area
h_b	handling time per bite	time
I	food intake rate (time frame unspecified), per unit of herbivore biomass	proportion/time
I_D	daily food intake rate, per unit of herbivore biomass	proportion/time
I_F	food intake rate obtained during foraging spells, per unit of herbivore biomass	proportion/time
I_P	food intake rate obtained within food patches, per unit of herbivore biomass	proportion/time
i_{max}	maximum intake rate (time frame unspecified)	proportion/time
k	saturation biomass level (carrying capacity), for either herbivores or vegetation; *or* first-order rate constant	mass/area
K	carrying capacity, as a variable	mass/area

M	metabolic + mortality losses combined, per unit of herbivore biomass; *or* body mass	proportion/time
M_P	physiological (metabolic) losses, per unit of herbivore biomass	proportion/time
M_Q	mortality losses, per unit of herbivore biomass	proportion/time
M_{P0}	resting or basal metabolic loss, per unit of herbivore biomass	proportion/time
P	physiological metabolic expenditure, as a state variable per unit of herbivore biomass	proportion/time
p	physiological metabolic losses, as a constant per unit of herbivore biomass	proportion/time
p_0	physiological metabolic expenditure for basal tissue maintenance	proportion/time
Q	mortality losses, as a variable per unit of herbivore biomass	proportion/time
Q_G	mortality rate from resource deficiencies, per unit of herbivore biomass	proportion/time
Q_0	mortality rate due to senescence, per unit of herbivore biomass	proportion/time
Q_A	additive mortality due to predation, parasitism and other accidents, per unit of herbivore biomass	proportion/time
q_0	minimum rate of mortality due to senescence and accidents, per unit of herbivore biomass	proportion/time
R	rate of resource renewal, per unit of biomass	proportion/time
r_m	maximum population growth rate	proportion/time
S	body size relative to adult body mass	proportion
s	search rate while foraging	area/time
T	time, as a variable	time
t_F	foraging time	time
t_{max}	maximum or terminal time horizon	time
t_R	resting time	time
t_s	searching time while foraging	time
t_0	starting time	time

V	vegetation biomass density	mass/area
V_d	dead or necromass component of vegetation	mass/area
V_g	green component of vegetation biomass	mass/area
$v_{1/2}$	half-saturation level for intake rate	mass/area
v_{max}	saturation biomass level of vegetation	mass/area
v_u	ungrazable level of vegetation biomass	mass/area
W	body mass (weight)	mass
z	constant determining how steeply mortality rises with diminishing resource gains	
z_0	constant determining mortality level when resource gain = metabolic requirement	

1 · *Conceptual origins: variability in time and space*

The motivation to write this book stems from a long-felt uneasiness with the theoretical models that throng the ecological literature. Having originally studied the physical sciences at university, I fully recognized the power of fundamental theory in science. But the mathematical models that I found in the ecological literature were discordant with the world that I encountered as a field biologist. Their conventional assumptions of uniform environments and near-equilibrium conditions seemed far divorced from the continually changing environments that I experienced in African savannas. Biological persistence seemed to be more a matter of coping with variability than balancing around some equilibrium state. The vast superabundance of green vegetation at one time of the year contrasted sharply with the sparse dry remnants at a later stage. Conditions were also spatially heterogeneous, with certain localities retaining green glades at times when only standing brown hay remained elsewhere. Furthermore, conditions fluctuated widely between years, some years being quite benign, others severely adverse. Animals persisted by responding to this variability in numerous ways: adjusting what they ate, the habitats they occupied, and when they reproduced. Conventional mathematical models omitted the basic features that distinguish biological from physical systems: temporal variability, spatial heterogeneity, and *adaptively* changing responses over different time scales.

Ecological theory has advanced considerably in recent years. Following May and Oster (1976), the susceptibility of systems governed by nonlinear relationships to cycles, and perhaps even chaotic oscillations, has become widely recognized. The effects of spatial structure have been accommodated into models of metapopulations (Hanski 1999), source–sink dynamics (Pulliam 1988), landscape processes (Hansson *et al.* 1995) and plant demography (Tilman and Kareiva 1997). Shifting patch mosaics have largely displaced the 'balance of nature' illusion (Botkin 1990; Wu and Loucks 1995). Behavioural ecology fundamentally addresses the adaptive responses of individual organisms to environmental heterogeneity (Krebs and McLeery 1984; Milinski and Parker 1991). Dynamic optimization approaches link individual behaviour to life history consequences (Mangel and Clark 1988; Clark and Mangel 1999; Houston and McNamara 1999). Individual-based population models (DeAngelis and Gross 1992) have grown into treatments of complex adaptive systems (Kaufmann 1995), taking into account how each organism responds to its immediate neighbourhood and the choices made by others.

Accelerating developments in computational technology offer the possibility to break loose from the strait-jacket of mathematical tractability. Undergraduate students can observe the transitions into chaos of nonlinear difference equations, and extract eigenvalues from projection matrices, during an afternoon's exercise with spreadsheet software. Although numerical outcomes lack the broad generality of analytic solutions, this reflects the contingent nature of ecological systems. Simple ecological principles rarely remain firm globally, being dependent upon the specific environmental context and particular species present.

Modelling philosophy

To be illuminating, models must seek a narrow window between the mirror of tautology (output merely reflects the assumptions) and the fog of unnecessary detail (uncertainty about how the output follows from the assumptions). It is helpful to recognize contrasting philosophies underlying different approaches to ecological modelling.

Mathematical models seek analytic solutions to abstract problems. Formal relationships between assumptions and their consequences are established, which may not be obvious *a priori*. Such models yield logical certitudes, but their connections with the real world are conjectural. Philosophically such models can be regarded as *parables*, offering irrefutable principles to those not inclined to query the assumptions. A good example is the logistic model of population growth, widely invoked as a starting approximation ('straw man') for much ecological theory. This model depicts some general consequences of density dependence in population growth rate, assuming that this exists. The chaotic dynamics that are generated for particular conditions and parameter values were initially a surprise. The potential for chaos cannot be doubted, but whether it is manifested for any real world examples is the subject of much recent debate. Such models are mainly of *heuristic* value.

At the opposite pole are simulation models numerically representing the dynamics of identified system components and their linkages. These are the kinds of ecological models associated with the IBP era, illustrated by networks of boxes, connecting lines and arrows of causation. Individual-based models constitute a currently fashionable elaboration, using computational power to integrate upwards from each individual in its specific context to the aggregate population dynamics (DeAngelis and Gross 1992). Potential bafflement about the relationship between the logically defensible design and parameter values, and the resultant output dynamics, tends to rise as model complexity increases. Nevertheless, simulation modelling remains widely employed, for example to represent forest ecosystem dynamics (Shugart 1998). Such models can be regarded as *narratives*, describing how specific causal relationships could lead to some general consequences. They may be *tactically* useful in specific contexts.

The middle road is represented by process-based models representing the mechanistic workings and functional inter-relationships among specific components of a clearly defined sub-system. Such models demonstrate cause–effect relationships that may have substantial generality, at least in the defined context. Philosophically they represent *allegories*, i.e. abstract stories presented in one medium to illustrate general principles that are applicable in other contexts. Practically they are of *strategic* value, providing a firm foundation for action. Because my aim is to develop strategic principles that are valid in a wide range of resource conservation situations, this modelling philosophy underlies this book.

The vision of such a middle path was opened by a meeting with Wayne Getz, visiting South Africa from his academic base at the University of California, Berkeley, in 1994. His metaphysiological modelling approach (Getz 1991, 1993) offered a simple, yet potentially powerful, framework for linking the dynamics of consumers to the interactive dynamics of their resource base across temporal and spatial scales. This led to the manifesto formulated for a workshop on modelling herbivore–vegetation interactions in semi-arid environments, held in conjunction with the 1995 meeting of the Resource Modeling Association:

The fundamental issue is how to maintain sustainability in the use of a resource that fluctuates widely in its supply. Analytic solutions to the optimal-yield problem become meaningless. New modelling approaches are needed that place greater emphasis on patterns of variance, in space as well as time, on resilience rather than stability, on rates of regrowth and regeneration, on reserve capacity for recovery, and on dynamic rather than static optimization. The feedback mechanisms that sustain return despite variability need to be identified, as well as processes governing return times.

Herbivores and vegetation

My perspective on ecology has been strongly influenced by the formative experience gained in studying two particular species of large mammalian herbivore in African savanna environments. Ideas developed during these studies permeate the models formulated in this book.

My doctoral research on the behavioural ecology of the white rhinoceros (*Ceratotherium simum*) (Fig. 1.1a), the largest extant grazing mammal, took place in what is now the Hluhluwe–Umfolozi Park in the KwaZulu–Natal (KZN) province of South Africa. I spent three-and-a-half years following these animals on foot as they went about their daily

a

Figure 1.1. Representative large mammalian herbivores. (a) Grazing white rhinoceros in the Hluhluwe–Umfolozi Park. (b) Browsing kudu in the Kruger National Park. (c) Grazing cattle in communal rangeland in South Africa.

c

Figure 1.1. (cont.)

activities, identifying each individually from variations in horn shapes. The management dilemma that provided the incentive for the study was a burgeoning population within the limited confines of a fenced protected area (Owen-Smith 1973, 1988). Grassland resources appeared severely overgrazed in places, while elsewhere abundant grass remained uneaten. Because white rhinos are effectively invulnerable to non-human predators as adults, lack of predation could not be invoked as the cause of the apparent overpopulation. I conducted a detailed investigation of social behaviour as a possible regulating mechanism (Owen-Smith 1975), only to reject it as having much influence. This left just the direct interaction between the consumer and food resources to contemplate. But how such a large and highly efficient grazer could reach any form of equilibrium with its food supply was difficult to discern. Eventually I invoked dispersal, now precluded by the fence, as being the only effective control in the short term, while recognizing that nutritionally induced changes in reproduction and survival must operate in the long term. This led to my recommendation that dispersal sinks, or rhino 'vacuum zones', be established within the fenced area (Owen-Smith 1974, 1983). This management policy is currently being implemented by the KZN conservation agency.

Subsequently, I embarked on a postdoctoral study of the greater kudu (*Tragelaphus strepsiceros*) (Fig. 1.1b), a large browsing antelope, in South Africa's Kruger National Park, which eventually spanned 10 years. Interestingly, kudu populations seemed well regulated, in the sense of becoming neither rare nor overabundant within protected areas, although not stable in density. In contrast to white rhinos, kudus appeared to be non-territorial, and occurred in cohesive social units. A special facility was being able to identify every individual from birth, from variations in stripe patterns and other marks. Despite the prevalence of predators in the park, findings revealed a close relationship between population change and rainfall, and hence food production (Owen-Smith 1990). Furthermore, for this browser on readily visible trees and shrubs, aspects of foraging behaviour could be quantified in a way not readily possible for grazers (Owen-Smith 1979). Foraging studies on kudus in the Kruger Park were augmented by more detailed investigations of the trophic ecology of kudus and impalas (*Aepyceros melampus*) in the Nylsvley Nature Reserve, as part of South Africa's Savanna Biome study (Scholes and Walker 1993). The latter employed habituated animals, enabling every woody plant and forb eaten during the course of a day to be recorded, as well as the detailed time allocation. Observations revealed how animals adjusted aspects of foraging behaviour to cope with the severe nutritional

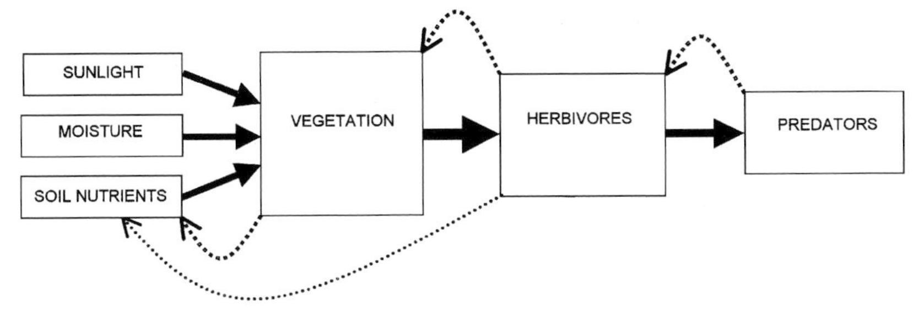

Figure 1.2. Schematic outline of an interactive herbivore–vegetation system.

bottleneck posed by the late dry season (Owen-Smith and Cooper 1987a, 1989; Owen-Smith 1994).

My personal studies have been focussed on wild herbivores, but the concepts developed in this book are applicable also to domestic ungulates managed under free-ranging conditions (Fig. 1.1c). Rangelands supporting livestock constitute a large portion of the Earth's land surface, and provide a livelihood for a substantial fraction of its people. Communally managed rangelands are widely believed to be severely overstocked, overgrazed and in danger of imminent collapse, yet continue to carry large numbers of domestic animals (Shackleton 1993). A perception of such systems as intrinsically disequilibrial and event-driven has emerged (Ellis and Swift 1988; Westoby *et al.* 1989; Behnke *et al.* 1993), but the implications of such a viewpoint for management remain contentious (Illius and O'Connor 1999).

The focus of the models developed in this book is on the herbivore–vegetation interface (Fig. 1.2). In particular I am concerned with how the dynamics of herbivores are governed by temporal and spatial variability in the food resource constituted by plants. The underlying resource fluxes contributing to vegetation growth are not considered explicitly. Herbivores can also influence plant growth by contributing to nutrient recycling, but this potentially important feedback is beyond the scope of this book (see DeAngelis 1992; Pastor *et al.* 1997). The top-down influence of predation on herbivore abundance will be considered, but not the cascading effects on vegetation that can arise under certain conditions (Hunter and Price 1992).

My aim is not an encyclopaedic treatise on herbivore–vegetation systems, but rather the development of a conceptual approach towards treating their dynamics that accommodates temporal and spatial variability. Nevertheless, I hope that these ideas will have wider applicability in theoretical assessments of consumer–resource dynamics, encompassing the plant–soil and predator–prey interfaces as well, and perhaps for harvesting theory involving humans as the consumers. Despite the huge scientific effort devoted to optimal harvest models (Walters 1986; Clark 1990), collapses of fish stocks have continued unabated. A fundamental problem is inadequate and unreliable information on changing fish populations in their oceanic environment. In contrast, the vegetation resources grazed or browsed by herbivores are rather more amenable to study.

Specifically, the herbivores I have in mind are large mammals, rather than rodents or insects. I believe that the basic concepts are applicable also to such small herbivores, although the temporal and spatial scales of relevance will be different. The time frame highlighted in this chapter and those that follow is the seasonal oscillation between the period of the year when vegetation is growing and thus a renewing resource, and the stage when vegetation is dormant and hence a diminishing resource. In savanna regions of Africa, this is represented by the annual or twice-annual alternation of wet and dry seasons; in temperate latitudes it is constituted by the annual oscillation between summer and winter. The spatial variability is constituted particularly by functional heterogeneity, in nutritional value and growth patterns, among the plant types forming the food resource, and their distribution within distinct habitat types.

The approach is orientated towards addressing three particular problems in applied herbivore ecology.

1. *Habitat suitability assessment.* Caughley (1994) emphasized that conservation biologists have paid much attention to the factors making small populations vulnerable to extinction, while largely ignoring the processes causing populations to become small. How do resources, in interaction with predation, shelter and other factors, limit the population size of a species that exists in a particular area?

2. *Coexistence among potentially competing species.* Species diversity may enhance overall ecological productivity, or at least stabilize production under conditions of environmental variability (Fritz and Duncan 1994; Tilman 1996). But how do multiple herbivore species coexist, despite substantial overlap in the vegetation resources they consume?

3. *Determinants of population instability.* Some populations of large mammalian herbivores appear relatively stable although not constant in density, whereas others show periodic instability, or even persistent irruptions and crashes (Fig. 1.3). To what extent do features of the resource base distinguish stable from unstable populations?

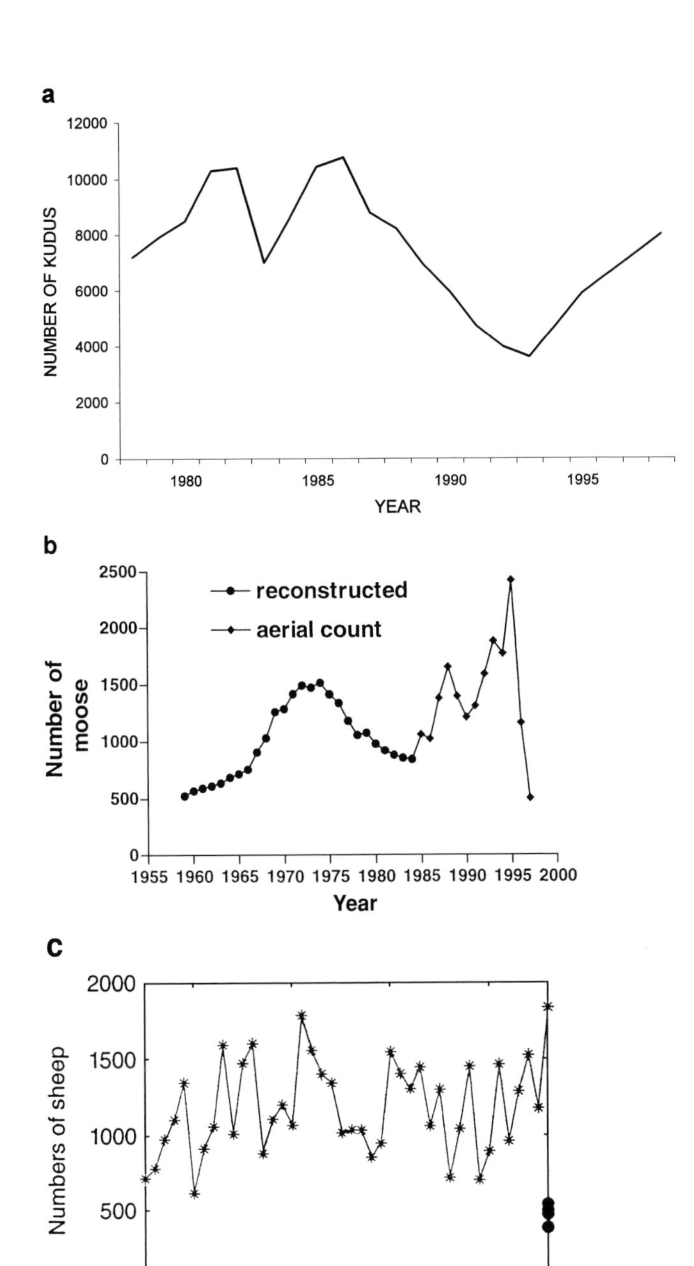

Figure 1.3. Patterns of population dynamics illustrated by representative large herbivore species. (a) Kudus in the Kruger National Park, showing regulated dynamics despite variable abundance (unpublished records, South African National Parks). (b) Moose in Isle Royale National Park, showing recent instability (from Peterson 1999). (c) Soay sheep on Hirta, St Kilda Islands, showing persistent instability (from Grenfell *et al.* 1998).

Modelling technology

Traditionally, ecological modelling has emphasized mathematical approaches, which seek analytic solutions for near-equilibrium conditions (Yodzis 1989). Ecological complexity may demand highly sophisticated mathematics, making models even more opaque to naive biologists than the ecosystems they study. This conflicts with the basic aim of modelling as a tool for aiding understanding through appropriate simplification (Starfield *et al.* 1990).

I approach modelling from a biological perspective, trying to capture the essential ecological processes through a clarifying theoretical formalism. Because we are concerned with temporal dynamics, functional relationships will usually be expressed as rates over time expressed as differential equations. Because functions are generally non-linear and incorporate time-varying parameters interacting with changing conditions, they cannot be solved analytically. Indeed, as we shall see, the equilibrium states sought by mathematicians are non-existent, or merely transient, in seasonally varying environments. Nevertheless, numerical solutions can easily be obtained with the aid of a personal computer.

The models developed in this book commonly entail a few equations iterated in a spreadsheet. When things become a little more complicated, particularly when conditional (IF . . . . . THEN) logic is involved, I resort to a friendly programming language called *True BASIC* (Kemeny and Kurtz 1990). My programming capabilities are self-taught, and so entail little sophistication. Most of these programs are so simple that readers conversant with basic spreadsheet or programming software could develop their own counterparts in the medium with which they are most familiar. I urge this response, rather than requests to borrow my programs, believing that models are most illuminating when users feel in full control over their workings. The symbol conventions that I have used widely in this book are listed in a preface, for ready reference.

Structure of the book

The structure of the book involves working hierarchically upwards from aspects of foraging behaviour towards processes occurring at higher levels of integration in populations and communities. This linking of adaptive behaviour to its consequences for population dynamics constitutes an *Adaptive Herbivore Ecology*, or, more generally, an *Adaptive Resource Ecology*.

Chapter 2 presents an outline of basic theory and its relation to models of consumer–resource dynamics. It leads to the conceptualization of the **GMM** model that is elaborated in the remainder of the book.

The succeeding four chapters cover the ecology of resource use by individual herbivores. Chapter 3 examines how the intake (or 'functional') response to resource availability changes over different time frames. Chapter 4 addresses consumer responses to patchy resource distributions at different spatial scales, taking into account local resource depletion. Chapter 5 considers how nutritional gains are influenced by heterogeneity in resource quality as well as abundance. Chapter 6 outlines how physiological processes including digestion, metabolic assimilation, and thermal tolerance, restrict resource gains over particular time frames.

The following block of four chapters establishes the links between resource gains and population dynamics. Chapter 7 is concerned with the dynamic allocation of resources to growth, storage or reproduction, over annual or longer periods. Chapter 8 outlines the regenerative responses of vegetation resources to herbivory in seasonally varying environments. Chapter 9 describes how within-population competition generates density dependence over the seasonal cycle. Chapter 10 considers the dependence of mortality on resource gains, including interactions with predation.

The next three chapters apply the **GMM** model to problems in resource conservation. Chapter 11 is concerned with habitat suitability assessment, i.e. the factors governing ecological or economic 'carrying capacity'. Chapter 12 assesses how distinctions among potentially competing herbivore species in morphology and physiology can lead to their coexistence. Chapter 13 evaluates resource features underlying instability in herbivore–vegetation interactions, and underlying tradeoffs between productivity and stability.

The final chapter assesses the vision of an *Adaptive Resource Ecology* that I hope the book has established. How much has been achieved? What problems remain to be addressed?

Further reading

Caughley and Sinclair (1994) provides a readable outline of basic principles of wildlife ecology and management, including applications of models. Chapters in Hodgson and Illius (1996) review current knowledge of herbivore–vegetation relations, covering both wild and domestic ungulates. The standard review of foraging theory is by Stephens and Krebs (1986). Sutherland (1996) explains consequences of behavioural ecology for population-level processes, with an emphasis on birds. Fryxell and Lundberg (1998) link models of individual behaviour to population or community processes, highlighting predation and habitat or territory occupation. Chapters in Tilman and Kareiva (1997) review current theory in spatial ecology. Milner-Gulland and Mace (1998) review ecological and economic models underlying resource harvesting, as well as the case histories of the few successes but many failures.

2 · *Consumer–resource models: theory and formulation*

This chapter outlines the theory, principles and concepts that underlie models of consumer–resource dynamics. We establish first the foundation in terms of the fundamental biological laws that apply to all organisms, and clarify the distinction between processes occurring in ecological versus evolutionary realms. This relates in particular to the currency to be used for the state variables in the models.

The kinds of consumers we have in mind in this book are large mammalian herbivores, with the resource constituted by vegetation. Starting from the simplest phenomenological models, we work upwards towards models of interactive herbivore–vegetation dynamics. Caughley's (1976a) classical model is elaborated further, drawing upon metaphysiological modelling concepts developed by Getz (1991, 1993). The three functional components contributing to herbivore biomass change – resource gains, metabolic attrition and mortality losses – are distinguished in formulating the **GMM** metaphysiological model of herbivore dynamics. The potential of this model to accommodate seasonal variability in resource production, as well as other environmental influences on herbivore population changes, will be outlined.

Some of the mathematical terminology and functions that will be adopted in the remainder of the book will be introduced. This sets the stage for the chapters that follow, which will explore in more detail the processes underlying population responses to spatial and temporal variability in environments.

Basic theory

Biophysical laws

All biological processes are governed by basic physical laws, and also by additional principles specific to the biological realm. The two classical laws of physical thermodynamics can be expressed as follows, from a biological perspective.

1. *Law of conservation of matter.* Matter cannot be created or destroyed, only transformed from one form into another. Thus mass balance must be maintained in all biological transactions.

2. *Law of entropy.* Continual energy inputs are needed to maintain ordered states, such as those characterizing biological organisms. Despite this, the ordered state inevitably disintegrates over time, such that organisms senesce and eventually die.

To these must be added two principles that are specifically biological.

3. *Law of regeneration.* To counterbalance losses through death and decay, organisms persistently regenerate new biomass of the same kind.

4. *Law of adaptation.* To cope with variability and change, biological entities adapt at various levels of integration.

Ecology versus evolution

Ecology and evolution are closely coupled, being viewed as different aspects of population biology (Wilson and Bossert 1971). This dialectic underlies an ambivalence in the currency used for demographic models (Ginzburg 1998; Berryman 1999).

Evolution concerns the replication of coded genetic information carried within individual organisms. Its perspective is intrinsically numerical, with regard to the replication of this information through the individuals that are its bearers. Nevertheless, the numerical magnitude of the replicates is inadequate alone as a guide to inter-generational persistence. Evolutionary success depends also on the survival chances of these progeny, to in turn produce more descendants. Numbers must be weighted by the expected likelihood of genetic persistence. In some circumstances fewer, larger offspring are more likely to be successful than many more smaller offspring, recognizing tradeoffs in the resources that can be allocated to reproduction.

A numerical currency has also become conventional in demographic modelling. This is largely a practical outcome of how population abundance is most readily assessed in the field. For descriptive models, it usually matters little whether the magnitude of the population is assessed numerically, or in terms of its total biomass. These two currencies are readily interchangeable, provided organisms do not vary too widely in mass, at least at the aggregate level. This does not hold for modular organisms like plants, where the choice of the individual becomes problematic. A plant population may grow in biomass even when the number of organisms constituting it diminishes, through processes of self-thinning (Harper 1977). A further anomaly arises for animal populations having births concentrated within a discrete period of the year. A stepwise change in numbers during the birth pulse is associated with no immediate increase in biomass, until the young survive and grow. This is conventionally sidestepped by invoking a discrete-time formulation with an annual time step. Hence, important ecological processes determining the survival and growth of these offspring are obscured.

Whether individual organisms survive and grow, and later reproduce, depends fundamentally on the material and energizing resources that they obtain from the environment. Gaining resources enables evolutionary persistence. The processes that govern the acquisition and allocation of these resources constitute the realm of ecology. The currency appropriate for interactive models, linking the dynamics of herbivores to changes in the resource base constituted by vegetation, is population biomass rather than numerical density.

Organisms adapt to changing environments not only evolutionarily, in their genetic information, but also at other levels of integration. Adjustments in behaviour, physiology, morphology and life history attributes can occur within the lifespans of individual organisms, i.e. over ecological time frames. Adaptive changes in these phenotypic features are central to the treatment in this book. From an ecological perspective, the consequences of such adaptation can be evaluated simply in terms of resultant gains in energy and nutrients, without any necessity to invoke genetic outcomes over generational time spans. Acquiring resources conveys the power to reproduce genetically.

Descriptive population models

Exponential growth

Despite its simplicity, the exponential growth equation encapsulates the most fundamental biological law, life begets like:

$$dN/dT = rN, \tag{2.1}$$

where dN represents the change in population abundance N over some very small time interval dT, and r represents the population growth rate as a proportion of the total population size. Following the convention adopted in this book, upper-case symbols represent state variables or rate functions, and lower-case symbols parameters that are effectively constant over the time span covered (see Acronym and symbol conventions, p. xiii). Growth is exponential because, integrating the above equation to determine the population size N_t after time t has elapsed, we obtain

$$N_t = N_0 e^{rt}, \tag{2.2}$$

where N_0 is the starting population size at time zero, and e is the basis of natural logarithms. Thus r appears in the exponent as a multiplier of time. The net growth rate r is usually interpreted as the numerical difference between the birth rate and death rate per capita, assuming that the population is closed to immigration and emigration. If births exceed deaths, the population grows; otherwise it declines.

Strictly, Equation 2.1 applies only if both natality and mortality occur continuously throughout the year. This does not hold when births are concentrated within a restricted season. In these circumstances we need to switch to the discrete-time formulation,

$$N_{t+1} - N_t = rN_t, \tag{2.3}$$

where N_t = abundance (numerical or biomass density) at the initial time t, N_{t+1} = abundance one time step later, and r becomes the proportional rate of population growth over the discrete time period. Growth is now geometric, rather than strictly exponential.

Notably, if population abundance N is measured as biomass rather than numerically, there is no need to adopt the discrete-time formulation. Biomass increase occurs continuously throughout the year, albeit at a somewhat variable rate.

The exponential growth model is unrealistic through ignoring any material restrictions on population increase. Eventually, however, there will be some space limit at which population growth ceases abruptly (Fig. 2.1a). This could result, for example, from territorial restrictions on space occupation.

Logistic growth

The logistic equation incorporates resource limitation via a decline in the proportional growth rate of the population as population density increases:

$$dN/dT = r_m(1 - N/k)N, \tag{2.4}$$

where r_m = proportional growth rate when density is approximately equal to zero, and k = density level at which the net population growth rate becomes zero (with k symbolized in lower case because it represents a

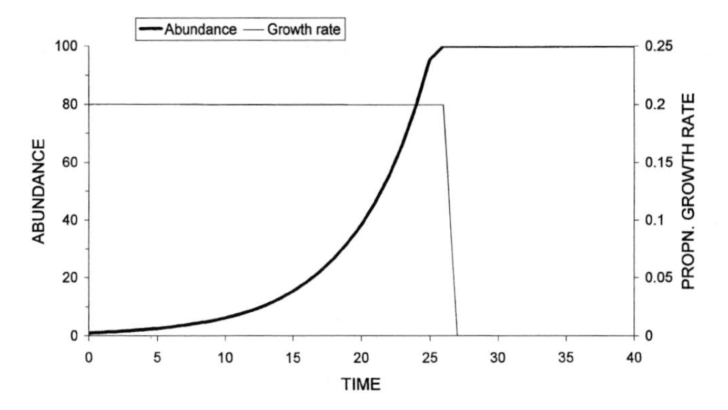

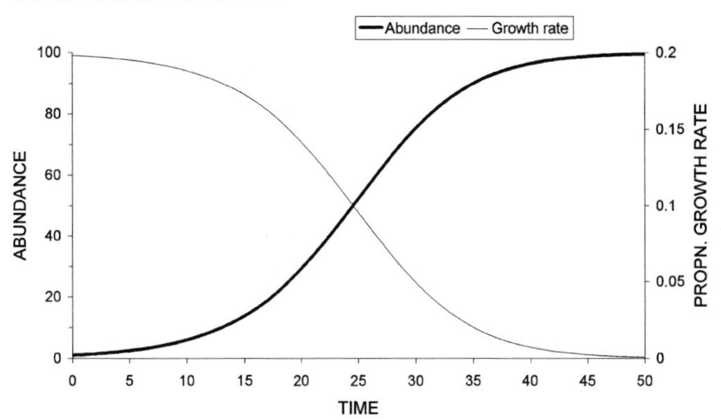

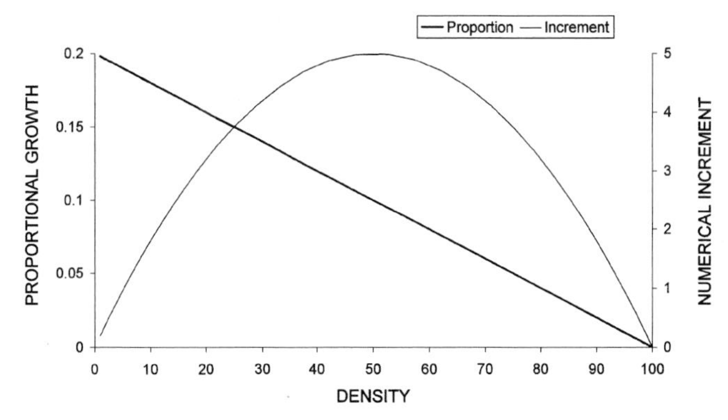

Figure 2.1. Output dynamics from descriptive population models. (a) Exponential population growth with growth rate $r = 0.2$ and upper limit to population size $N = 200$. (b) Logistic population growth, with intrinsic growth rate $r_m = 0.2$ and equilibrium population size $K = 100$. (c) Density dependence in proportional growth rate and numerical growth increment generated by a logistic model with the above parameter values.

fixed parameter rather than a state variable). Starting from an initially low density, population abundance increases sigmoidally over time towards its upper limit where the net growth rate reaches zero (Fig. 2.1b). More specifically, the proportional growth rate declines linearly as the population density increases because, rearranging Equation 2.4, we obtain

$$(1/N)\,\mathrm{d}N/\mathrm{d}T = r_m - (r_m/k)N, \tag{2.5}$$

which is the equation for a straight-line relation with N, with intercept r_m and negative slope r_m/k (Fig. 2.1c). The asymptotic density k is conventionally termed the 'carrying capacity' of the environment for the population. Paraphrasing Kooijman (1990), when $N = k$ the population produces only faeces, corpses and gases.

Equation 2.2 is a non-linear differential equation because, rearranging again, we obtain

$$\mathrm{d}N/\mathrm{d}T = r_m N - (r_m/k)N^2, \tag{2.6}$$

which involves subtraction of a squared term in N. Accordingly, the incremental rate of population change is a parabolic function of N with a maximum when $N = k/2$ (Fig. 2.1c). The proportional rate of population growth could itself be made non-linear by incorporating a power coefficient into Equation 2.5 (following Maynard Smith and Slatkin 1973):

$$(1/N)\,\mathrm{d}N/\mathrm{d}T = r_m\{1 - (N/k)^z\}. \tag{2.7}$$

If the power coefficient $z > 1$, the decline in relative growth rate accelerates as population density increases (Fig. 2.2a); if $z < 1$ it decelerates (Fig. 2.2b). The former pattern has been termed 'overcompensation' and the latter 'undercompensation' (Bellows 1981). These contrasting forms of non-linearity substantially alter the maximum growth increment and hence the harvest quota that can potentially be obtained (Fig. 2.2). Overcompensating density dependence seems to be typical of most large mammal populations (Fowler 1981, 1987), whereas undercompensation has been found in some insect populations (Hassell *et al.* 1976).

The constant k in Equation 2.4 may be interpreted as having some vague relation with resource availability. The more productive the environment, the greater the population level attained before the rate of increase reaches zero. Density dependence arises because each individual receives a smaller share of the fixed resource supply as the population size grows. However, for herbivores resource production is likely to vary from year to year, since vegetation growth will be influenced by variable rainfall conditions, temperature regimes and other factors. To represent

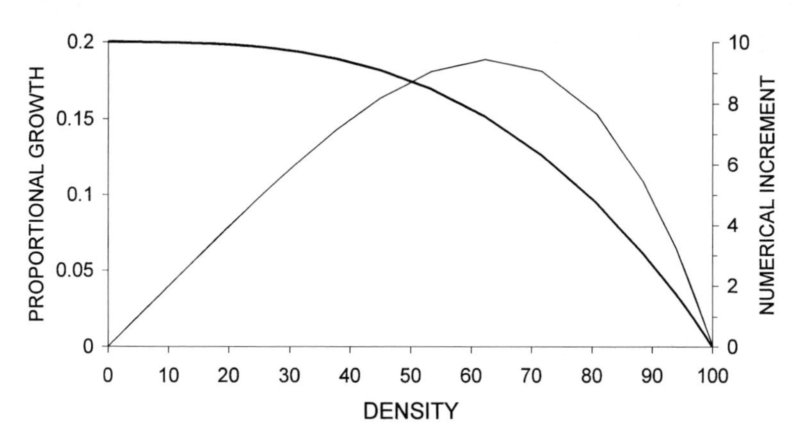

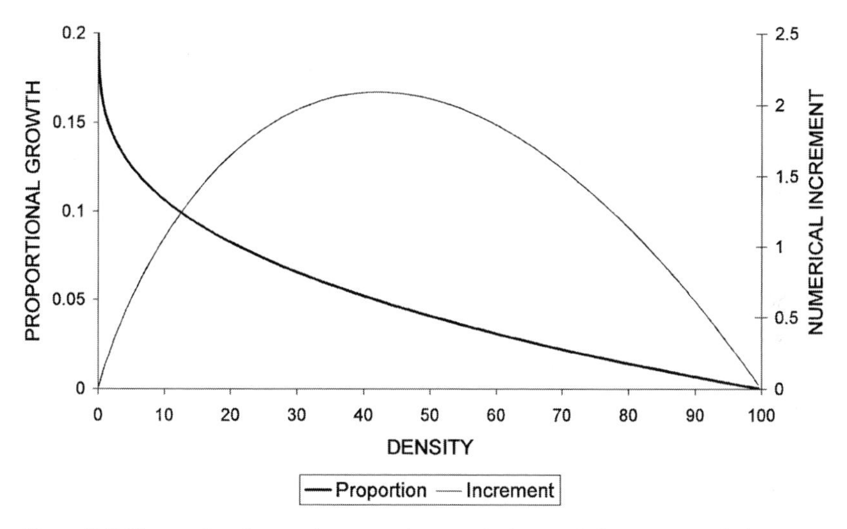

Figure 2.2. Proportional growth rate and numerical growth increment relative to density from a modified logistic model allowing for non-linear density dependence. (a) Overcompensation, with intrinsic growth rate $r_m = 0.2$, equilibrium population size $K = 100$ and power factor $z = 3$. (b) Undercompensation, with $r_m = 0.2$, $K = 100$ and $z = 0.33$.

changes in the environmental capacity to support herbivores, k could be replaced by an environmental state variable K, dependent for example on rainfall. The modelled population then tracks the changing zero growth level in some kind of dynamic equilibrium. If K varies widely while the population responds slowly, density will fluctuate somewhat below the equilibrium level that might be reached in a constant environment. This recognition led Ellis and Swift (1988) to emphasize the persistent disequilibrium inherent in livestock dynamics in arid rangelands.

Transforming from continuous to discrete time, Equation 2.4 becomes

$$N_{t+1} - N_t = r_m N_t (1 - N_t / k). \tag{2.8}$$

This formulation differs subtly, but importantly, because an automatic time lag between population change and the density to which it responds is built in. Specifically, the population increment is governed by the population density at the start of the time interval, and is unaffected by changes in density occurring during the interval. This can allow the population size to rise above its equilibrium level k before density dependence kicks in and draws the population size down. Under these circumstances, oscillations rather than a smooth approach to equilibrium can be generated, especially if r_m is large, or density dependence somewhat overcompensating in form. If the proportion by which the population grows per unit time step is sufficiently large, the oscillations may be sustained as a limit cycle, or even become chaotic in form (May 1981).

To a large extent, the tendency of difference equations towards non-equilibrium dynamics is an artefact of the time step adopted. With an annual time interval, the assumption is that offspring produced during the year grow automatically towards adulthood, without experiencing any effect of resource insufficiency until the end of the annual cycle. Then mortality takes hold when the inadequacy of resources is suddenly encountered. In reality, resource limitations would take hold somewhat sooner than this, with offspring incurring reduced growth and elevated mortality during the course of the year. A lag in the onset of density dependence could be specified explicitly in the differential equation model, by relating the change in population size to the density prevailing at some previous time. In this case, the appropriate delay could be made less than a year, if appropriate.

Computer algorithms automatically operate in discrete time. Hence, an important consideration in formulating a computational model is the time interval chosen for iteration.

Demographically structured population models

Structured population models partition the total population abundance N among the age, sex or other classes constituting it. Age- or stage-specific rates of survival and fecundity are used to project the resultant population dynamics in discrete time, using matrix algebra. Conventionally, the vital rates are incorporated as constant parameters, so that the resultant pattern of population change is exponential growth or decline. Moreover, the currency is almost invariably numeric.

In reality, the vital rates are unlikely to be constant, being influenced by density as well as by other environmental factors. Age-class-specific rates of survival and fecundity could be formulated as functions rather than constants, dependent upon such factors. Solutions must then be obtained computationally, because matrix algebra cannot handle variable coefficients of transition between classes. Age-structured models incorporating environmental dependence as well as density dependence were developed for white rhinos by Owen-Smith (1988), and for kudus by Owen-Smith (2000). In these examples, both biomass density and numerical density were accommodated, by multiplying the numbers in each age class by the specific masses. An annual iteration was used for both models.

Age or stage classes could be sub-divided further to take into account the likely differences in survival chances and reproductive outputs among particular sets of individuals within the population. This could take into account the specific size reached at a particular age, extent of stored body reserves, reproductive history, social status, neighbourhood context, and other factors that might be relevant. Such individual-based models compute the aggregate population dynamics from the specific circumstances of each organism constituting the population (DeAngelis and Gross 1992).

Interactive consumer–resource models

Interactive consumer–resource models relate the dynamics of the consumer population directly to the dynamics of the food resource base, and allow also for the impact of consumption on resource dynamics. The prototype took the form of the coupled differential equations formulated independently by Lotka (1925) and Volterra (1926), with predator–prey systems in mind. Expressed most generally, we have

$$\mathrm{d}P/\mathrm{d}T = R(P) - I(P, C); \tag{2.9}$$

$$\mathrm{d}C/\mathrm{d}T = G(C) - M(C), \tag{2.10}$$

where P = prey abundance, and C = predator (carnivore) abundance. R is the intrinsic rate of production of prey, expressed as some function of prey abundance. I is the loss to the prey population as a result of consumption (intake) by predators, dependent on both predator and prey abundance. The function G transforms prey consumed into predator population gain. M takes account of the intrinsic shrinkage of the predator population through mortality losses and metabolic attrition. Note that the growth rate of the consumer population is governed by the net difference between resource gains and inherent attrition, rather than by the balance between birth and death rates.

For simplicity, Lotka and Volterra assumed that the prey increase at a constant exponential rate r in the absence of predation. Each contact between predator and prey results in a constant proportion of the prey being consumed. A fixed proportion of the food consumed becomes incorporated into the consumer population. Finally, losses from the consumer population through mortality and other causes occur at a constant rate. Hence, we have

$$\mathrm{d}P/\mathrm{d}T = rP - aPC; \tag{2.11}$$

$$\mathrm{d}C/\mathrm{d}T = caPC - mC, \tag{2.12}$$

where a indexes how encounters become transformed into consumption, c the fractional conversion of consumed material into the consumer population, and m the constant proportional loss from the consumer population.

For constant prey biomass, P, the predator population grows exponentially if $caP > m$, and declines exponentially if $caP < m$. Hence, whether the predator population expands initially depends upon the prevailing prey abundance when predators are introduced. However, an expanding predator population increasingly depresses its prey base. At some stage this forces the predator population into decline, eventually allowing the prey population to recover. The reciprocal interaction generates sustained oscillations in predator and prey abundance, even though no explicit time lag is specified in the differential equations. If the predator population grows, it depresses prey abundance through increasing consumption. The outcome is reciprocal oscillations in herbivore and vegetation abundance. Since neither the predator nor the prey population experiences intrinsic density dependence, these oscillations are neutrally stable, i.e. they persist at whatever amplitude is generated initially (May 1981).

Caughley's model

Caughley (1976a; see also Caughley and Lawton 1981) modified the Lotka–Volterra equations to apply to herbivore–vegetation systems. In simple overview, we have (adjusting Equations 2.11 and 2.12)

$$dV/dT = RV - IH; \tag{2.13}$$

$$dH/dT = GH - MH. \tag{2.14}$$

Notably, the functions R, I, G and M are all rates expressed relative to the corresponding state variables, with V representing vegetation abundance and H herbivore abundance. Brackets indicating the functional dependencies have been omitted for simplicity.

Caughley modified the rate functions to allow for non-linear saturating effects. The relative growth rate of vegetation, R, was represented by a logistic function:

$$R = r_V(1 - V/k_V), \tag{2.15}$$

where r_V = maximum relative growth rate, and k_V = density of vegetation at which net growth becomes zero, assumed constant. As was noted by Caughley, an underlying assumption is that plants do not influence the supply rates of their trophic resources. This may be valid for sunlight and rainfall, but is questionable for soil nutrients.

Caughley assumed that the intake rate per herbivore, I, as a function of vegetation abundance, was exponentially saturating:

$$I = i_m(1 - e^{-bV}), \tag{2.16}$$

where i_m = maximum relative intake rate at high food abundance, and b indexes how strongly the feeding efficiency of the herbivores declines towards low vegetation biomass (Fig. 2.3). However, the hyperbolically saturating Michaelis–Menten function has become more widely adopted, because the controlling parameter is more easily interpreted:

$$I = i_m V/(v_{1/2} + V), \tag{2.17}$$

where $v_{1/2}$ = vegetation biomass at which the intake rate is reduced to half of its maximum (Fig. 2.3).

The gain in herbivore abundance as a consequence of food consumed was assumed to be directly proportional to the food intake rate, so that, transforming Equation 2.17, we get

$$G = c\, i_m V/(v_{1/2} + V), \tag{2.18}$$

where c indexes the proportional conversion efficiency. Caughley assumed that the intrinsic rate of attrition in herbivore abundance, through mortality or whatever source, remained constant, as in the Lotka–Volterra formulation, i.e. $M = m$.

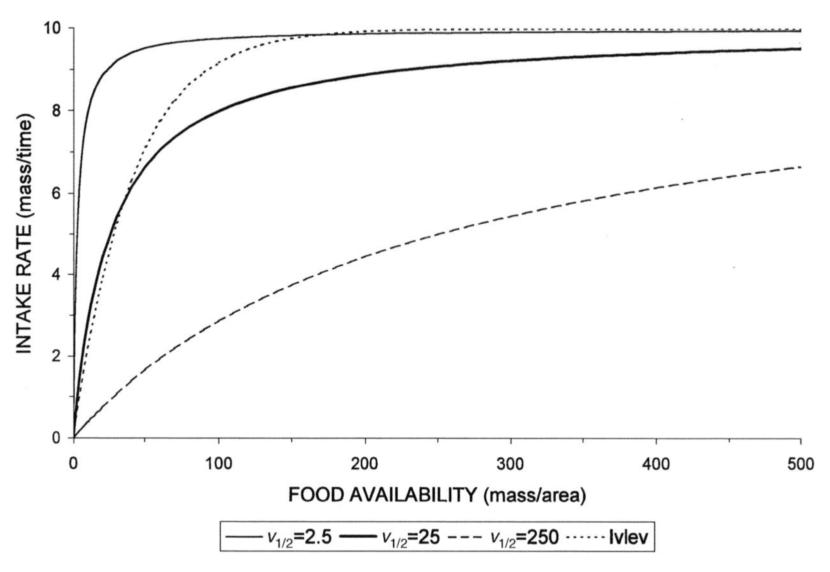

Figure 2.3. Form of the intake response to changing food availability generated by the exponentially saturating Ivlev function ($I = a(1 - e^{-bV})$) compared with the Michaelis–Menten function ($I = aV(v_{1/2} + V)$) for different values of $v_{1/2}$.

This model is likewise prone to instability in the output dynamics, arising largely from the non-linearities in the response functions. Caughley (1976a) used it to exemplify the observed propensity of introduced herbivore populations to 'irrupt' initially to a high density, before crashing and then eventually stabilizing at a lower density in a degraded vegetation. Whether oscillations are dampened or persistent depends largely on the value of the half-saturation parameter $v_{1/2}$ (Fig. 2.4). To reveal this algebraically, we establish the situation associated with stability in the herbivore population, i.e. where $dH/dT = 0$. Incorporating Equation 2.18 into Equation 2.14, we obtain

$$V^* = mv_{1/2}/(ci_m - m), \tag{2.19}$$

where V^* = vegetation abundance associated with zero herbivore growth. The smaller $v_{1/2}$, the less the vegetation amount remaining at this stage. Lowering metabolic or mortality losses m from the herbivore population likewise reduces this vegetation amount, as also does raising the conversion coefficient c or the maximum eating rate i_m. Correspondingly, the herbivore population level H^* associated with a

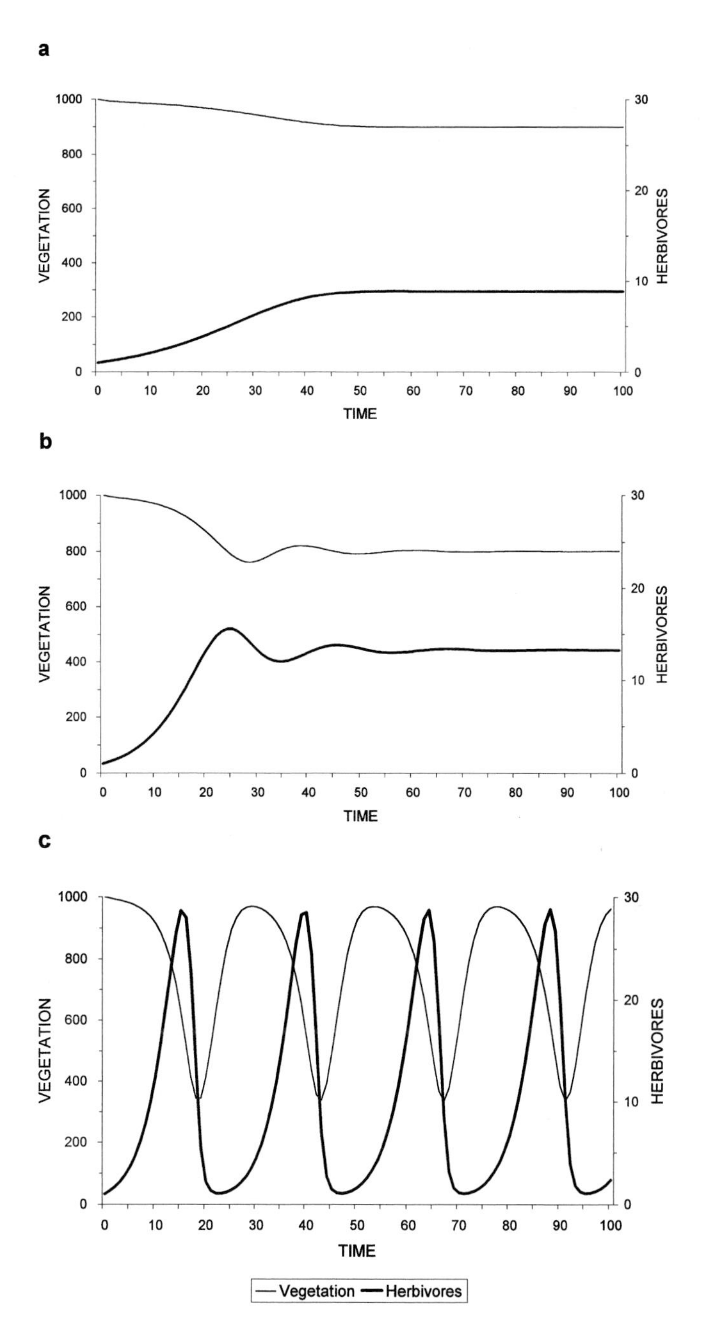

Figure 2.4. Output generated by Caughley's herbivore–vegetation model for different values of the half-saturation level $v_{1/2}$, with other parameters set as follows: vegetation growth rate $r_V = 0.5$, conversion coefficient $c = 0.5$, mortality loss rate $m = 4$. (a) $v_{1/2} = 225$, showing smooth approach to equilibrium. (b) $v_{1/2} = 200$, showing damped oscillations. (c) $v_{1/2} = 162$, showing sustained limit cycle.

zero rate of growth in vegetation is given by

$$H^* = (r_V/i_m)(1 - V/k_V)(v_{1/2} + V), \tag{2.20}$$

which generates a humped curve peaking at some intermediate value of V. Where the two zero growth isoclines intersect defines the point of joint equilibrium, which may be stable or unstable. Persistent cycling by both populations tends to develop when V^* is somewhat less than $k_V/2$; in other words, when grazing reduces vegetation abundance to less than half of the level that it could have attained in an ungrazed state, before the herbivore population stops increasing. An interesting corollary is that a sufficient increase in k_V, leaving $v_{1/2}$ and other parameters unchanged, can destabilize a previously stable equilibrium. This phenomenon has been called the 'paradox of enrichment' (Rosenzweig 1971). Notably, enriching the nutritional quality of the vegetation, by elevating the conversion coefficient c, tends likewise to be destabilizing (cf. Equation 2.19).

Caughley (1976a) was equivocal about the units used to express vegetation and herbivore abundance, because his model remained somewhat abstract. In fact, Equations 2.13 and 2.14 make sense, from a mass balance perspective, only if the abundance of both populations is expressed as biomass rather than numerical density (see Ginzburg 1998; Berryman 1999).

Further elaboration

(a) Density-dependent interference

In Caughley's (1976a) formulation, the density feedback counteracting unlimited growth by the herbivore population arises solely through the impact of consumption on resource abundance, i.e. through exploitation competition. However, because of the non-linear dependence of herbivore intake on food abundance, the form of density dependence that it generates tends to be overcompensating, especially if $v_{1/2}$ is small (Fig. 2.5). This helps promote population oscillations, evident from a spiralling in the growth rate of the herbivore population towards equilibrium.

Competition may also occur through direct interference with feeding. This is manifested as a progressive reduction in food intake rate as the density of competitors increases. Interference competition can be incorporated by inserting a term dependent upon H into the intake function.

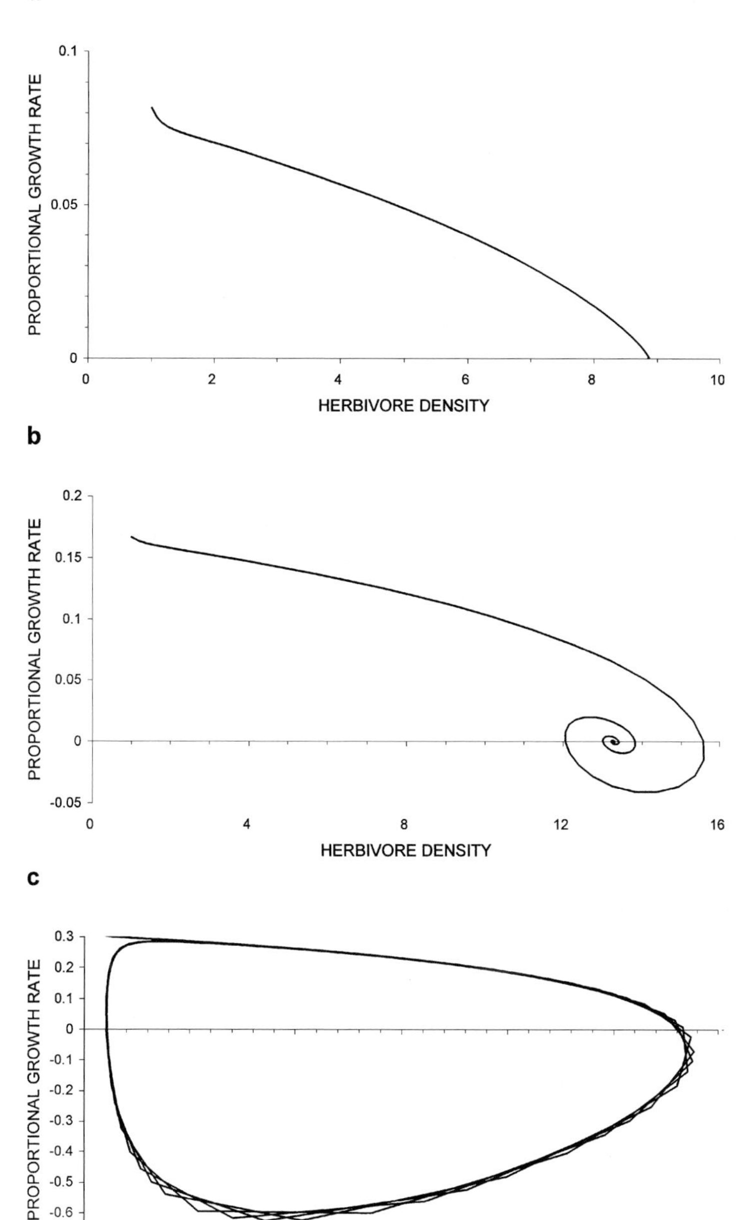

Figure 2.5. Form of density dependence in herbivore growth rate generated by Caughley's model. (a) Parameters as in Fig. 2.4a. (b) Parameters as in Fig. 2.4b. (c) Parameters as in Fig. 2.4c.

For example, following DeAngelis *et al.* (1975), Equation 2.17 could become

$$I = i_m V/(v_{1/2} + V + \delta H) \quad (2.21)$$

where δ = the interference coefficient. The higher the value of δ, the stronger the impact of each additional herbivore on the food intake rate of other herbivores.

If competition, whether through interference (e.g. through strict spatial exclusion) or as a result of exploitation, is sufficiently strong, the resource share available to each consumer is effectively the ratio V/H. In these circumstances, we have

$$I = i_m(V/H)/\{v_{1/2} + (V/H)\}. \quad (2.22)$$

This *ratio-dependent* intake (or 'functional') response has been advocated by several workers (Getz 1984; Arditi & Ginzburg 1989; Akcakaya 1992; Berryman 1992), but challenged by others (e.g. Abrams 1994; Gleeson 1995). In the short term, the food intake rate is controlled by the prevailing food availability, with perhaps some effect of interference competition. However, in the long run the amount of food available for consumption by each consumer depends on the annual production of food relative to the density of the consumer population. The problem confronted is how to assess the longer term outcomes of short term processes, especially when the underlying relations are non-linear. While resource impacts remain sufficiently low, food intake is hardly affected. Beyond some level of resource depression, the effect on food intake accelerates rapidly.

Even fairly weak interference, as represented in Equation 2.21, tends to stabilize the consumer–resource interaction. However, aggressive interactions over food are rarely manifested by large mammalian herbivores, except in confined situations. Because food is widely dispersed, animals can forage along parallel or even interweaving pathways without experiencing much effect on food intake rates even while moving as a group. Only when herd size becomes quite large do the back-markers in the group encounter diminished resource access.

(b) Mechanisms of vegetation change

Caughley's (1976a) formulation of vegetation dynamics did not distinguish between annual growth in vegetation biomass, and changes in the plant populations producing this biomass. His chosen model parameters imply an annual iteration, and hence demographic responses by the plant population following losses to herbivory. However, the amount of forage produced to feed herbivores depends on the annual biomass growth by these plants. If consumption is excessive, some of these plants die, so that the capacity of the vegetation to produce forage the next year is reduced. Following this mortality, the space opened up can be re-colonized either vegetatively, or via seed germination.

Accordingly, in applying Caughley's model to the white rhino – grassland interaction, I distinguished between these components of vegetation change (Owen-Smith 1988). The annual amount of grass produced for consumption by white rhinos depended on the prior rainfall total. If the aggregate consumption by the white rhino population and other herbivores exceeded some threshold level, an index of the grass population at the start of the next year was reduced, and accordingly the forage production potential. However, the grass population was able to recover to some extent, following such shrinkage, through its logistic growth potential.

The stability of this model was strongly dependent on the intrinsic rate of increase of the grass population (i.e. r_V). If this was sufficiently rapid, overgrazing was transient. If too slow, the grass population declined to a baseline level, leading to a crash in the white rhino population. This crash could be averted if white rhinos were able to disperse before critical thresholds were exceeded. The model demonstrated the susceptibility of the herbivore–vegetation system to instability, especially under conditions of variable rainfall and hence variable grass production.

The modelling exercise also raised questions about the mechanisms governing grassland recovery. In fact, a grassland could be grazed almost bare, and yet completely re-vegetate the area over a growing season from a persistent seed bank. Vegetative expansion of surviving plants could help achieve this re-establishment. A distinction between demographic increase and biomass production did not seem that meaningful. Although vegetation could thus recover very rapidly, the problem for herbivores would be to find sufficient food for support during the period of re-growth. Their grazing pressure could indeed prevent the grass from re-growing.

Metaphysiological population models

Getz's formulation

Getz (1991, 1993) approached the formulation of linked consumer–resource models from a different perspective. He was concerned with resolving anomalies in classical models coupling species in food webs

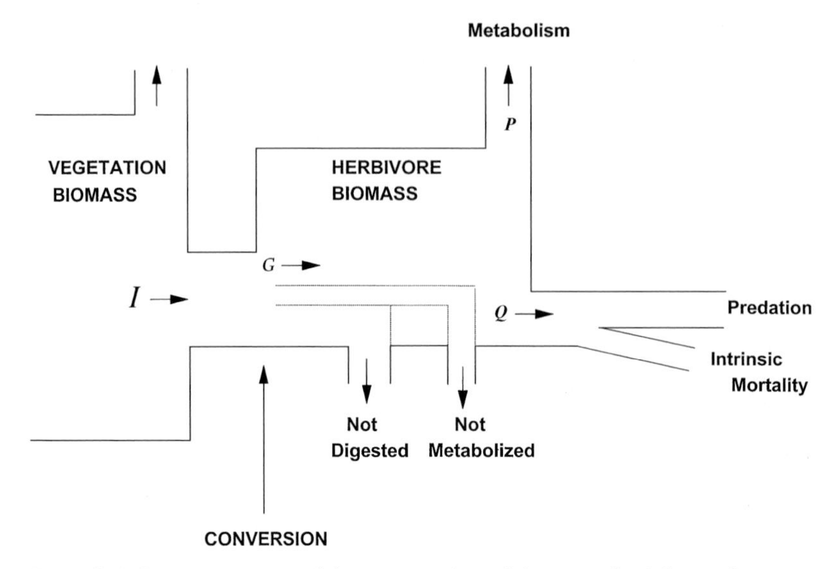

Figure 2.6. Compartment model representation of the metaphysiology of a herbivore population. I = food intake from vegetation, G = gain in herbivore biomass from food consumed, P = biomass losses through metabolic attrition, Q = biomass losses through mortality.

via demographic accounting of births and deaths. He proposed that the consumer population could be conceptualized as a meta-organism, gaining material from its food resources, and losing biomass through metabolic attrition as well as via consumption by other organisms (Fig. 2.6). In overview, this perspective resembles the classical compartment models describing energy fluxes through trophic levels within ecosystems. However, Getz took things a stage further, by focussing attention on the processes governing the extraction of resources and their conversion into consumer biomass. His special concern was how the form of the functional relations emerged from the underlying mechanisms. He showed how this approach enabled the parameters r and K in the classical logistic model to be dissected into their underlying components. He revealed further how this perspective provided insights into the evolutionary tradeoffs associated with the somewhat discredited notion of r versus K selection. A summary overview is provided by Getz (1994).

From a metaphysiological perspective, populations grow only if the net biomass gained from resources consumed exceeds losses to metabolism and mortality. How this biomass is partitioned numerically is of secondary consideration. Reproduction itself generates no new biomass; a fraction of parental biomass is merely budded off as offspring. Whereas numerically the maximum rate of population growth depends on the difference between reproductive potential and inevitable mortality, in biomass terms it is governed by potential rates of food intake and conversion, relative to metabolic and mortality losses. To see this, set V very large in Equation 2.18 to estimate maximum G, and insert this into Equation 2.14, to obtain

$$(1/H)(\mathrm{d}H/\mathrm{dT}) = c\,i_m - m. \qquad (2.23)$$

Generating many offspring creates the potential for rapid population growth. For this growth to be realized, these offspring must have the capacity to eat fast and convert food consumed efficiently, plus the resources to achieve this potential. In environments where such resources are frequently available, species are likely to evolve the ability to exploit them effectively, through appropriate adaptations in physiological capacity as well as in demography.

For zero net population growth, we must have $G = M$, from Equation 2.14. In the absence of interference competition, the equilibrium herbivore density H^* is undefined, except through the interaction with vegetation. Allowing interference of the form specified in Equation 2.21, and provided a stable biomass of vegetation V^* pertains, we obtain

$$H^* = [\{(c\,i_m/m) - 1\}\,V^* - v_{1/2}]/\delta. \qquad (2.24)$$

With the intake response dependent on the resource ratio, i.e. Equation 2.22, the equilibrium biomass of herbivores is given by

$$H^* = (c\,i_m - 1)\,V^*/(m\,v_{1/2}). \qquad (2.25)$$

Hence, the herbivore abundance attained also depends on the maximum intake rate and food conversion efficiency, relative to intrinsic population attrition. This reveals the interdependence between the parameters r_m and K in the logistic equation. However, the herbivore density at equilibrium is influenced additionally by the half-saturation parameter $v_{1/2}$ and by the strength of interference competition. In crowded environments, consumers must also be efficient at exploiting sparse resources, and in coping with competitive interference.

The form of the intake ('functional') response, and its influence on population stability, has received much theoretical consideration, but little past attention has been given to other components of the 'numerical' response of the consumer population, represented by Equation 2.14. Getz (1991, 1993) suggested that, rather than being constant, mortality losses would be inversely related to resource gains. With no food, herbivores die infinitely fast, ignoring the temporary lag imposed by stored

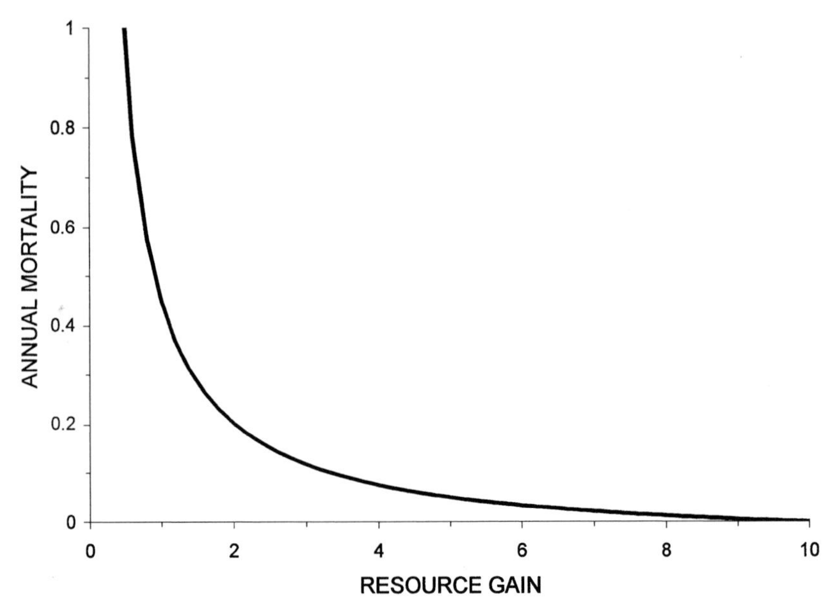

Figure 2.7. Hyperbolic dependence of mortality rate on the effective rate of resource gain in units of metabolizable energy.

body reserves. Increasing food gains alleviate this mortality, until a level is reached at which further increases in food supply have little effect, because food is no longer limiting. This form of relation can be represented by a negative hyperbolic function (Fig. 2.7):

$$M = z_0 + \{z/G\}, \tag{2.26}$$

where z_0 sets the background level of mortality when the resource gain G is high, and z determines how steeply mortality rises with diminishing gains. If food is sufficiently plentiful, animals die only from accidents, chance predation and eventually old age. With diminishing resource gains, malnutrition increases susceptibility to starvation, predation and disease. The nutritionally dependent component of mortality tends to be stabilizing, because it reduces the population growth rate by an increasing proportion as the resource gain per unit of consumer biomass diminishes (Fig. 2.8).

The *GMM* model

Getz's (1991, 1993) metaphysiological modelling approach provided the foundation for this book. It presented a conceptual framework for

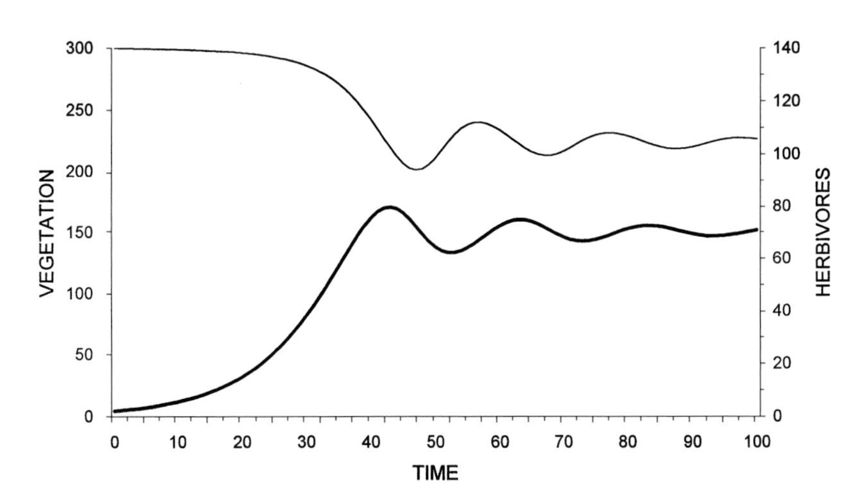

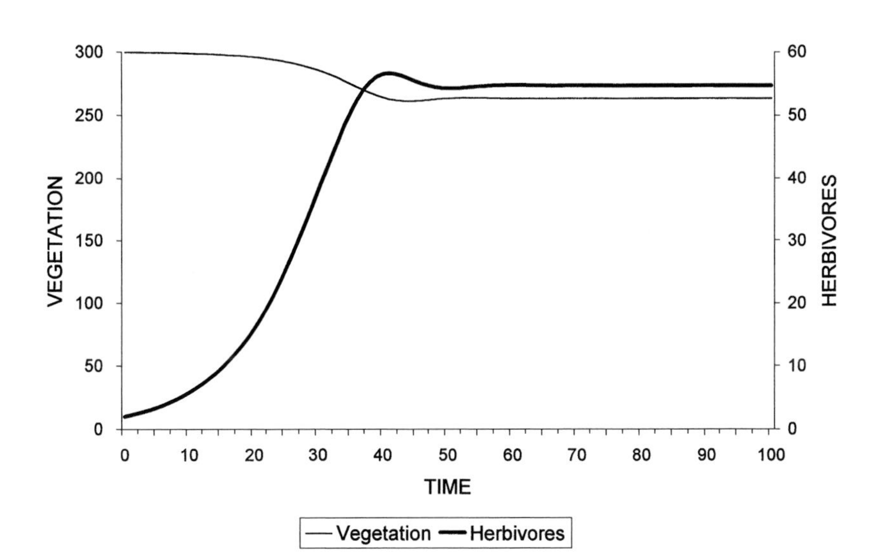

Figure 2.8. Comparison between output of Caughley's model assuming a constant mortality loss, and that of a modified model allowing for resource-dependent mortality as in Fig. 2.7. Model parameters: vegetation growth rate $r_V = 0.8$, maximum vegetation biomass $k_V = 300$, half-saturation level $v_{1/2} = 120$, maximum intake rate $i_m = 1.2$, conversion coefficient $c = 1.3$. (a) Constant rate of mortality loss $m = 1.1$. (b) Mortality coefficients: $z_0 = 0.6$, $z = 1.1$.

integrating upwards from the adaptive responses of individual organisms to changing resource availability, towards consequent outcomes for populations, and ultimately communities and ecosystems. It allowed a consistent currency of biomass flux, or its energetic equivalent, to be adopted. The obscuring of within-year dynamics associated with demographic models employing a numerical currency could be avoided, because biomass change occurs continuously, albeit at varying rates. Although time change is necessarily discrete in computational models, the time step could be made as short as is appropriate.

Despite the applicability of continuous differential equations, mathematicians cannot provide analytic solutions in circumstances where conditions are variable, and functional relations non-linear. Moreover, seeking equilibrium outcomes is inappropriate in seasonal environments. Resource availability and other conditions can change drastically during the course of an annual cycle, with resource renewal generally ceasing for a portion of the year. Any equilibrium between production and consumption would be no more than transient. Consumer biomass expands during times of the year when resources are abundant, and contracts when these resources diminish. The consumer–resource interaction is intrinsically disequilibrial, but entrained to daily, annual and perhaps longer environmental cycles. The output dynamics have to be obtained computationally, despite inevitable numerical contingencies.

The remaining chapters of this book will be concerned with examining how the form of various functional relationships emerges from underlying mechanisms, and changes with expanding temporal and spatial scales. A modification needs to be made to Equation 2.14, defining consumer biomass dynamics, before we proceed further. The two components contributing to the relative rate of consumer biomass loss, M, need to be distinguished, because they are likely to differ in functional form. Physiological attrition, M_P, via energy metabolism for tissue maintenance and activity costs, occurs continuously at a fairly steady rate. Mortality losses, M_Q (with Q borrowed from life table convention), arise through various causes, including starvation, senescence, predation, disease and various accidents, and may be strongly dependent on resource gains.

The minimum level of mortality is set by deaths through senescence. If all individuals survive to their maximum lifespan, then die suddenly, the loss rate is the reciprocal of the life expectancy. Thus if the potential lifespan is 10 years, on average about 10% of the population is lost each year. However, other sources of mortality pre-empt deaths through senescence. Moreover, the susceptibility of animals to dying as a result of malnutrition will depend on resource gains relative to physiological costs. If metabolic costs are low, animals can survive on little food, and vice versa. Finally, allowance needs to be made for deaths through predation, unrelated to malnutrition, but dependent on predator abundance. Thus Equation 2.26 requires elaboration, as follows:

$$M = M_P + z_0 + \{zM_P/G\} + f(H, C), \qquad (2.27)$$

where $C =$ carnivore biomass. The function $f(H, C)$ represents the intake response of the carnivores to changes in their food abundance, as represented by herbivore biomass.

Accordingly, the fundamental equation governing herbivore biomass dynamics takes this form:

$$(1/H)\mathrm{d}H/\mathrm{d}T = G - M_P - M_Q. \qquad (2.28)$$

Equation 2.28 defines the **GMM** metaphysiological model, labelled mnemonically for the components **G**rowth, **M**etabolism and **M**ortality (although strictly growth is the difference between G and M_P). Notably, each component in the formula relates to a potential environmental influence. The rate of biomass gain, G, depends directly on resource availability. Metabolic costs, M_P, may be affected by environmental conditions such as temperature, although the resource supply can influence the activity component. The mortality loss, M_Q, depends largely on the risk of predation, in interaction with resources and senescence. Overall, the resource dependency of all components of consumer biomass change is captured.

Adoption of the metaphysiological formulation does not preclude taking into account the demographic structure of the population. This can be brought in as a secondary refinement, as I will show in final chapters. Ultimately, we could recognize differences in the physiology and spatial context of each animal or plant, thereby ending up with an individual-based, physiologically explicit, model, if this is appropriate. However, this would be at the expense of the conceptual simplicity of the **GMM** model. The important value of the metaphysiological approach is to direct attention to the environmental determinants of population dynamics: towards environmental structure, rather than demographic structure.

Metaphysiological population models are fully concordant with the basic laws of physical thermodynamics, as well as with biological principles. The material basis for population growth is explicit. The thermodynamically open nature of biological systems, dissipating energy to maintain their ordered states, is fully recognized (Prigogine and Wiame 1946; see also Nicolis and Prigogine, 1977; Weber *et al.* 1988). Such systems are intrinsically disequilibrial, tending towards dynamic rather than static stability – remaining in place in a flowing stream, rather than resting in some unperturbed recess on the bottom.

To find the source of any semblance of stability, we must turn to the most basically biological law, the law of adaptation. To persist in a dynamically changing world, organisms must continually adapt, in behaviour, in physiology, in demography and in genotype, over varying time frames. In the following chapters of this book, the consequences of these adaptive responses for the functional forms of the processes governing herbivore biomass dynamics will be assessed.

Overview

This chapter outlines the principles, concepts and modelling approaches that provide the foundation for this book. Numerical population models have been developed from descriptive equations to coupled demographic equations in matrix notation, and ultimately towards individual-based models. The consequent population dynamics have become largely decoupled from environmental influences, except through some vague, and misleadingly constant, K carrying capacity. The alternative biomass-based approach leads from coupled differential equations linking consumer populations directly to the dynamics of their resource base, towards metaphysiological formulations accommodating the functional outcomes of various environmental influences on population dynamics. It provides a framework for integrating upwards from the adaptive responses of individual organisms to environmental variability, towards consequent outcomes for population dynamics, food web interactions, and ultimately ecosystem processes. Metaphysiological models are fully concordant with basic physical and biological laws, and capture the intrinsically disequilibrial dynamics of biological systems.

Further reading

Useful introductions to population ecology theory and models are provided by Gotelli (1995), Hastings (1997), Gurney and Nisbet (1998) and Case (2000). Crawley (1983) gives a comprehensive evaluation of consumer–resource models in the context of herbivore–plant interactions. For the original development of the metaphysiological modelling approach, see Getz (1991, 1993, 1994). This approach may be contrasted with physiologically structured treatments of population dynamics by Metz and Dieckmann (1986) and Kooijman (1993).

3 · *Resource abundance: intake response and time frames*

For herbivores, food resources change in amount both intrinsically through vegetation growth and decay, and as a result of consumption. Such changes in food abundance influence the rate of food intake obtained by each individual herbivore. The relationship between food intake rate and resource abundance has conventionally been termed the 'functional response' (following Solomon 1949). However, because this is just one of several response functions considered in this book, I will call it specifically the *intake response*.

Not all of the standing vegetation biomass that a botanist would measure is effectively available for consumption by a specific herbivore. Some plant material is *inaccessible*, e.g. underground plant parts for most antelope and deer. Inaccessibility may be temporary; for example, leaves high in tree canopies may later be shed. A portion of plant biomass is *inedible*, e.g. woody trunks of trees for most ungulates. But inedibility is relative; vegetation components rejected as food at one time may later be eaten when little else is available. In effect, different plant species and parts vary over time in their *acceptability* to consumers. I will use the term *potential food* to encompass all vegetation components that are eaten at some time or other. Currently edible, accessible and acceptable material constitutes *available food*.

The food intake rate may be measured over different time frames: (a) the short-term ('instantaneous') rate of food ingestion during periods while animals are engaged solely in feeding, i.e. prehending, biting, chewing and swallowing; (b) the average intake rate over the duration of a foraging bout, encompassing searching and moving as well as feeding; and (c) the daily intake, allowing for time devoted to activities unrelated to food-seeking. Food intake could additionally be aggregated over longer periods, such as a season, year or lifespan. Table 3.1 outlines how different temporal scales relate to corresponding spatial scales, behavioural processes and vegetation units. The intake response may change somewhat in its form as the temporal frame is expanded.

Table 3.1. *Hierarchical set of temporal and spatial scales associated with food intake by large mammalian herbivores*

Temporal scale	Spatial scale	Defining behaviour	Vegetation unit
1–2 s	bite	plucking, chewing and swallowing	plant part
2 s – 2 min	feeding station	moving head, prehending, biting	plant (grass tuft, shrub)
0.5–30 min	food patch	feeding (eating), stepping	clump of plants
1–4 h	foraging area	feeding, walking, standing alert	plant species association
12–24 h	daily range	foraging, travelling, drinking, ruminating, resting	landscape unit
3–12 months	seasonal range	growth, reproduction, mortality	landscape type
several years	lifetime range	survivorship, fecundity, dispersal	geographical region

Source: Adapted from Bailey *et al.* (1996).

The food intake rate is commonly assessed *per capita*, i.e. per individual animal. However, for standardization it is more appropriate to express it as a mass-specific (proportional or relative) rate. For some purposes it is more revealing for the intake rate to be related to the metabolic mass equivalent, calculated as $M^{0.75}$ where M = body mass.

Basic theory

Holling's response types

Holling (1959, 1965) distinguished between three forms of intake response, labelled Types I, II and III. For Type I, the intake rate increases linearly with increasing food availability, below some satiation limit. For Type II, the intake rate decelerates asymptotically towards its upper limit. For Type III, the initial increase is retarded at low food abundance (Fig. 3.1).

The Type I response is supposedly typical of situations where food comes in fine particles that can be ingested immediately, e.g. suspended organic matter for filter feeders like barnacles. According to Holling (1965), a Type II response arises where predators must pause to handle (capture, kill and consume) individual prey before resuming searching for

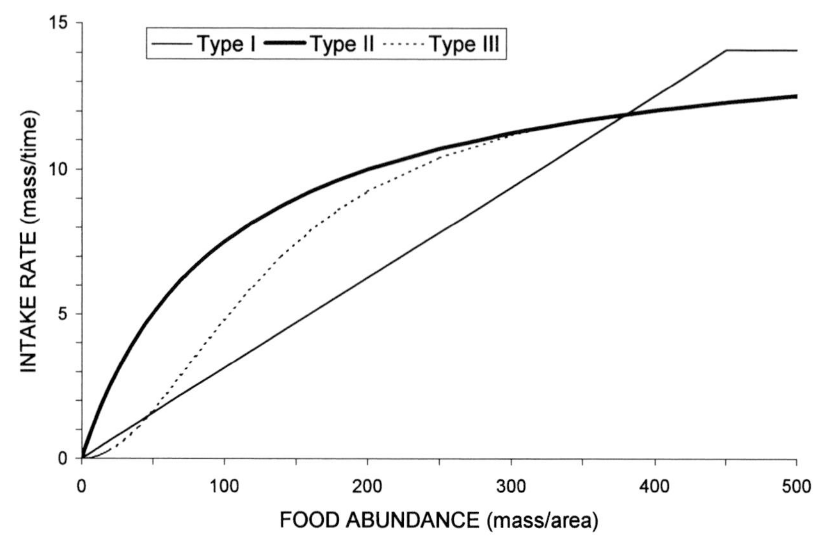

Figure 3.1. Classical Types I, II and III intake (or functional) response patterns, from Holling (1959, 1965).

the next food item. He mimicked the basic mechanisms experimentally by having a blindfolded assistant transfer sandpaper discs (food items) from a table (the environment) into a receptacle (the consumer's gut). Searching is first needed to locate a food item. Searching activity is then suspended while the prey is prehended and transferred to the receptacle. Thereafter, searching is resumed. If prey are rare, most time is spent searching. When prey are abundant, time is devoted largely to handling the prey captured. This leads to Holling's 'disc' equation:

$$N = sDT/(1 + s\,h\,D) \tag{3.1}$$

where $N =$ number of 'prey' consumed during a time interval T, s = area searched per unit time, $D =$ the prey density per unit area, and h = handling time per prey item.

If handling time is zero, the prey capture rate N/T is simply sD, so that intake rate rises linearly with prey density. This is the 'Type I' response, below the satiation limit.

Equation 3.1 is functionally identical to the Michaelis–Menten equation, which expressed in general form is

$$\mathrm{d}X/\mathrm{d}T = a\,X/(b + X) \tag{3.2}$$

where a is the maximum rate and b the half-saturation level, i.e. the value of X at which dX/dT reaches half of this maximum. To see this, Equation 3.1 can be rearranged as follows:

$$N/T = (1/h)D/\{(1/sh) + D\}. \qquad (3.3)$$

Hence the maximum intake rate when D is large is inversely related to the handling time per unit prey $1/h$. How quickly intake rate approaches its maximum as food density increases is dependent on the product of area searched per unit time and the handling time.

The Type III response arises where some of the prey become effectively unavailable toward low prey abundance. This could arise when the prey mostly occupy safe refuges when rare, or when predators shift to alternative habitats or prey types when the normally selected prey becomes uncommon. Expressed relative to prey availability, the form of the intake response remains Type II; but relative to prey abundance there is a decline in the predation impact per capita on the prey. This helps prevent the prey population from being completely eliminated.

Modification for herbivores

For large mammalian herbivores, food takes the form of bite-sized clusters of plant parts. The time required to prehend, pluck, chew and swallow a leaf or shoot tip can be quite brief (a matter of seconds). These food items are concentrated within the patches constituted by a tree, shrub or extended herbaceous sward. While animals feed in such patches, the search time required to locate the next bite is negligible, overlapping mostly with the time taken to chew the previous bite. Animals may move a step or two to re-position themselves within the patch without interrupting feeding. Hence, the rate of food intake is governed largely by the time taken to handle (pluck and chew) each bite. This is 'Process 3' foraging in the terminology of Spalinger and Hobbs (1992), or handling-limited foraging for Farnsworth and Illius (1998).

Eventually, edible and accessible food items become depleted and herbivores need to transfer to a new patch. Spalinger and Hobbs (1992) distinguished situations where patches are readily apparent, so that animals can proceed straight to them ('Process 1'), from those where animals must search for the next patch ('Process 2'). This widens the time frame to encompass searching movements between food patches. If patches are sufficiently sparse, the rate at which patches are encountered can become the controlling influence on intake rate (Farnsworth and Illius 1998). When the time frame is broadened even further to the daily cycle, interruptions of foraging by other activities, such as drinking, resting or ruminating, as well as for travelling between the sites of these activities, must also be taken into account.

Within-patch feeding

The food intake rate obtained while feeding within food-concentrated patches is effectively the product of bite mass and biting rate. This *within-patch* intake rate will be symbolized I_P. Following Spalinger and Hobbs (1992), let b_m represent the bite mass, and h_b the handling time per bite (the reciprocal of the biting rate). Thus, $I_P = b_m/h_b$. The time per bite includes two components: the time required to manipulate and pluck each bite, h_a, and the time required to chew and swallow the mass obtained, h_c. If animals cannot chew while plucking, these two actions compete for time, so that $I_P = b_m/(h_a + h_c)$. It is helpful to substitute h_c by h_m, the handling time per unit mass of food, i.e. $h_m = h_c/b_m$. We then obtain

$$I_P = b_m/(h_a + h_m b_m). \tag{3.4}$$

This is a Michaelis–Menten equation, describing a situation in which intake rate increases asymptotically with increasing bite mass. The latter is governed by the bulk density of plant parts within a bite-sized volume. When bite mass is very small, so that chewing and swallowing are momentary, the eating rate is determined directly by bite size relative to the minimum plucking time per bite: $I_P = b_m/h_a$. If bites are very large, so that animals spend most of their time handling the ingested mass, $I_P = 1/h_m$, i.e. the maximum eating rate is governed by the oral processing time per unit mass of food.

For some herbivores plucking and chewing can overlap in time, using the same jaw movements (Laca *et al.* 1994b; Ginnett and Demment 1995). Hence, control over the eating rate is exerted by bite mass, until a bite size is reached at which additional chewing movements become necessary before the next bite can be taken. Larger bites may also take longer to pluck because of the extra manipulation needed to gather material. For the *ca.*750 kg steers studied by Laca *et al.* (1994a), the time per bite remained constant at 2 s until the bite mass exceeded 1.3 g (equivalent to 1.8 mg per kg live mass). As a result, there was initially a linear increase in intake rate with increasing bite mass, until the threshold bite mass was exceeded (Fig. 3.2). When the bite mass reached around 4 g, further increases in bite size did not increase the intake rate, being fully

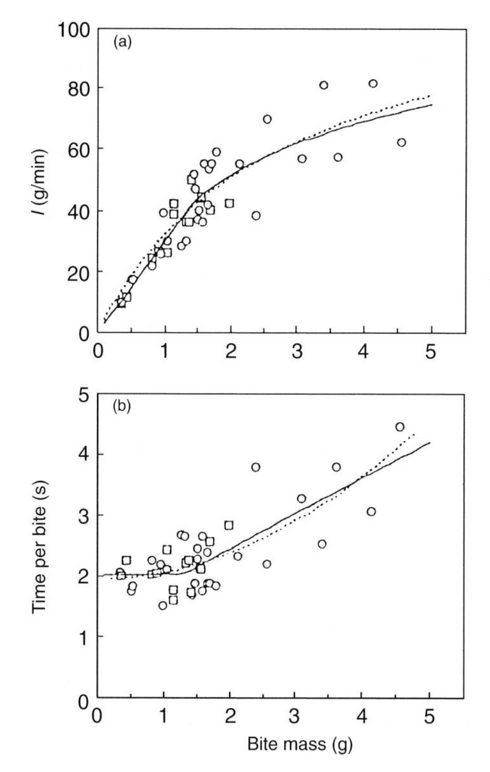

Figure 3.2. Discontinuous relation between within-patch intake rate, handling time per bite and bite mass for 750 kg steers consuming hand-constructed alfalfa and dallisgrass swards. (a) Change in intake rate with increasing bite mass, showing an initially linear segment, then a curved region tending towards the maximum intake rate. (b) Underlying change in time per bite, showing the threshold bite mass at which additional chewing movements become necessary. (From Laca *et al.* 1994b. Reprinted with permission from Elsevier Science).

compensated by the longer handling time required per bite. Spalinger and Hobbs (1992) recognized that, rather than being strictly asymptotic, the 'instantaneous' intake rate could reach some upper limit at a finite resource density.

Observations confirm that the short-term intake rate obtained by various ungulates tends to increase almost linearly initially with increasing bite mass, towards some upper limit (Fig. 3.3). However, observed eating rates are often less than the potential maximum for a particular bite size. Either animals chose to feed below the maximum rate, or some external factor restricted the biting rate below its maximum. Potential influences on plucking rate include the presence of thorns, affecting leaf prehension,

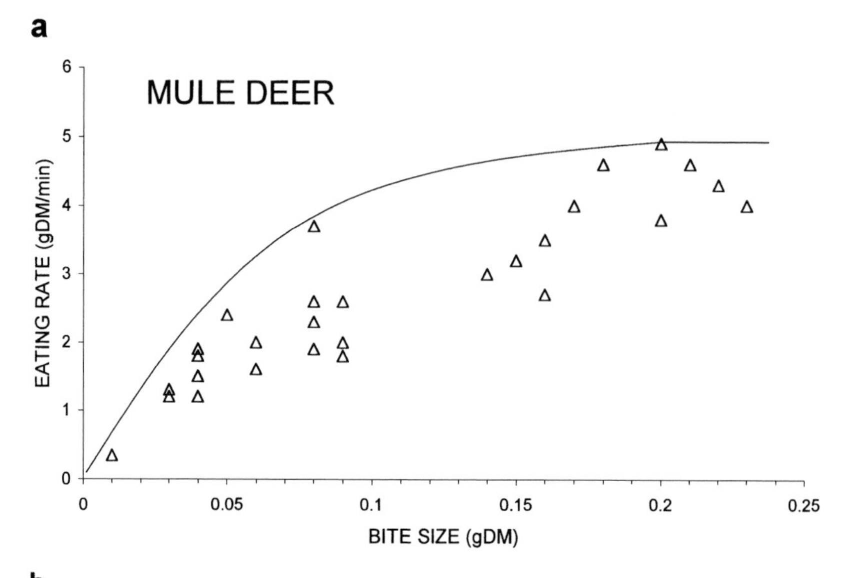

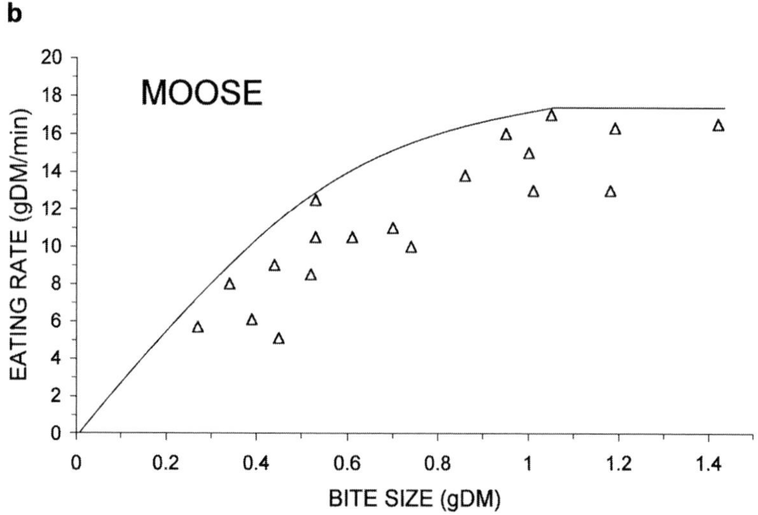

Figure 3.3. Relation between within-patch intake rate and bite size for various ungulates, showing an outline of the upper envelope to the distribution. (a) Browsing mule deer, mass 32–52 kg (from Spalinger and Hobbs 1992, based on data from Wickstrom *et al.* 1984). (b) Browsing moose, mass *ca.* 350 kg (from Spalinger and Hobbs 1992, based on data from Risenhoover 1987). (c) Browsing kudus, mass 100 kg, distinguishing spinescent (triangles) from unarmed (squares) woody plant species (from Cooper 1985; Cooper and Owen-Smith 1986). (d) Grazing North American elk, mass *ca.* 250 kg, distinguishing grazing (squares) from shrub (asterisks) browsing (from Hudson and Watkins 1986, © 1992 by the University of Chicago. All rights reserved).

c

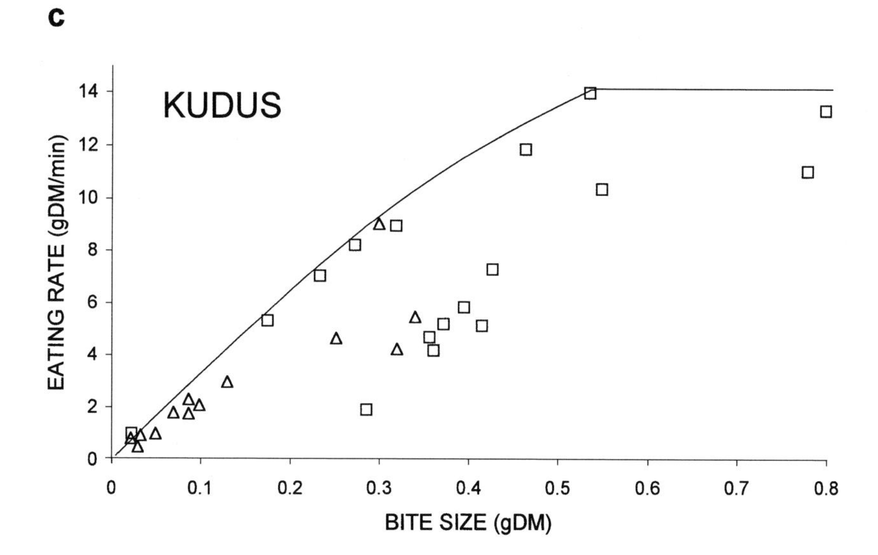

d

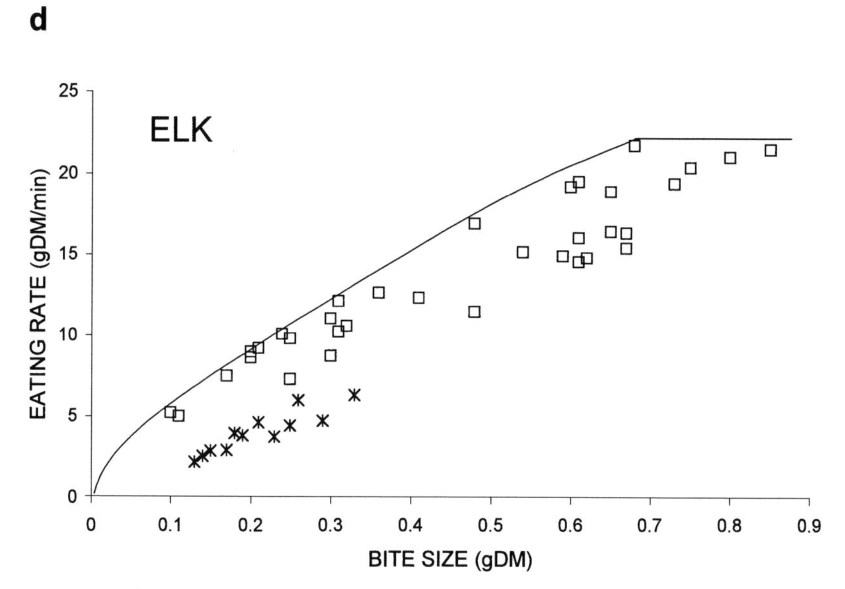

Figure 3.3. (cont.)

or high tensile strength of stems or petioles, slowing leaf abscission. Mastication time could be lengthened by a high fibre content in the material ingested. The maximum eating rates observed under field conditions seem to be 15–30% less than the theoretical maximum predicted by the composite scaling model of Shipley *et al.* (1994), which was based on hand-constructed alfalfa swards.

Elk exhibited a substantially reduced eating rate when feeding on shrubs growing in the forest understory, compared with grass swards (Hudson and Watkins 1986) (Fig. 3.3d). Browsing kudus, in contrast, obtained a slow intake rate when consuming grass (Cooper 1985). The maximum biting rate is about 25/min for moose and 45/min for kudus and deer (extracted from Fig. 3.3), increasing to about 60/min for impala (Cooper and Owen-Smith 1986) and sheep (O'Reagain and Owen-Smith 1996), and 78/min for Thomson's gazelle (Bradbury *et al.* 1996). Hence the minimum time per bite ranges between 0.8 and 2.4 s. Shipley *et al.* (1994) suggest that the biting rate is constrained allometrically by the period required for pendulum movements of the lower jaw.

Despite being theoretically controlled by mouth volume (effectively bite width × gape × bite depth), maximum bite mass scales across herbivore species with a body mass exponent of 0.6 for natural vegetation, and 0.7 for artificially constructed patches (Shipley *et al.* 1994). This suggests that either the surface area covered by the bite, or the molar tooth area, exerts the underlying control on bite size.

For browsing antelope, the bite mass obtained from woody plant species is determined largely by the leaf mass, although large browsers like kudus commonly bite off shoot tips as well as plucking single leaves (Cooper and Owen-Smith 1986). However, the presence of spines and other deterrent structures on stems inhibits stem browsing, and this feature is commonly associated with small leaf size. For grazing ungulates, bite mass is dependent largely on grass height. Bite depth increases linearly with increasing sward height, up to some maximum level (Burlison *et al.* 1991; Laca *et al.* 1992; Jiang and Hudson 1994; Edwards *et al.* 1995). The corresponding increase in bite mass is counteracted to some degree by a decrease in the bulk density of grass tissues with increasing sward height. Bite mass may be reduced below its potential for a particular grass height through preferential selection for the green leaf component. Notably, ungulates can increase their eating rates substantially above the norm following fasting (Greenwood and Demment 1988), or on cold, wet days (O'Reagain *et al.* 1996).

In summary, the within-patch intake rate is restricted largely by the bite size obtained, with handling time per bite less variable. Above some level the biting rate decreases with further increases in bite mass, because of the additional chewing time required. The maximum bite mass is governed largely by sward height and leaf bulk density for grazers, and by leaf or twig dimensions for browsers. An upper limit to the eating rate is set by the processing time required for chewing and swallowing the ingested material. Often the bite mass observed is less than the maximum possible, through selection for particular plant tissues. The range of variation in eating rate may be as wide as 15-fold, as shown for kudus in Fig. 3.3.

Foraging spells

Foraging activity encompasses bouts of feeding within food patches, interrupted by movements between food patches, standing looking around, grooming actions, and perhaps brief social interactions. Because the time per step is typically similar to the time per bite (1–2 s), walking interrupts feeding only if more than 1–2 steps are taken in sequence. Besides movements in search of food, walking may also take place to keep up with herd companions. From analysis of the distribution of intervals between feeding events, foraging was considered to have ended when more than one minute elapsed without feeding for kudus (Owen-Smith 1993a), or more than two minutes for deer (Gillingham *et al.* 1997). Foraging bouts interrupted only briefly by extended walking and other non-feeding activities can be amalgamated into foraging spells, which commonly span a few hours. The area traversed during a foraging spell represents the foraging area. Foraging spells, interspersed with spells of socializing, travelling to water, or moving to a new foraging area, can in turn be amalgamated into activity sessions. Hence, the daily activity cycle consists of an alternation of active and resting sessions, the latter including stationary activities such as ruminating.

The rate of food intake obtained while foraging, I_F, will be lower than the within-patch eating rate, I_P, because of the additional time spent moving between food patches, or diverted to minor interruptions. Assume that an amount of food F_P is offered by each food patch, of which a fraction a is accepted, i.e. consumed. This allows for the likelihood that some potentially edible material will remain behind when patches are abandoned. The time entailed feeding with a food patch is thus aF/I_P. Let the search time per patch be t_S, and the additional time occupied by other activities be t_O. Thus, overall we have

$$I_F = aF_P/(t_S + t_O + aF_P/I_P). \qquad (3.5)$$

This equation is more useful if we replace F_P with the biomass density of food per unit area F, and correspondingly replace the search time t_S by the area searched per unit time s. The time devoted to activities besides feeding and moving is most conveniently expressed as a proportion t'_O of the total foraging time. With these changes Equation 3.5 becomes

$$I_F = \{saF/(1 + saF/I_P)\}\{1 - t'_O\}. \quad (3.6)$$

This is a modified form of Holling's disc equation appropriate for large herbivores. It can be rearranged into the Michaelis–Menten format, thus:

$$I_F = [I_P F/\{(I_P/sa) + F\}][1 - t'_O]. \quad (3.7)$$

The asymptotic intake rate while foraging is set by the eating rate obtained within food patches. The half-saturation level depends on the relation between this eating rate and the product of searching rate and acceptance fraction. The intake rate actually obtained depends on the proportion of foraging time spent feeding relative to searching (i.e. the factor saF/I_P in the denominator of Equation 3.6), as well as on the time diverted to other activities. Notably, search time does not include movements within food patches, which by definition do not interrupt feeding. Food abundance has two components: the density of food patches per unit area, and the density of acceptable food items within these patches. The handling time per unit amount of food is the inverse of the eating rate within a patch. Following the convention of this book, all of the factors symbolized in upper case are recognised to be variables rather than constants.

For large mammalian herbivores the time entailed moving between food patches amounts to only a small fraction of total foraging time. Feeding to walking time ratios of between 3:1 and 6:1 seem typical of large browsers (Owen-Smith 1979; Renecker and Hudson 1986; Ginnett and Demment 1997). For grazing antelope species or mixed grazer–browsers, feeding to walking time ratios range from 6:1 up to 12:1 (Dunham 1982; Underwood 1983). However, for most of these assessments single steps within extended patches were enumerated as well as sequences of steps between patches. For kudus, almost half of the steps taken while foraging constituted within-patch movements (from Owen-Smith and Novellie 1982).

The feeding time obtained per unit step taken while foraging is an index of effective food availability, assuming that the eating rate and acceptance coefficient remain fairly constant. For kudus, this measure varied from 5.6 s per step in the wet season to 8.2 s per step in the early dry season, declining further to 2.6 s per step by the late dry season, in the favoured habitat type (Owen-Smith 1979). The feeding time per step obtained from a less favourable habitat was less than half that obtained from the preferred habitat during the wet season, with the former habitat type being abandoned in the dry season after this index had fallen below 2.0 s per step. Since each step taken by a kudu takes just over a second, the proportion of foraging time spent walking increased correspondingly during the dry season.

The proportion of foraging time taken up by miscellaneous activities varied seasonally between 7 and 15% for kudus (Owen-Smith 1979). For black-tailed deer the proportion was constant at about 8% of active time (Gillingham *et al.* 1997). For various grazing ungulates vigilance (i.e. standing alert) occupied between 3% and 11% of foraging time (Underwood 1982). Accordingly, the overall proportion of foraging time spent feeding typically ranges between 65 and 80% for browsers like kudus and deer, compared with 80–90% for grazing ungulates. As a consequence, quite substantial changes in food availability, and consequently in the feeding:walking time ratio, have a relatively small effect on the food intake rate obtained while foraging.

Measurements on grazing ungulates, usually made with monospecific swards that are progressively depleted, commonly show a decelerating increase in food intake rate with increasing food abundance (Fig. 3.4). However, the resemblance to a 'Type II' intake response is misleading. Other functional forms could be fitted, e.g. an abrupt 'broken stick' transition from an increasing phase to a maximum level. Spalinger and Hobbs (1992) explained the decline in intake rate towards low food abundance as governed mostly by the reduction in bite size consequent upon diminished sward height, rather than via increases in search time relative to feeding. For bison, the effect of sward height on bite size weakened when the stem proportion in the sward was increased (Bergman *et al.* 2000). Effectively only the leaf component constituted available food, with stem consumed incidentally. The height effect is generally less apparent in heterogeneous grass swards, where herbivores may forage selectively among grass species differing in leaf density and leaf:stem ratio (Forbes 1988). When swards become very tall, the intake rate may decline owing to reduced leaf accessibility and consequent decline in biting rate (documented for sheep by O'Reagain *et al.* 1996).

For browsers, in contrast, bite size declines little with diminishing forage biomass, so that the food intake rate remains fairly constant until very little food is left (Fig. 3.5). A further problem in comparing intake

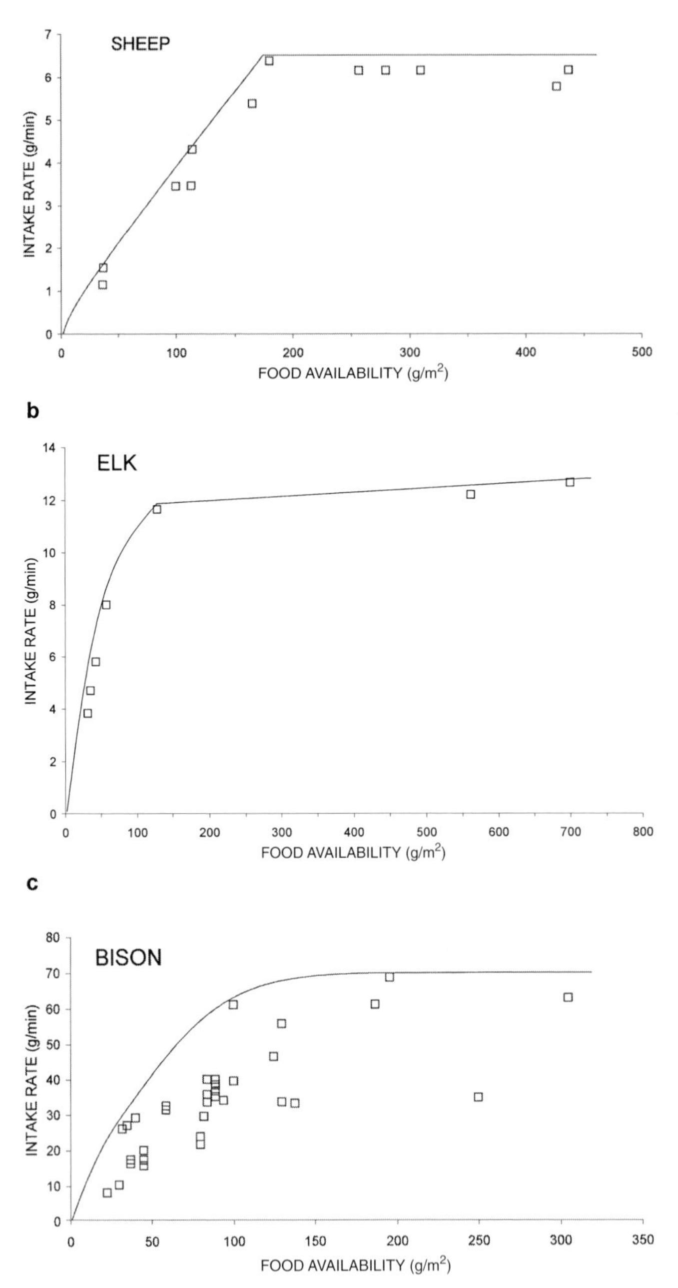

Figure 3.4. Relation between food intake rate while foraging and food abundance for various grazing ungulates. (a) Sheep, body mass *ca.* 40 kg (from Allden and Whittaker 1970). (b) North American elk, mass 140–190 kg (from Wickstrom *et al.* 1984). (c) Bison, mass *ca.* 400 kg (from Hudson and Frank 1987).

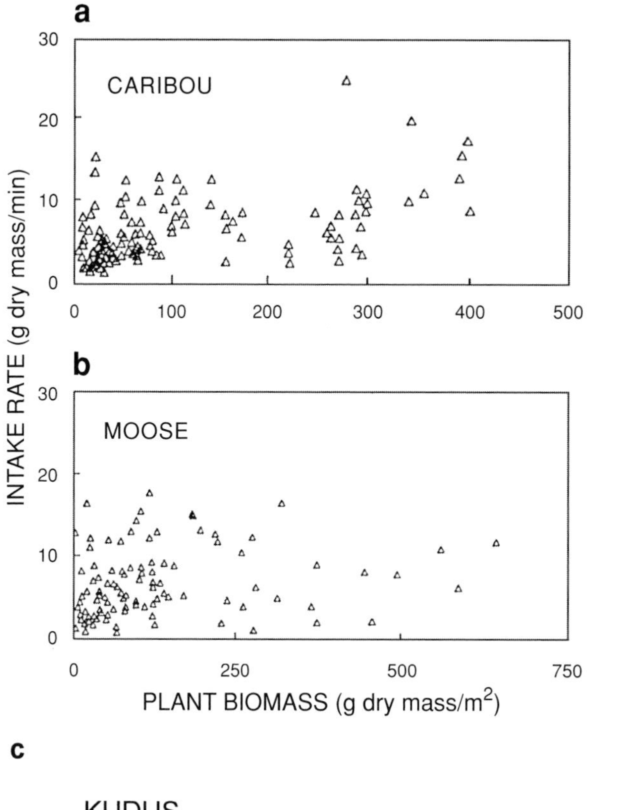

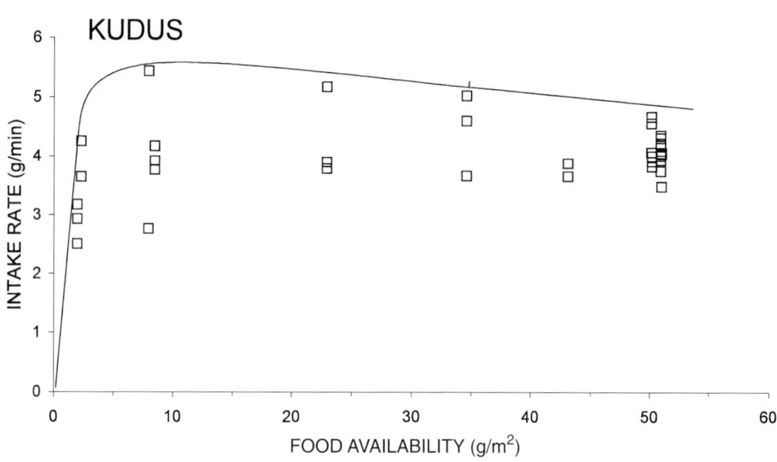

Figure 3.5. Relation between food intake rate while foraging and food abundance, for various browsing ungulates. (a) Caribou, mass 70 kg (from Spalinger and Hobbs 1992, based on data from Trudell and White 1981, © 1992 University of Chicago Press). (b) Moose, mass *ca.* 350 kg (from Spalinger and Hobbs 1992, based on data from Risenhoover 1987, © 1992 University of Chicago Press). (c) Kudus, mass 100 kg (from Owen-Smith 1994 and unpublished; data are averages over a day, line indicates upper envelope).

responses across species arises in the assessment of food availability. For grazers, this is generally estimated as total standing biomass in homogeneous swards clipped close to ground level. For browsers like kudus, generally only dicotyledonous foliage constitutes acceptable food, although small amounts of grass are eaten on occasions. Moose browse leafless twigs as well as foliage. Caribou consume graminoids as well as browse and lichens. Thus, the vegetation components contributing to food availability can differ quite widely.

In summary, the food intake rate obtained while foraging remains strongly dependent upon the within-patch eating rate, because search time spent moving between patches occupies only a minor proportion of total time. Accordingly, changes in bite size continue to have an overriding effect on the intake rate. The amount of food effectively available for consumption varies widely among herbivore species.

Daily food intake

Over the daily cycle, spells of foraging activity alternate with periods of resting, travelling, ruminating, socializing and other miscellaneous activities. The daily rate of food intake, I_D, is accordingly influenced by the proportion of the day engaged in non-foraging activities, so that we have

$$I_D = I_F(1 - t_R - t_{mis}), \tag{3.8}$$

where t_R = proportion of the day spent inactive (resting), and t_{mis} = proportion of the day engaged in miscellaneous activities. Notably, time allocation now becomes a control variable. For ruminants, ruminating takes place mostly during resting spells, while animals remain immobile. However, because ruminating time is influenced by the amount of food consumed, t_R can depend on I_F. This influences the daily food intake only if the ruminating time exceeds the basic resting time.

For large mammalian herbivores, foraging typically occupies 30–50% of the 24 h day–night (diel) cycle (Owen-Smith 1988). The time spent actually feeding will be somewhat less than this, because of the fraction of foraging time taken up by moving and other activities. However, a distinction between feeding time and foraging time is not usually made for grazers, for which feeding occupies 90% or more of foraging time. The proportion of the daylight hours devoted to foraging is usually higher than the 24 h proportion for medium-sized ungulates, which spend less time foraging at night than during the day. This does not apply to very large species like elephants and rhinos, which are equally active night and day.

Certain small ungulates are more active nocturnally than during daylight. The maximum daily foraging time seems to be about 13–14 h (i.e. about 55% of the 24 h cycle) for cattle, even under starvation conditions, for reasons that are unclear. For kudus, foraging time amounted to 91% of daylight hours on a cool rainy day (Owen-Smith 1998a).

Because foraging occupies such a large proportion of daily time, the capacity for herbivores to compensate for a reduced foraging intake by increasing daily foraging time is limited. Accordingly, the form of the daily intake response corresponds closely with that of the intake rate obtained while foraging (Fig. 3.6). Nevertheless, an asymptotically saturating hyperbola is not the only function that could be fitted to these data. Recognizing that much of the scatter in the data is due not to random error in measurement, but rather to day-to-day variability in daily food intake, it may be more revealing to outline the upper envelope enclosing the data, as shown in Fig. 3.6. This suggests how the maximum daily intake for a particular vegetation biomass changes with increasing food abundance. Presumably, on some days animals ate less than the maximum amount, for various reasons. A 'broken-stick' pattern is suggested, with maximum daily intake increasing rapidly with increasing food abundance when food is sparse, then somewhat more slowly at intermediate vegetation biomass levels.

Towards high vegetation biomass, the maximum daily intake may even tend to decline, as is indicated for kangaroos and kudus in Fig. 3.6. For Thomson's gazelles, there was no significant relation between sward height and intake rate once grass height exceeded 15 cm, corresponding with a sward biomass of about 25 g/m^2 (Wilmshurst *et al.* 1999a). Above this limit, digestion rate limited the daily food intake. Because the digestive processing capacity depends on food quality, it will be considered in more detail in Chapters 5 and 6.

Half-saturation level for intake

From Fig. 3.4, the half-saturation level for foraging intake rate varies from 11 g/m^2 for sheep to 60 g/m^2 for bison. For kudus, the foraging intake rate did not drop below half of its maximum until less than 3 g/m^2 of food remained; data for other browsers vaguely indicate a half-saturation level no higher than this (Fig. 3.5). For daily food intake (Fig. 3.6), a half-saturation level under 10 g/m^2 is apparent for kangaroos and sheep, and as low as 2 g/m^2 for kudus. Diverse estimates tabulated by Wilmshurst *et al.* (2000) indicate values for the half-saturation biomass

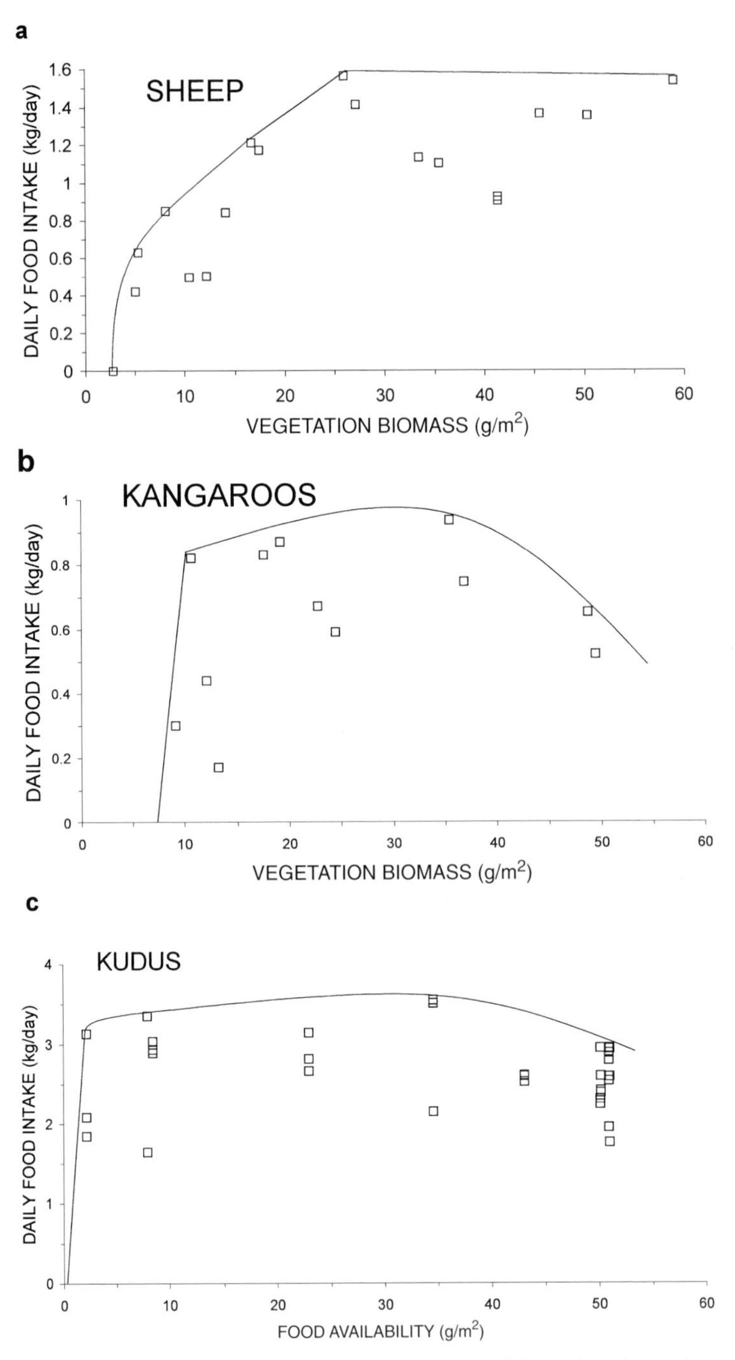

Figure 3.6. Relation between daily food intake and food abundance for various herbivores. (a) Sheep, mass 57 kg (from Short 1985). (b) Kangaroos, mass 26 kg (from Short 1985). (c) Kudus, mass 100 kg (from Owen-Smith 1994).

for daily intake ranging from 2 g/m^2 for mule deer to 99 g/m^2 for bison, with a couple of outliers even higher. Their estimate of this measure for Thomson's gazelle was 16.6 g/m^2, based on artificially constructed swards (Wilmshurst *et al.* 1999a). Data reported by O'Reagain *et al.* (1996) indicate a half-saturation biomass of around 40 g/m^2 for cattle grazing a heterogeneous grass sward.

The level at which the foraging intake rate reaches half of its maximum could also be determined from Equation 3.7, using estimates for the relevant parameters. For kudus, walking speed was estimated to be 20 m/min, and the path width scanned on either side of the body midline to be 0.5 m (Owen-Smith and Novellie 1982), yielding a search rate of 20 m^2/min. The maximum sustained eating rate may be as high as 10 g/min. The fraction of plants accepted along the foraging pathway did not exceed 0.5 (Owen-Smith 1994). This yields a half-saturation biomass estimate of 1 g/m^2. These parameter values should apply to any similar-sized ungulate, so much higher values for the half-saturation level can arise only if a considerably smaller fraction of the measured vegetation biomass is accepted during a foraging traverse.

The effective value of the acceptance coefficient a can be estimated from the ratio between feeding time and moving time while foraging, which are the two components of the denominator of Equation 3.6, using estimates for s, I_P and F. For kudus only 0.03–0.06 of available food biomass was accepted during the wet season, increasing to about 0.25 by the late dry season, based on the data presented by Owen-Smith (1993b, 1994). For grazers the feeding time per step is typically twice that for a browser (see above), but so also is the standing food biomass. Hence the range of variation in the acceptance coefficient should not differ much from that estimated for kudus.

Allometry of food intake

The daily food intake as a proportion of body mass declines from about 3.5% (dry mass/live mass) in small antelope weighing around 10 kg, to about 1% in the African elephant (Fig. 3.7). Typical values for cattle and medium-sized antelope are around 2%. Non-ruminants like horses and zebras require a substantially higher food intake than ruminants of similar size, amounting to about 4% of body mass per day, to compensate for their lower digestive efficiency (Bell 1971). Daily food intakes falling much below these levels will result in starvation.

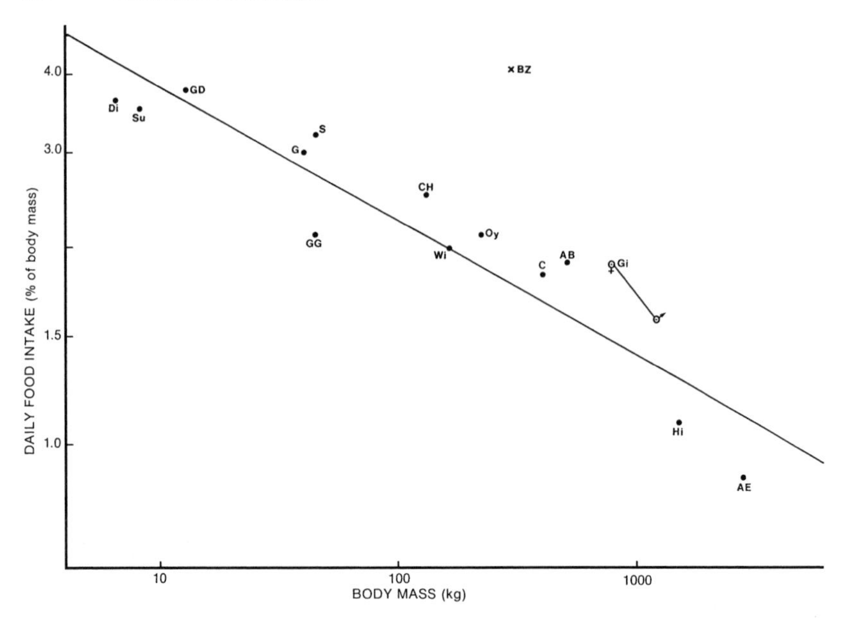

Figure 3.7. Relation between daily food intake and body mass for various ungulate species (from Owen-Smith 1988). AB, African buffalo; AE, African elephant; BZ, Burchell's zebra; C, cattle; CH, Coke's hartebeest; Di, dikdik; G, goat; GD, grey duiker; GG, Grant's gazelle; Gi, giraffe (separate points for males and females); Hi, hippopotamus; Oy, oryx; S, sheep; Su, suni; Wi, wildebeest.

Extrapolating to an annual time frame, a food intake amounting to eight times live mass over the course of the year is needed by a typical medium-sized ruminant. This ratio sets an upper limit to the relation between vegetation production (in dry mass units) and herbivore production (as live biomass). If both plants and herbivores are expressed equivalently in dry mass units, this ratio increases to around 24:1, assuming a dry to wet mass ratio of 0.33 for animal tissues. In other words, secondary production could not exceed about 4% of primary production even if all vegetation was utilized, and in practice is limited to below around 2%.

Influences on the form of the intake response

Because the half-saturation level for intake is small relative to the potential food abundance, the foraging intake response is intrinsically steeply saturating for large mammalian herbivores. This is because food abundance is such that search time occupies only a small fraction of foraging time. All else being equal, the foraging intake response becomes more gradually asymptotic as the range in food abundance spanned is narrowed (Fig. 3.8a–c). Moreover, in response to a surfeit of potential food, much of inferior nutritional value, herbivores may consume only a small fraction of the material encountered while foraging. As food abundance declines, the proportion accepted widens. Through this mechanism the food intake rate could be held fairly constant around some target level over a wide range in vegetation biomass, until food abundance reaches the critical threshold where almost all of the potential food is consumed (Fig. 3.8a–c). As a result the intake response will appear truncated rather than asymptotically saturating towards higher vegetation biomass levels. Adjustments in the daily time allocation to foraging as well as in the proportion of foraging time diverted to miscellaneous activities could also contribute to a levelling out of the intake rate when food is abundant. Truncation could arise additionally through the operation of physiological constraints, as covered in Chapter 5.

Over all foraging time scales, the intake rate remains highly sensitive to changes in the eating rate obtained within food patches. Where bite size is a function of food biomass, through the shared influence of grass height, a gradually saturating intake response results (Fig. 3.8d), as is commonly documented for grazers (Fig. 3.4). For browsers feeding upon leaves and twig tips, this mechanism does not apply, so that the intake response remains steeply saturating (Fig. 3.5).

As outlined in Chapter 2, a situation in which the intake rate of consumers remains high until food abundance is reduced well below half of its potential biomass in an ungrazed state generates instability in the consumer–resource interaction. Why, then, are all large herbivore populations not cyclic or even chaotic in their dynamics? Some possible answers will be developed in the chapters to follow.

Overview

This chapter outlined the processes that determine the form of the intake (or 'functional') response of herbivores to changing food availability, distinguishing among time frames. Within food patches, food intake rate is controlled largely by ingestive processes, specifically factors determining bite size and biting rate. Over the duration of foraging spells, movements between food patches influence intake rate, more so for browsers than for grazers. Through the daily cycle, food intake is influenced additionally by time apportionment between food-seeking and non-foraging activities. Nevertheless, bite mass remains a basic determinant at all scales.

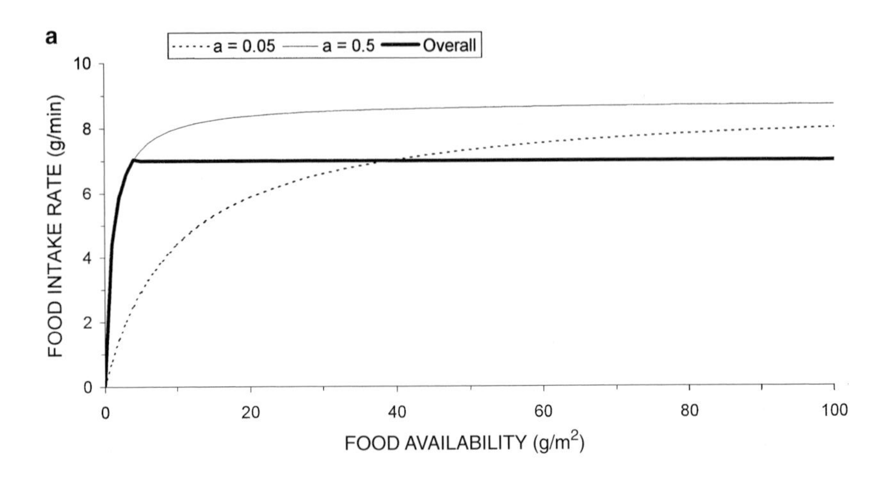

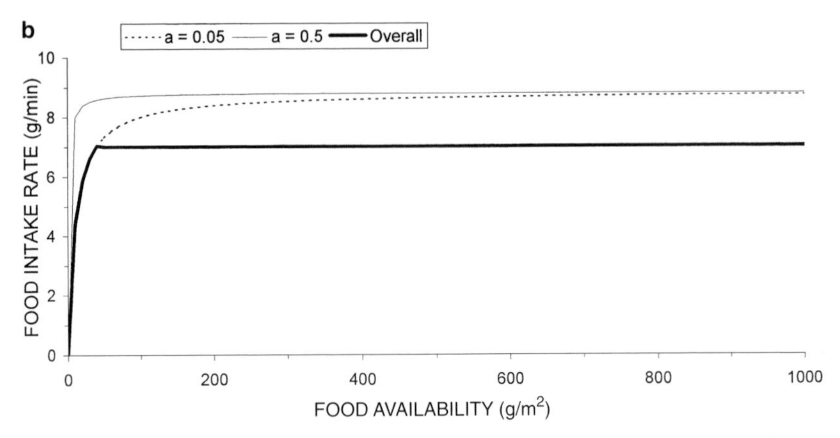

Figure 3.8. Effects of parameter adjustments on the foraging intake response of a hypothetical 200 kg ungulate, calculated by using Equation 3.6 with $s = 20\ \mathrm{m^2/min}$ and maximum $I_P = 10$ g/min; heavy line indicates overall outcome for a target intake rate of 7 g/min. (a) An order of magnitude difference in the acceptance fraction a. (b) An order of magnitude increase in the maximum food availability F. (c) An order of magnitude reduction in the maximum food availability F. (d) Maximum eating rate I_P dependent on height of grass sward, specifically proportional to $F^{1/3}$.

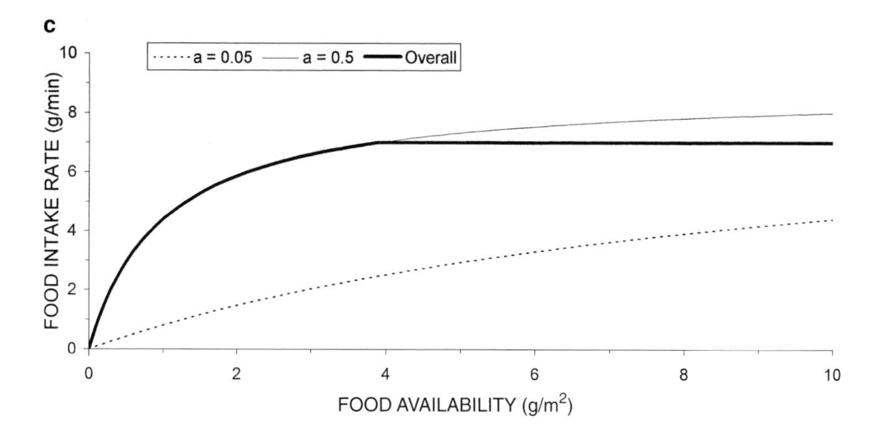

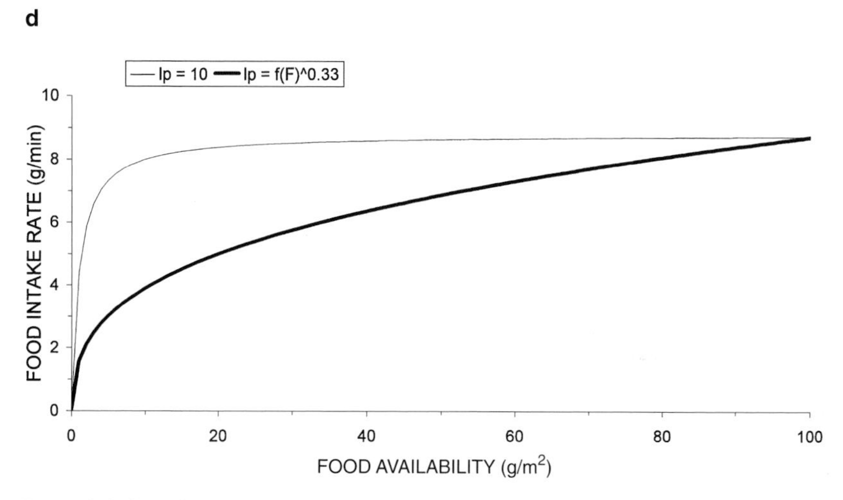

Figure 3.8. (cont.)

Across all time frames, the intake response is steeply rather than gradually asymptotic, or even truncated, because of the superabundance of potentially edible food confronting large herbivores. In response to this abundance, animals may accept only a small fraction of the edible and accessible plant biomass presented along foraging pathways, broadening this proportion as food availability declines. Changes in bite mass with diminishing height, and hence standing biomass, of grass may generate an intake response that is more gradually saturating for grazing ungulates feeding in homogeneous swards.

Further reading

A basic outline of functional response concepts is presented by Real (1977). The evaluation of their application to herbivores by Spalinger and Hobbs (1992) is focussed especially on short time frames. Farnsworth and Illius (1996, 1998) extend this approach to consider variability in the proportion of handling time that is exclusive of searching. Schmitz *et al.* (1997) describe how various adaptive responses by herbivores can generate functional response forms that are radically different from those classically presented in textbooks.

4 · *Resource distribution: patch scales and depletion*

In the preceding chapter, we recognized a distinction between within-patch feeding and between-patch searching. However, patches may be identified at a diversity of scales. To a large extent they are generated by the feeding responses of herbivores to particular forms of heterogeneity in the vegetation. Plants occur underfoot almost everywhere, except in desert regions. Herbivores selectively consume certain plant parts and species, and forage preferentially in particular vegetation types and landscape regions (Senft *et al.* 1987; Bailey *et al.* 1996). By rejecting the intervening plant material as food, herbivores define the gaps separating patches, thereby accentuating what might be somewhat fuzzy patterns in plant distribution (Arditi and Dacorogna 1988). Even in a uniform grass sward, herbivores create a patch structure with the first bite that they take, depleting food in the spot where they fed relative to the surrounding matrix.

The handling time required to deal with food that is patchily distributed determines the asymptotic form of the classical 'Type II' intake response (Holling 1959, 1965). If food arrived in the form of fine particles that could be swallowed without pause, as for filter feeders like barnacles or whales, the intake rate would rise linearly with food density until digestive capacity was satiated. The clustering of plant parts into bite-sized entities causes herbivores to divert their attention from searching, even if only momentarily, to deal with the material plucked. They halt their foraging movements to feed intensively within the larger patches constituted by individual woody plants or monospecific clumps of grasses and forbs. Over several hours or days they may forage circuitously within a particular vegetation association or landscape region, before sallying off to seek another foraging area.

The existence of a prior patch structure in vegetation presents opportunities for selective responses at various relevant scales. Having already dealt with bite size effects in Chapter 3, we turn our attention now to the consequences of patchiness in feeding sites, foraging areas and habitat types. These labels, defined from the herbivore perspective, correspond with particular vegetation units and temporal frames, as summarized in Table 3.1.

Besides patch selection, we are concerned with the consequences of food depletion within patches, through consumption, on the duration of patch occupation. Patches are generally abandoned before all the potential food they contain has been consumed, so that food acceptance within them is partial. Vegetation components passed over at one time may later be consumed, so that acceptance can be manifested as a sliding proportion over lengthening time scales. Plants over time regenerate the parts removed by herbivores, reconstituting the patches. This depends on their growth potential, and hence the season of the year. At times plant resources are non-renewing, so that patches are progressively depleted.

Changes in patch selection and abandonment affect the proportion of foraging time spent walking relative to feeding, and hence the form of the food intake response to changing plant biomass. Such processes are the subject of the present chapter. We consider here only variability in food amount within patches. The consequences of changing food quality will be addressed in Chapters 5 and 6.

Feeding sites

Feeding sites are equivalent to the food concentrated patches recognized by Spalinger and Hobbs (1992). They are constituted by a set of adjacent *feeding stations*, presenting forage that an animal can reach without moving its feet (Goddard 1968; Jiang and Hudson 1993). A potential feeding station is rejected if an animal moves an additional step without pausing to feed, so that sets of two or more steps in sequence identify the gaps between extended feeding sites (Novellie 1978). Foraging activity thus consists of feeding within food patches, interspersed with movements through regions where no acceptable food is encountered.

For browsing ungulates feeding largely on woody plants, food patches can be equated with individual trees or shrubs, or clumps thereof. However, many browsers also consume certain forbs (non-graminoid herbs) and shrublets growing in the spaces between trees. To identify objectively a gap criterion for kudus, Owen-Smith and Novellie (1982) analysed the distribution of step sets by using log-survivor plots. An abrupt break in the slope indicated that the probability of encountering a feeding station diminished after more than three steps had been taken in sequence.

This suggested that kudus tended to move beyond the bounds of a feeding site following a movement sequence of the corresponding distance. Direct observations showed that kudus could take 2–3 steps in succession while repositioning themselves at a particular bush. Almost half of the steps taken by kudus during foraging spells represented within-patch movements constituted by step sets of three or less.

Grazing ungulates can potentially meander through extended grass swards for prolonged periods, so that the limits to feeding sites are inherently more nebulous. Underwood (1983) found no distinct inflexion in the log-survivor distribution of step-sets for the grazing antelope species that he studied. Hence, he regarded all movements of two or more steps in sequence as constituting transfers between patches. He found that 60–85% of steps taken by these grazers were single steps representing movements within extended food patches, varying among animal species and with environmental conditions.

Having identified the feeding-site scale, our concern is next with the factors governing (a) the initial acceptance of the patch, as indicated by commencement of feeding, and (b) patch abandonment, as shown by movements encompassing several steps without feeding.

Patch acceptance

Theory

A potential feeding station may be regarded as *accepted* if one or more bites are taken before the next step is made. According to optimality theory, a food item should be accepted if the benefit obtained from consuming it outweighs the opportunity cost of searching for and ingesting a more profitable food item within the time entailed (Stephens and Krebs 1986). Ignoring differences in food quality (i.e. protein or digestible energy content), the benefit of feeding is simply the within-patch intake I_P, i.e. the quotient of bite mass b_m and handling time per bite h_b (Chapter 3). If potential food within the patch is rejected, an animal must search for an additional time t_s, dependent upon food abundance, before encountering the next food patch. Since t_s is a random variable, consumers must decide on the acceptance or rejection of a feeding station based on some expected value of t_s.

Restricting consideration to just two patch types, choice depends on whether

$$b_{m2}/h_{b2} \leq b_{m1}/(t_{s1} + h_{b1}) \qquad (4.1)$$

where the subscript 2 symbolizes an inferior patch, and subscript 1 a superior patch potentially available elsewhere in the environment. If the search time required to find a better patch is sufficiently long, consumers should feed in both patch types. Otherwise they should selectively seek out only the best patches.

Equation 4.1 can be generalized to any number of patch types, $i = 1, \ldots, n$, ranked in order of value, so that Equation 4.1 becomes

$$b_{mi}/h_{bi} \leq \boldsymbol{b_m}/(\boldsymbol{t_s}+\boldsymbol{h_b}) \tag{4.2}$$

where the right-hand-side of the expression, in bold, represents the gain expected if only patch types ranked higher in value than i are consumed, based on averaged values for their parameters. Hence a lower-ranked patch should be consumed only when better patches are so sparse that search time for them is lengthy. While the highest-yielding patches remain sufficiently abundant, herbivores should restrict their foraging to them.

Observations

Goats that were experimentally offered a choice between patches constituted by single grass species preferentially selected the patches yielding highest intake rates, irrespective of the grass species involved (Illius *et al.* 1999). Nevertheless, the lower-yielding alternative was also grazed in most trials. This conflicts with the prediction of the optimality model outlined above that the acceptance of a food type should be all or nothing ('one-zero'). Illius *et al.* (1999) suggest that the discrepancy shown by their goats might have been due to discrimination error in identifying the better patches. Numerous other hypotheses have been advanced to explain the 'partial preferences' that are widely shown by consumers, in contradiction of the absolute choices predicted by optimality theory (McNamara and Houston 1987a). Among them are notably the costs of acquiring the information needed to distinguish between patch benefits, and random variability in encounter rates with patches.

Cattle likewise favoured food patches where grass was taller and hence intake rate higher than elsewhere (Distel *et al.* 1995). Grass species acceptability to cattle was found to be positively related to leaf height (O'Reagain and Mentis 1989a), the main factor determining bite size (Laca *et al.* 1992). In contrast, sheep preferred somewhat shorter grass species, suggesting an overriding influence of quality (O'Reagain 1993). Among browsing ungulates, moose preferentially selected the largest shoots of adequate quality presented by trees during winter (Vivas *et al.* 1991; Molvar *et al.* 1993). For kudus, bite size varied widely among woody plant species, being influenced primarily by leaf size, especially for spinescent species (Cooper and Owen-Smith 1986). Under wet season conditions the acceptance frequencies shown by kudus for various species were uninfluenced by eating rate, unless bite size and hence intake rate was very low (Fig. 4.1a). In the dry season when foliage was less abundant, a correlation between patch acceptance and eating rate emerged for the relatively palatable subset of species (Fig. 4.1b). Impalas and goats, being somewhat smaller than kudus, were less influenced by bite size restrictions imposed by leaf size (Cooper and Owen-Smith 1986).

Overall, the patch choices made by herbivores appear to be strongly related to the rates of food intake that these patches yield, whether patches are represented by grasses of varying height, woody shoots of varying mass, or trees differing in leaf size. However, rather than the one–zero acceptance predicted by foraging theory, preferences appear to be graded, with inferior patches being accepted less frequently rather than being completely rejected. Larger herbivores appear to be more responsive to intake rate restrictions than smaller species.

Patch departure

Theory

Although patch acceptance may be influenced by the high intake rate obtained initially, in the process of feeding herbivores deplete the food items supporting this rate. This generates a situation of diminishing returns, through lengthened time to find each bite, decreased bite mass or extended handling time. The benefits of continuing to feed for longer in the patch need to be weighed against the potential gain that could be obtained by seeking another patch. The optimality analysis invokes 'marginal value' principles (Charnov 1976). Accordingly, a patch should be abandoned when the rate of food intake that it yields declines to the level expected from a new patch, devalued by the time required to reach this patch.

The optimal departure time is most easily assessed from a plot of the cumulative intake, $\sum I$, against feeding time. This tends asymptotically towards the limit set by the total amount of food offered within the patch. Astrom *et al.* (1990) and Laca *et al.* (1994a) proposed that the cumulative intake could be represented by a hyperbolic function

$$\sum I = i_0 t_p/(1 + k_{1/2} t_p), \qquad (4.3)$$

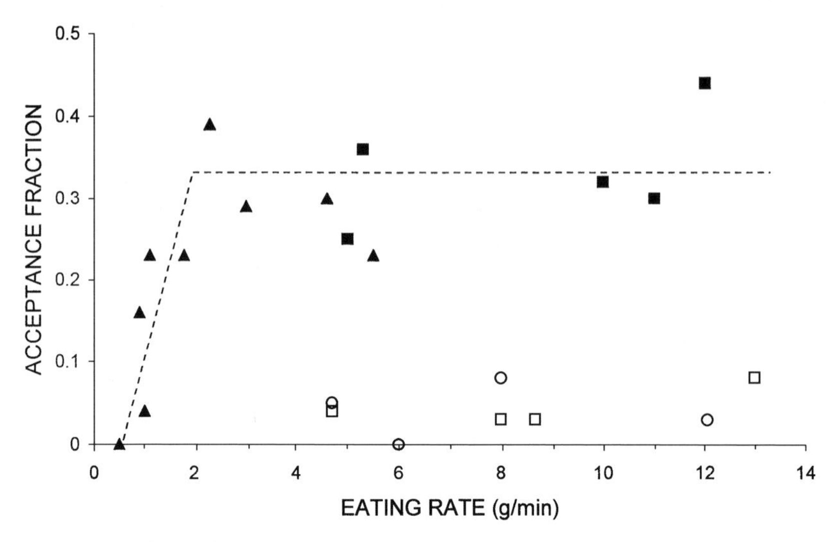

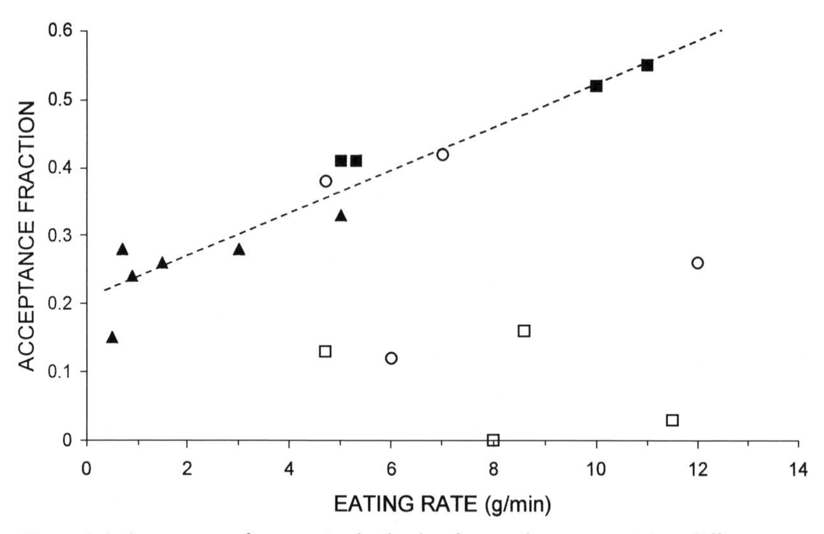

Figure 4.1. Acceptance frequencies by kudus for patches representing different woody species, in relation to the eating rates obtained. Symbols indicate palatability categories: solid squares, palatable deciduous; solid triangles, palatable spinescent; open squares, unpalatable deciduous; open circles, evergreens. (a) Wet season; line indicates threshold decline in acceptance of palatable species for eating rates <2 g min^{-1}. (b) Dry season; line indicates correlation between acceptance and eating rate for palatable deciduous and evergreen species. (From Cooper 1985; Cooper and Owen-Smith 1986; Owen-Smith and Cooper 1987a).

where t_p = feeding time within the patch, i_0 = initial intake rate, and $k_{1/2}$ = reciprocal of the time taken to remove half of the available food within the patch. The ratio $i_0/k_{1/2}$ gives the total amount of food offered by the patch. To work out the optimal feeding duration, we have to consider also the expected search time to find a new patch, t_s. The maximum rate of food intake while foraging is indicated by the steepest line from the origin tangential to the cumulative intake curve, spanning the expected search time t_s (Fig. 4.2a). If patches are sparse, so that travel time between them tends to be lengthy, consumers should stay feeding for longer than under conditions where patches are closely spaced. Patches should be abandoned at the same marginal rate of gain, irrespective of varying initial yield or total amount of food presented. Since large herbivores typically spend only 10–20% of their foraging time walking between food patches (Chapter 3), a decline in I_P by this small proportion should prompt patch departure.

A hyperbolic $\sum I$ function is generated by an intake rate that declines rapidly at first and then more and more slowly as food is depleted towards lower levels (Fig. 4.2b). The corresponding intake rate response to changing food availability is concave-upwards in shape (Fig. 4.2c). This suggests a situation in which patches offer just a few large bites, which are quickly depleted. Under these conditions, consumers should remain within patches for relatively brief periods.

If the intake rate response itself is gradually asymptotic (cf. Fig. 3.8c), the slope of the cumulative intake curve changes little until a substantial fraction of the potential food in the patch has been eaten (Fig. 4.3a). For more steeply asymptotic or even truncated forms of intake response (cf. Fig. 3.8a), the effectively linear region of the cumulative intake function is extended (Fig. 4.3b). This exemplifies a situation where patches offer an abundance of similar-sized bites, so that eating rate declines appreciably only after most of these bites have been consumed. Under such conditions, feeding durations within patches should be quite prolonged. If the bite sizes offered within patches are stratified into discrete size categories, abrupt declines in intake rate should occur at the stages when these bite classes become mostly depleted, generating a 'broken stick' pattern of response (Fig. 4.3c).

For grazers, each feeding station can be regarded as a mini-patch, with depletion of the food easily accessible within neck reach prompting a step to access a new area. Eventually the limits of the extended feeding site are reached when the vegetation presented does not offer any acceptable forage.

a

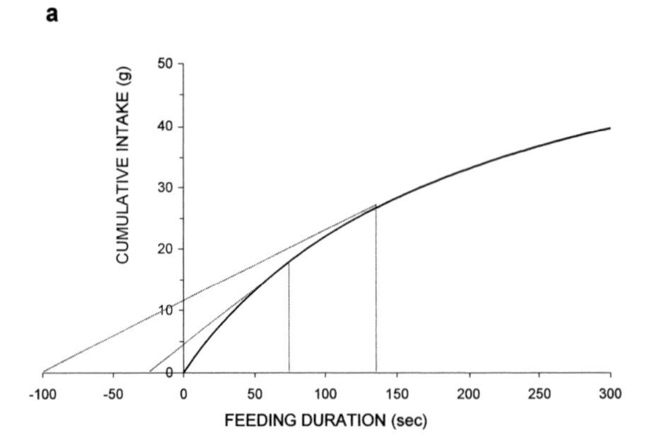

b

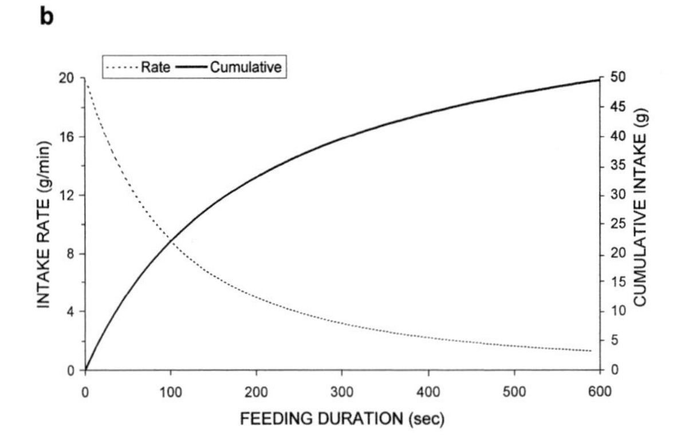

c

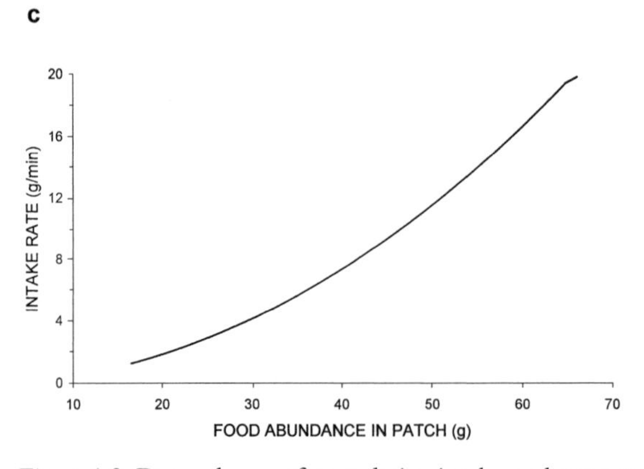

Figure 4.2. Dependence of cumulative intake and corresponding intake rate on feeding duration within a patch, assuming a hyperbolic cumulative function. (a) Optimal feeding durations indicated by lines tangential to the cumulative intake curve for short (20 s) and long (100 s) travel times between patches (represented by negative feeding times). (b) Change in intake rate with increasing feeding duration relative to the cumulative intake curve. (c) Corresponding intake rate response to changing food abundance.

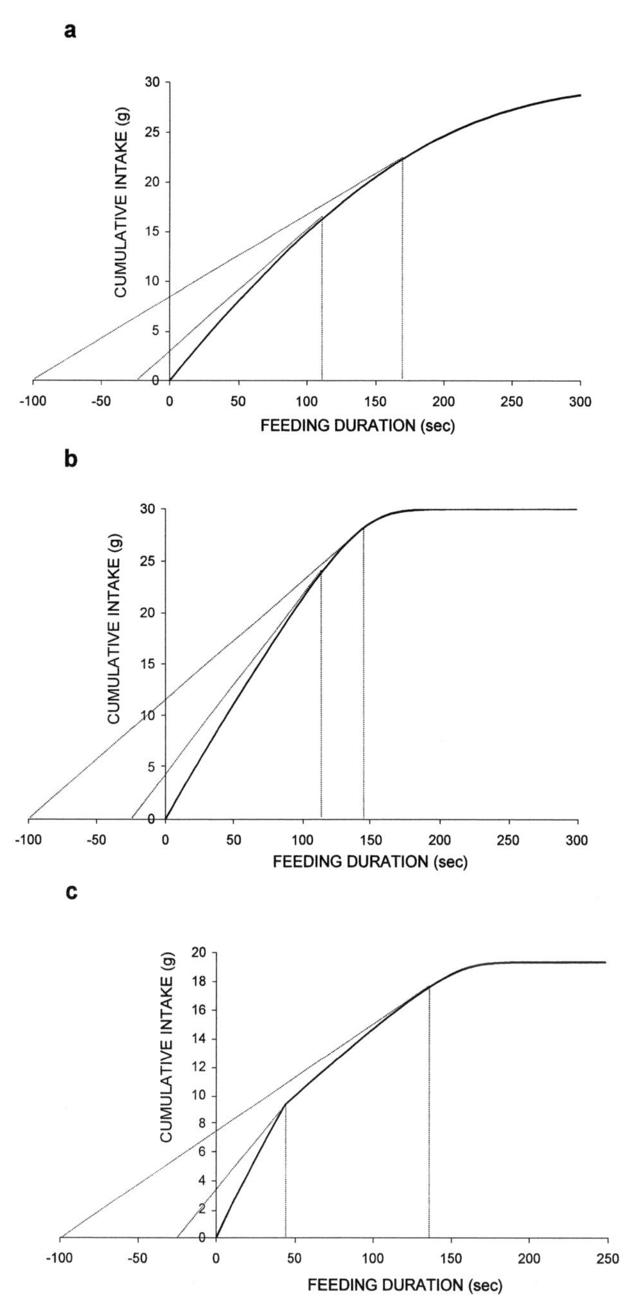

Figure 4.3. Influence of alternative intake rate responses on the optimal feeding duration within a patch. (a) Intake response gradually asymptotic. (b) Intake rate response steeply asymptotic. (c) Abrupt changes in intake rate generated by bite size changes when particular sward horizons become depleted.

Observations on grazers

Elk grazing semi-natural pastures fed on average for 15 s per feeding station, taking around 8 bites during this time, without seasonal variation (Jiang and Hudson 1993). Animals moved to new feeding stations when their lateral neck angle reached a critical aspect, without any detectable change in bite rate (bite size could not be measured under the study conditions). Extended feeding sites encompassed on average 31 consecutive feeding stations in early summer, declining to 12 by early autumn. During early summer, elk moved on to seek a new feeding site after the bite rate they obtained dropped below the seasonal mean. By autumn, patches were abandoned only after this had occurred at two consecutive stations, with the magnitude of the decline being about 10%.

For various wild ungulate species grazing African savanna grasslands, feeding typically lasted 6–12 s per feeding station, during which time 4–10 bites were taken (Underwood 1983). Feeding time per station peaked during the mid dry season, possibly because of the additional manipulation required to extract green leaves from obstructive stems and dry leaves (see also Novellie 1978). The extent of contiguous feeding stations tended to be greater in the wet season than in the dry season, as also did that of the gaps between patches. This suggests a finer-scale mosaic of accepted feeding sites in the dry season, without much change in the overall proportion of stations accepted for feeding.

Ungar *et al.* (1992) suggested that the bite sizes offered within grass swards could differ substantially between grazing horizons (layers within the sward), and that intake rate could change sharply when consumers shifted their feeding from one horizon to the next. The response of cattle to such a situation was investigated by Laca *et al.* (1994a; see also Ginnett *et al.* 1999) using hand-constructed swards. With tall (20 cm height) swards, a substantial decline in bite mass was recorded after a set number of bites had been consumed, but not with short (10 cm height) swards (Fig. 4.4a). Time per bite varied relatively little. Correspondingly, for tall swards the cumulative intake curve showed an abrupt change in slope corresponding with the stage when bite size changed sharply (Fig. 4.4b). However, a linear rather than discontinuous decline in bite mass was found when patches were constituted by natural swards, even when these were quite tall (Laca *et al.* 1994a; WallisDeVries *et al.* 1998) (Fig. 4.4c). This was because overlap between bites caused bite dimensions to decline progressively even within sward horizons. As a result, the slope of the cumulative intake curve decelerated quite slowly (Fig. 4.4d). Cattle grazing natural grassland took on average seven bites

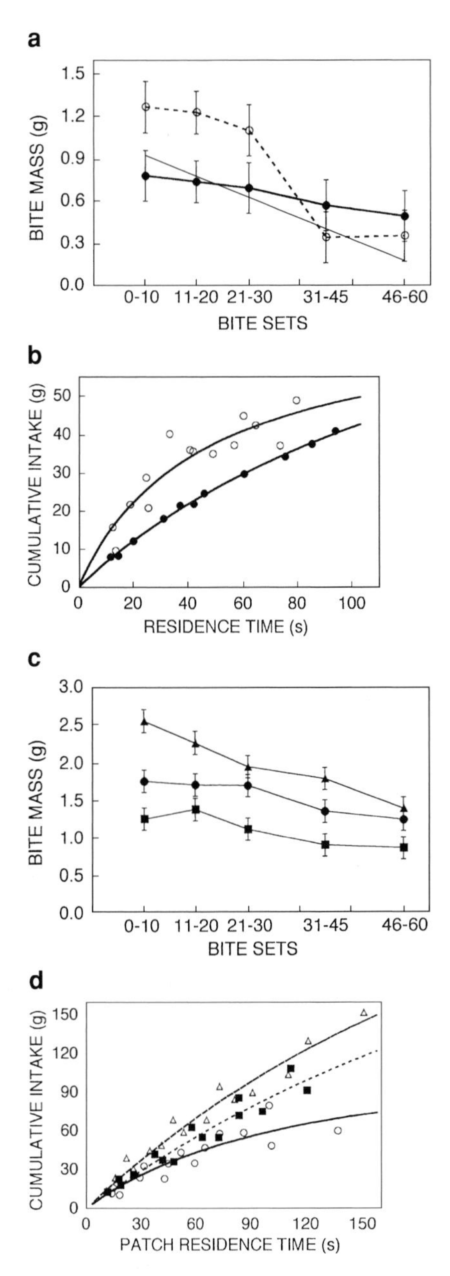

Figure 4.4. Observed changes in bite size and corresponding cumulative intake patterns of cattle related to the depletion of grass sward horizons within patches. (a) Bite mass changes within hand-constructed patches, for two grass heights: 20 cm (open circles) and 10 cm (closed circles). (b) Cumulative intake curves associated with (a). (c) Bite mass changes within grass patches created in the field, for three grass heights: 18 cm (triangles), 12.5 cm (circles) and 7 cm (squares). (d) Cumulative intake curves associated with (c). (From Laca *et al.* 1994a.)

per feeding station, and had not exhausted the top sward horizon when they moved on (WallisDeVries *et al.* 1998). When the spacing between patches was increased experimentally by mowing intervening grass, cattle lengthened their feeding time within patches, qualitatively in accordance with marginal value principles (Laca *et al.* 1993).

Observations on browsers

For browsers, the food patches constituted by trees or shrubs are discretely spaced, so that animals must travel several steps between them. Moose presented with experimental patches formed by cut trees fed typically for about 40–120 s per tree (Astrom *et al.* 1990) (Fig. 4.5a). Feeding was typically terminated after about one third of edible twigs had been consumed (maximum 70%), by which stage intake rate had declined by roughly 30%. Since the travel time between trees was only about 5 s, observed feeding durations were substantially longer than expected from marginal-value principles. The feeding duration nevertheless lengthened with increases in the food biomass on offer.

In a similar experiment, Shipley and Spalinger (1995) presented young moose and white-tailed deer with patches assembled from dormant maple stems in clusters of varying size and spacing. Their data are suggestive of abrupt changes in intake rate, and hence in the slope of the cumulative intake curve, at certain stages of depletion, at least in the upper envelope to the wide scatter of points (Fig. 4.5 b–e). Feeding duration increased with increasing patch spacing, but was much longer than expected relative to travel time. Although the twigs offered a constant food quality and structure, both moose and deer increased their bite sizes as patch size and density declined, in some cases by over 50%. The deer took consistently smaller bites than did the larger moose.

For free-ranging kudus, feeding times at the patches formed by palatable trees and shrubs averaged 50–60 s, but were longer for spinescent species yielding reduced eating rates (Owen-Smith and Cooper 1987b; Owen-Smith 1994). Within this time about 5 g of foliage was consumed, representing roughly 2% of the leaf biomass presented by a shrub 2.5 m tall with most of its canopy within reach. The mean feeding duration per tree tended to increase with increasing tree size, owing largely to the truncation of extended durations for smaller trees (author's personal observations, unpublished). As leaf availability declined over the course of the dry season, feeding times for palatable deciduous species remained relatively constant, while those for evergreen trees lengthened markedly,

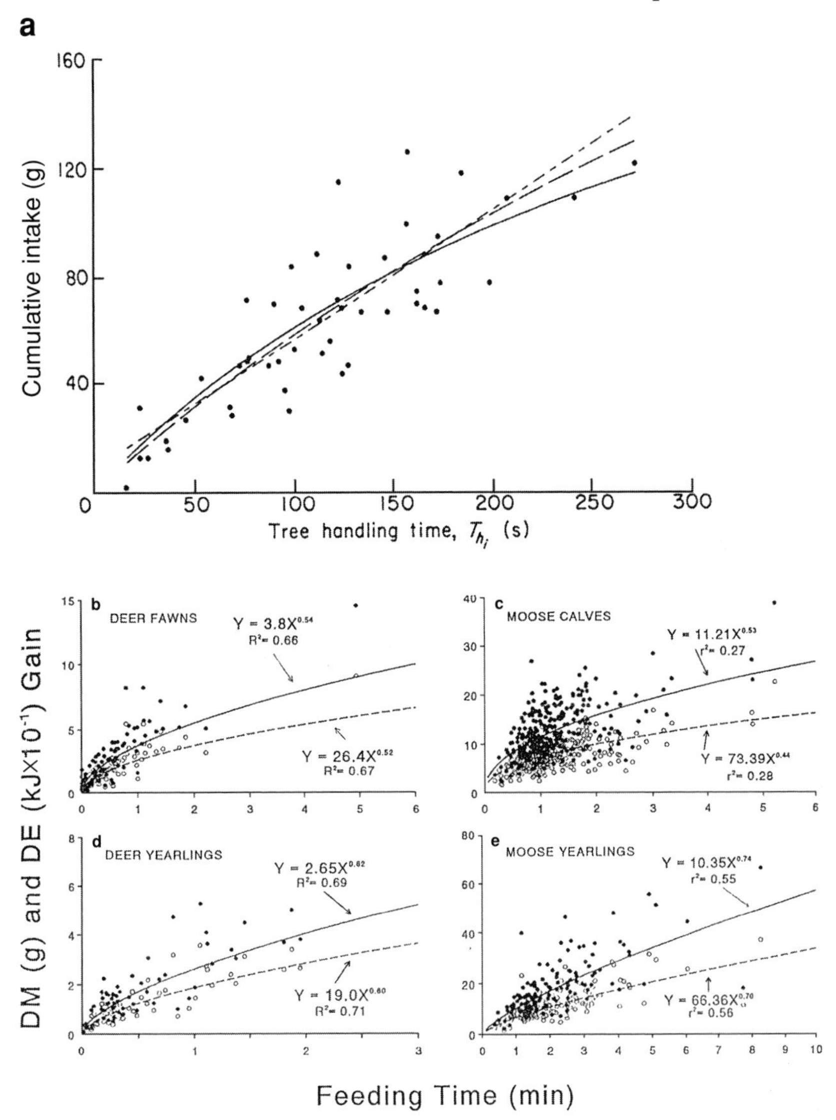

Figure 4.5. Cumulative food intake patterns of browsing ungulates feeding on artificially constructed food patches. (a) Moose feeding on cut trees offering 121–170 g total twig biomass; solid line indicates fitted hyperbolic function, dashed line a linear regression through the data points (from Astrom *et al.* 1990, © Blackwell Science). (b–e). Moose and white-tailed deer feeding on clusters of dormant maple stems (nine per patch) (from Shipley and Spalinger 1995, with permission © 1995 by Springer-Verlag).

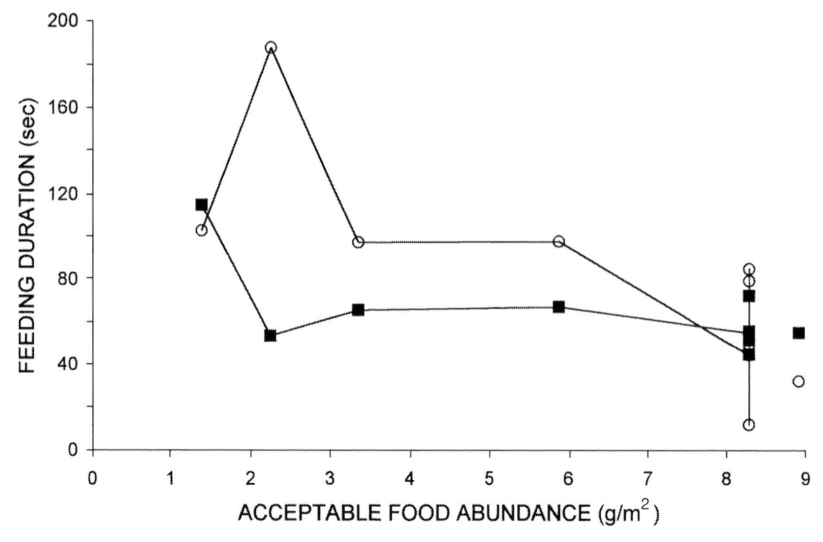

Figure 4.6. Feeding duration per tree by kudus in relation to seasonal changes in acceptable food availability (i.e., leaf biomass remaining on those species included in the diet), comparing palatable evergreen (open circles) with palatable deciduous species (closed squares) (from Owen-Smith 1994).

then declined once their foliage had also become depleted (Fig. 4.6) (Owen-Smith 1994). The pattern indicates a general increase in the fraction of leaf biomass consumed per patch as food abundance declined.

Giraffe exhibited sex differences in patch use, with males feeding at higher levels and on taller trees than females (Ginnett and Demment 1997). Males showed longer feeding durations per patch than females, associated with the greater amount of food persisting at higher canopy levels, but similar travel times between patches. The bite sizes of males were larger than those of females (particularly in the wet season), whereas males exhibited shorter handling times per gram of intake. As a result, male giraffe achieved higher intake rates than females, in accordance with their larger size and hence needs, but at the cost of reduced food quality.

In summary, there is widespread evidence that both grazing and browsing herbivores vary their feeding durations within feeding sites in response to changes in both the amount of food presented within these patches, and the spacing between them. The quantity of food consumed consistently represents only a small fraction of the total amount of potential forage offered within patches. Animals tend to feed in patches for somewhat longer than predicted from simple optimality principles considering time allocation alone.

Foraging areas

Theory

The decision when to leave a depleting patch depends on expectations of potential patch intake rates and travel times between patches in some surrounding region. This will be referred to as the *foraging area,* following Bailey *et al.* (1996) (see also Table 3.1). Some cognitive mechanism is needed to update expectations based on recent rates of intake, and devalue expectations from regions visited in the more distant past. Bailey *et al.* (1996) suggest that animals should periodically sample various sites within the foraging area, to assess the potential returns that are available.

For domestic ungulates, the foraging area may be restricted to the camp or paddock within which animals are confined. Feeding sites within such areas could potentially be exploited systematically, until all have been depleted. However, if foraging movements are somewhat random, an increasing proportion of the patches encountered will have been previously depleted. For objects that are randomly distributed, spacing is inversely proportional to the square root of density (Pielou 1969). Accordingly, the mean distance between ungrazed grass patches increases little until 80–90% of patches have been grazed, but thereafter rises rapidly. At this stage, herbivores should either move to a new foraging area, or widen their patch acceptance. Confined domestic ungulates may simply stand at the gate, waiting to be moved to a new paddock.

Resource depression within foraging areas over broader time scales can likewise be analysed by using principles of marginal value, but without invoking a cumulative intake function. The choice between a current foraging area, labelled 1, and an alternative area, labelled 2, can be evaluated by comparing the immediate gain expected from each, taking into account the travel time required to move between areas:

$$a_1 F_1/(t_{p1} + t_{s1}) \leq a_2 F_2/(t_{p2} + t_{s2} + t_t) \quad (4.4)$$

where F = amount of food offered in acceptable patches, a = proportion of this food consumed before patch departure, t_p = feeding duration per food patch, t_s = time spent moving between food patches, and t_t = expected travel time to reach a new foraging area. Note that the effective intake rate I_P within patches is represented by the ratio between the amount of food consumed per patch ($a\,F$) and the time entailed to consume it (t_p). Because bite size and biting rate tend to decline as feeding time is prolonged, I_P is a variable. However, the magnitude of the reduction in I_P is quite small, and so can be ignored.

If alternative foraging areas are adjacent, travel time between them will be short, and transfers frequent. If a prolonged journey is required to move to a new feeding area, the current area must be depleted quite severely before it becomes advantageous to abandon it. This analysis does not consider the competitive interactions that arise within populations when animals need to weigh up also the consequences of foraging decisions made by others.

Observations

Cattle are rarely found in the same section of large paddocks on successive mornings (Bailey *et al.* 1990). Over periods spanning several days or weeks, animals may shift among sites in the same general region, or move between a set of disjunct foraging areas. By the time all sections have been exploited, sufficient time may have elapsed for vegetation to have regenerated in the localities previously visited, while conditions are favourable for plant growth.

This grazing rotation may be mimicked by livestock ranchers who establish a set of small fenced camps and transfer livestock between these areas at regular intervals. Cattle within such camps began re-grazing tillers of preferred grass species after 60% of the tillers of these species had been cropped, and at the same time increased their acceptance of species of lower palatability (O'Reagain and Grau 1995). After 80–90% of tillers of both preferred and intermediate species had been grazed, they began eating even the least-preferred species. The bite sizes obtained from re-grazed tillers were substantially lower than those obtained at first grazing. Sheep were more selective, grazing the least-preferred species at a later stage than did cattle.

Observations on broad-scale foraging movements by free-ranging wild animals are sparse, so a single example must suffice. Joos-Vandewalle (2000) related the movements of zebra between two regions of their summer range to weekly changes in sward height presented by a favoured annual grass species. During the early growing season, zebra concentrated in the Savuti area, and through their grazing pressure suppressed the growth in height of this grass species (Fig. 4.7). After nine weeks they transferred to the Mababe area, where the grass had grown relatively tall in the absence of much grazing pressure. Over four weeks they depressed the mean grass height from 30 cm to 20 cm, and then returned to Savuti where the grass had been left to grow taller. After having grazed the grass in Savuti down to a height of 20 cm, the zebra moved off to their dry-season range.

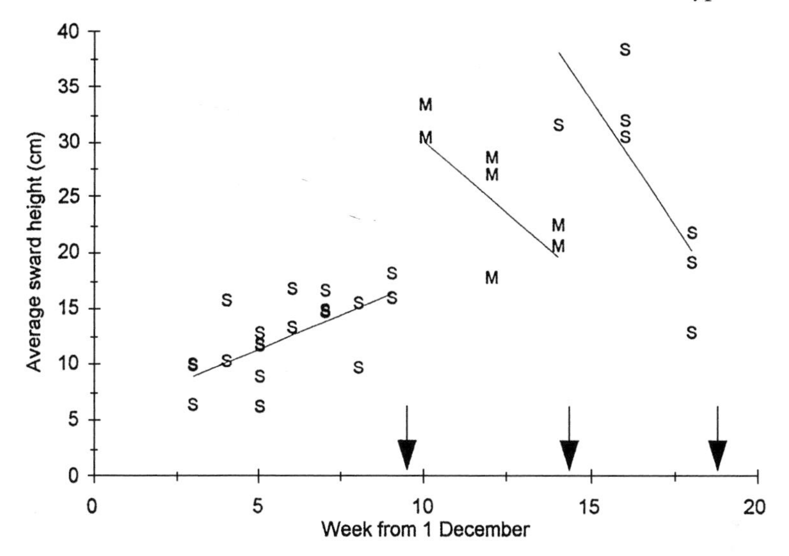

Figure 4.7. Weekly changes in the height of a favoured annual grass species in relation to the movements of zebra between two feeding areas in northern Botswana during the wet season. S, Savuti; M, Mababe; arrows indicate the times of movements between these two areas, and finally the migration away to the dry season range. (From Joos-Vandewalle 2000.)

Habitat types

Over longer time periods, herbivores may shift between *habitat types* that are fundamentally distinct in the resource attributes they present. For example, grazing ungulates in the western Serengeti region move between short grass on uplands during the rainy season and taller grass in bottomlands during the dry season, in a sequence governed largely by relative body size (Bell 1970). Within the Serengeti ecosystem, other populations of these same species migrate between semi-arid short grass plains in the south-east during the wet season and much taller grass growing in the north where the rainfall is substantially higher during the dry season. White rhinos in the Umfolozi Game Reserve shifted their grazing location from the short-grass grasslands favoured in the wet season, towards taller grasslands offering higher food biomass during the dry season (Owen-Smith 1988). By the late dry season, when much of the taller grassland on flatter areas had been grazed down, white rhinos moved up into hilly areas retaining tall grass reserves.

Habitat types may be distinguished also by differences in woody canopy cover or species composition of the vegetation. Observations in the

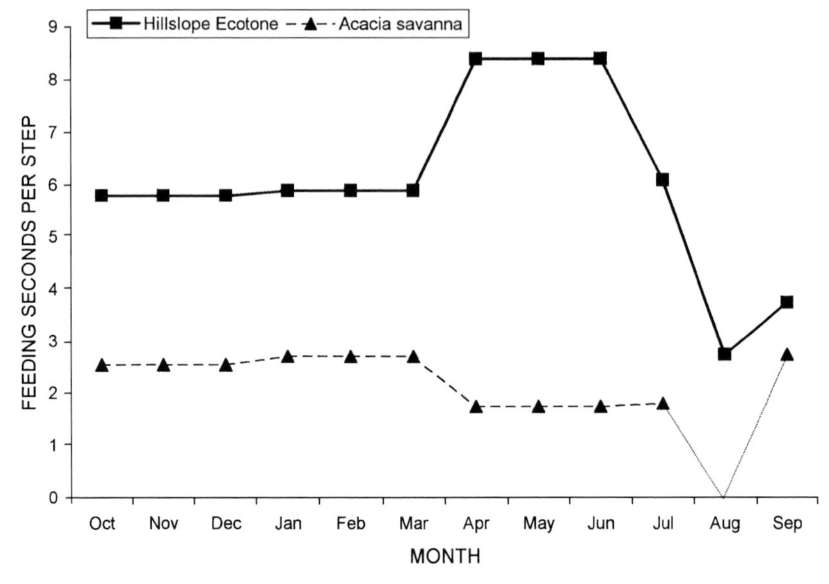

Figure 4.8. Seasonal changes in acceptable food abundance, assessed by the feeding time obtained per step made while foraging, in two habitat types used by kudus in the Kruger National Park. (From Owen-Smith 1979.)

Kruger Park showed that the hill base ecotone consistently yielded higher foraging returns for kudus than the adjoining plains, as indexed by the feeding time obtained per step taken while foraging (Owen-Smith 1979) (Fig. 4.8). Nevertheless, during the wet season the kudus foraged fairly evenly through both habitats, possibly owing to the abundance of high-quality forbs and creepers in the herbaceous layer of the plains habitat. During the dry season, when the feeding time per step in the plains habitat declined to under 2 s, and hence the proportion of foraging time spent feeding to less than 60%, this habitat was largely abandoned. Notably, the feeding time per step in the hill base ecotone rose substantially during the early dry season, largely because the kudus widened their diet to include a higher proportion of the woody species still retaining foliage in this region.

For kudus observed at Nylsvley, *Acacia*-dominated regions yielded lower rates of food intake from the woody vegetation, but higher rates of intake from the forb component of the herbaceous layer, than the predominant *Burkea* savanna (Owen-Smith 1993a). The *Acacia* habitat was favoured in the wet season, but used little during the dry season after most forbs had withered away (Owen-Smith 1993b, 1994).

Although the focus of this chapter is on the effects of food distribution, it must be recognized that habitat use can be influenced additionally by many other factors. These include predation risk, proximity to water, and protection from weather extremes. A broader consideration will be developed in Chapter 11, including also the effects of competitive feeding.

Waterpoint restrictions

Besides food, a major restriction on the distribution of many ungulate species in arid and semi-arid regions is water availability for drinking. Surface water is generally widely available during the rainy season, but becomes restricted to major rivers, scattered pools along seasonal watercourses and other pools that may persist in drainage depressions. Particularly for domestic ungulates, but increasingly also for wild ungulates, natural water sources may be augmented by artificial waterpoints in the form of dams or boreholes supplying drinking troughs. Grazers tend to be strongly dependent on drinking water, but browsers less so because tree foliage and succulents tend to retain higher moisture contents than grasses.

The dependency of animals on access to water for drinking during the dry season restricts their use of food resources. Vegetation growing close to water becomes depressed in abundance and transformed in composition as a result of the grazing, trampling and fertilizing influences of animal concentrations. This radially concentrated impact around waterpoints has been termed the piosphere effect (Lange 1969; Andrew 1988). The consequences for vegetation pattern have been reviewed by James *et al.* (1999) for arid Australia, and more generally by Thrash and Derry (1999). Dry-season distribution patterns of ungulates relative to water were documented for the Amboseli region of Kenya by Western (1975). Some 80–90% of the populations of water-dependent grazers occurred within 4 km of water.

Where water is restricted to isolated points in the landscape, a situation that has been termed central-place foraging (Orians and Pearson 1979; Stephens and Krebs 1986) is generated. This has been widely studied for birds provisioning young in the nest, and for rodents sallying out from refuge burrows. Consumers must trade the reduced travel costs of foraging close to the central point against the greater benefits offered by food that is less depleted further away from this locality. The situation is essentially one of a choice between foraging areas, where the travel costs depend on location relative to the central place. Superimposed on this is a situation of progressive depletion of food resources concentrated radially towards the central point. Hence, the formal structure of the choice situation is essentially that represented by Equation 4.4; only the context differs.

Owen-Smith (1996a) offered some guiding principles for artificial waterpoint provision in the context of wildlife management within extensive protected areas. However, I am not aware of any studies that have quantified the costs and benefits to herbivores of foraging at different distances from water over the course of the dry season.

Selectivity in foraging

Acceptance coefficient

As outlined above, large mammalian herbivores selectively exploit spatial heterogeneity in vegetation at various scales, thereby enhancing the rate of food intake that they obtain. As a consequence, the amount of food effectively available for consumption may be substantially less than the standing crop of potential food in the environment, at least in the short term. Over longer time periods selectivity tends to be widened, so that an increasing fraction of the edible vegetation biomass becomes consumed. Such selectivity transforms the standing crop of potentially edible and accessible food, F, into the effectively available food biomass, aF, where a is the acceptance coefficient. The vegetation components rejected as food do not influence the effective rate of food intake.

Browsing kudus commenced feeding on only about one third of the specimens of acceptable plant species that they encountered within neck reach while foraging (see Fig. 4.1). They abandoned these patches having consumed as little as 2% of the foliage available within their height reach, suggesting an effective acceptance coefficient of less than 0.01. However, a somewhat higher estimate is obtained if the amount of food consumed per foraging step (the product of feeding time per step and eating rate) is related to the standing biomass of acceptable foliage within the foraging habitat During the wet season, the effective acceptance coefficient for kudus calculated in this way averaged 0.03 for woody plant foliage and 0.06 for herbaceous browse (forbs), considering only movements between food patches (Owen-Smith 1993a). In the dry season, when fewer leaves remained, the corresponding acceptance coefficient for woody foliage increased to around 0.1 (Owen-Smith 1994). The discrepancy between the alternative estimates can be explained partly by the additional feeding movements that take place within food patches (i.e. tree environs), and partly by the elevated encounter rates with favoured food species as a result of directed orientation of foraging movements.

A medium–large grazing ungulate moving about 0.5 m per step, and scanning a path width extending about 0.5 m on either side of its feet, searches an area of approximately 0.5 m^2 per step. This area would present a standing crop of available food amounting to somewhere between 50 and 500 g during the growing season (cf. Figs. 3.4 and 3.6). With bite mass averaging around 2.0 g for a 700 kg steer, and typically around 5–10 bites taken per feeding station, about 0.04–0.2 of the available forage would be removed per feeding station by a cow, and somewhat less by a smaller ungulate. Allowing also for potential feeding stations that are rejected as food, the effective acceptance coefficient would be reduced even further.

The above estimates are short-term acceptances applicable over the duration of a foraging spell. Over extended time periods when vegetation is non-renewing, and patches become re-visited, the cumulative fraction of vegetation consumed will expand progressively. By the end of the dormant (winter or dry) season, most of the edible and accessible material may have been consumed.

The consequence of the narrowing in acceptance with increasing food availability (and vice versa) is to extend the plateau region of the intake rate response (Fig. 3.8). Indeed, both bite size and time apportionment within and between food patches may be adjusted so as to maintain the overall intake rate around some target level. However, this target rate may vary somewhat with changing food quality, as will be outlined in Chapter 5.

Optimality approximations

Observations show that herbivores generally prefer to feed in those patches that offer higher eating rates, and feed for longer in patches that are larger or more widely spaced. In this sense their behaviour is qualitatively in accordance with optimality principles. However, in contrast to the predictions of simple optimality models, herbivores consume food types offering less favourable intake rates to some extent, and tend to feed for more prolonged periods in patches than expected from marginal-value accounting. A wider range of habitats may be used than just those that are best in terms of feeding returns. This is a widespread finding: animals are generally less narrowly selective than predicted from naive optimality theory.

Numerous reasons have been advanced for the discrepancy between optimality predictions and observed behaviour. Animals lack information to make precise choices, have to cope with random variability in time and space, must take into account other influences than those considered in the model, or are governed by outcomes over longer temporal scales than those considered. Optimality analysis identifies the upper limit to potential performance, given the assumptions made in the model. Real animals fall short of this ideal to varying degrees, but nevertheless perform better than would a naive consumer unresponsive to environmental heterogeneity. Hence, whatever their shortcoming, optimality principles provide a useful guideline for identifying both the upper bound to performance, and the direction of the adaptive responses that are likely in particular circumstances.

An alternative concept borrowed from psychology theory is that animals respond not by narrowly optimizing, but rather by *matching* their pattern of use to the rewards provided (Staddon 1983). Thus options that offer lower rewards are adopted less frequently, in a proportion corresponding approximately with the relative value of the rewards obtained. A reward-matching pattern could simply be an outcome of momentary maximizing of gains (Stephens and Krebs 1986). However, a graded correspondence between the effective intake rates offered by alternative food patches and the frequency with which these patches were exploited was reported for red deer by Langvatn and Hanley (1993). A matching rule may be useful for describing the outcome of selection in complexly varying environments, although not capturing the mechanisms underlying this outcome.

Body size influences

Because of the metabolic rate – body mass relation (see Chapter 7), smaller herbivores should be more narrowly selective for specific plant parts, whereas large herbivores should be broadly tolerant of a wide range in forage quality. Jarman (1974) proposed that, as a consequence, small antelope species should spend much of their foraging time travelling between widely separated feeding sites, whereas large grazing bovines in particular should generally feed almost continuously. Moreover, smaller animals take longer to travel the same distance than larger species because their step length tends to be shorter. Accordingly, the ratio of feeding to moving time should increase with increasing body mass.

Comparative observations on the feeding behaviour of various ungulate species by Underwood (1983) provided inconsistent support for this expectation. Steenbok, a small mixed feeder, fed for little time per feeding station, rejected many potential stations, and thus spent much time searching relative to feeding. In contrast, reedbuck, the smallest strict grazer observed, fed intensively for long periods within particular localities with little movement. Buffalo, the largest grazer, rejected few feeding stations, fed for a high proportion of foraging time, and appeared to be selective mainly for habitat type; but so also did tsessebe, a medium-sized antelope less than half the size of buffalo. Species patterns are evidently dependent on the fraction of the available vegetation that constitutes acceptable food in particular patches and habitats.

Overview

Spatial heterogeneity in resource distribution creates opportunities for selection at four spatial scales: (1) bite-sized clusters of plant parts; (2) feeding sites, formed by sets of adjacent feeding stations; (3) foraging areas representing regional aggregations of food patches; and (4) habitat types, constituted by vegetation that is structurally or compositionally distinct in resource attributes and other features. Heterogeneity in bite sizes offered may be the main influence on selection, provided differences in food quality are small. Feeding sites may be selected if they offer larger-than-average bites, and are abandoned once intake rate declines below some level as a consequence of local food depletion. Variation in the form of the cumulative intake curve may influence the duration of feeding within a patch. Because movements between food patches occupy only a small fraction of the foraging time of large herbivores, a relatively minor decrease in the eating rate obtained within a patch could be sufficient to prompt patch departure. Small chance effects on bite sizes or biting rates could underlie the observed wide variability in feeding durations within patches.

Grazers experience a relatively even patch distribution, extended in a horizontal plane, and browsers a more aggregated food distribution with a substantial vertical dimension. Accordingly, movements between food patches occupy a higher proportion of the foraging time of browsers than of grazers. Larger ungulates experience a more even food distribution at both patch and habitat scales than smaller species, unless food abundance is so strongly depleted that quantity rather than quality is limiting. Relatively little attention has been paid to factors governing the movements of herbivores between foraging areas, over periods when resources are effectively non-renewing. Transfers by animals between habitat types may be influenced by additional factors besides resource gains.

As a consequence of selective feeding responses to various scales of patchiness, only a fraction of the available food is accepted, and hence contributes to intake rate, during foraging spells. Nevertheless, over the course of time the accepted fraction may be expanded to compensate for declining food abundance. Such adaptive responses accentuate the plateau region of the intake response to changing food availability.

Further reading

The standard reference on foraging theory is the book by Stephens and Krebs (1986). Useful reviews of the consequences of spatial heterogeneity for herbivore foraging behaviour are provided by Coughenour (1991), Laca and Demment (1991, 1993), Bailey *et al.* (1996) and O'Reagain (2001).

5 · *Resource quality: nutritional gain and diet choice*

For herbivores, the nutritional gain depends not only on the rate of food intake, but also on the *quality* of the material consumed. Accordingly, in this chapter the focus shifts from the intake response to the biomass gain response, represented by the function G in Equation 2.14, and hence to factors determining the value of the conversion coefficient c from plant to herbivore biomass.

The nutritional quality of herbage consumed is governed by its composition, firstly in terms of the proportion of cell wall fibre relative to cell contents, and secondly by concentrations of protein, soluble carbohydrates, mineral elements and other nutrients in the cell contents. These factors, as well as the composition of the cell wall, particularly its degree of lignification, determine the potential nutritional yield in the form of metabolizable energy and material substrates. However, the cell wall content also slows down the rate of digestive processing, restricting the rate at which these energizing and material nutrients become available, particularly for ruminants. In addition, many plant species contain various secondary chemicals, which may function as potential toxins, or impede digestion in various ways, reducing nutrient availability. Diet selection thus becomes largely a matter of balancing different nutritional benefits against various costs and rate restrictions. Immediate gains may need to be balanced against longer-term consequences. Quality as governed by nutritional yields must be traded against quantity as determined by the rate of intake obtained, over different time frames.

Herbivores are confronted by a wide variety of potential food items in the form of various plant species and parts. These *food types* are distinct in the nutritional properties that they present, as well as in the rates of food intake that they yield. A single feeding station may present more than one food type, among which animals must choose. For modelling to be tractable, the multiplicity of plant species and tissues needs to be aggregated into a more limited set of distinct food classes. Categoric distinctions in nutritional properties exist between grasses and dicotyledonous plants, and between herbaceous and woody browse in the latter category, as well as between plant parts such as leaves, stems, fruits and flowers. Within these categories, some groups of plant species appear quite similar in their nutritional value, as revealed not only by chemical analyses but also through the responses of herbivores to them. This enables species to be grouped into palatability classes, based on the likelihood that they will be consumed when encountered by a foraging herbivore, under different conditions of food availability (Owen-Smith and Cooper 1987a).

In this chapter, we consider how diet selection models based on optimality models need to be modified to take into account differences in nutritional value among the plant types and parts potentially available as food for large herbivores. A central issue is how immediate rates of gain during foraging spells need to be counterbalanced against daily gains influenced additionally by digestive constraints. The problem faced by a hungry herbivore is how narrowly selective, or broadly tolerant, to be when faced with a superabundance of potential food, most of which is of low nutritional quality. The inherent nutritional quality of a food type has to be weighted against its effective value in terms of the rate of biomass gain that it yields. Additional factors have also to be considered, such as what other food might be available, and the changes in food quality as well as food abundance that take place over the seasonal cycle of plant growth.

Diet breadth model

Basic contingency model

The classical 'contingency' or *diet breadth* model of optimal foraging theory considers how the effective nutritional gain changes as the range of food types included in the diet is widened (Stephens and Krebs 1986). To identify the optimal diet, these food types must be ranked from best to worst in terms of their effective value. The immediate or short-term value is governed by the nutritional yield (as digestible energy, or any other target nutrient), relative to the handling time entailed. From the perspective of the nutritional gain function, G, this is effectively the within-patch eating rate, I_P, multiplied by the conversion coefficient, c, from food to consumer biomass (cf. Equations 2.17 and 2.18). In turn, I_P is the quotient of the bite mass b_m and handling time per bite h_b (cf. Equation 3.4). Hence, for any particular food type i, we have

$$G_i \propto c_i b_{mi} / h_{bi}. \tag{5.1}$$

Additional handling time besides that involved in food ingestion is ignored. For generality, G should be expressed as a relative rate per unit of herbivore biomass, meaning that the bite mass b_m should be scaled as a proportion of herbivore biomass.

Notably, the conversion coefficient c is a composite measure of various influences restricting the fraction of the food consumed that becomes transformed into consumer biomass. It takes into account the fraction of the ingested material that is digestible, biomass losses via added secretions and inevitable attrition of gut lining during food passage, and the fraction of absorbed products assimilated into body tissues. In practice, it also accommodates the conversion from the dry mass units in which plant biomass is usually measured, into live herbivore biomass.

Applying the principle of opportunity cost, the gain G must be compared with the potential gain that could have been obtained had the same amount of time been spent seeking out and consuming an alternative food type. Furthermore, it is assumed that food types are interspersed, so that search time is shared in common. The choice is thus whether to commence feeding on a food type immediately available for consumption, or move on to seek out a better food type. Obviously it is advantageous to consume the best food type whenever it is encountered, so the question becomes how many additional food types should be eaten. Restricting consideration to two food types for simplicity, choice depends on assessing the inequality

$$c_2 b_{m2}/h_{b2} \geq c_1 b_{m1}/(t_{s1} + h_{b1}), \tag{5.2}$$

where t_{s1} = expected search time to find the better food type, with food types subscripted in rank order of their effective value. If the higher-value food is sufficiently sparse, i.e. search time to find it is sufficiently lengthy, it becomes worthwhile for the lower-value food type to be included in the diet.

Equation 5.2 can be generalized to any number of food types, by comparing the nutritional gain from a particular food type with the averaged rate of gain obtained by consuming all superior food types:

$$c_i b_{mi}/h_{bi} \geq \underline{c}\,\underline{b}_m/(\underline{t} + \underline{h}_b), \tag{5.3}$$

where the underscores denote average values, and i indexes the rank order of value. Note that a lower-quality food type may be favoured over a higher-quality one if the eating rate that the former offers is sufficiently high.

The averaged rate of gain from all food types consumed, allowing for search time, is the effective rate of biomass gain achieved during foraging spells, G_F. Adapting Equation 3.6, for any particular diet breadth $i = 1 \ldots r$ we have

$$G_F \propto s \sum_{i=1..r} c_i a_i F_i \bigg/ \left(1 + s \sum_{i=1..r} a_i F_i / I_P\right), \tag{5.4}$$

where s = search rate (area searched per unit time), F_i = food density (mass per unit area) for food type i, a_i = acceptance fraction for food type i, and other parameters are as defined for Equation 5.1.

The corresponding food intake rate, I_F, is obtained by omitting the conversion coefficient c_i, i.e.

$$I_F \propto s \sum_{i=1..r} a_i F_i \bigg/ \left(1 + s \sum_{i=1..r} a_i F_i h_{bi}\right). \tag{5.5}$$

The food intake rate increases asymptotically with widening diet breadth r, because the accepted food biomass increases correspondingly (Fig. 5.1).

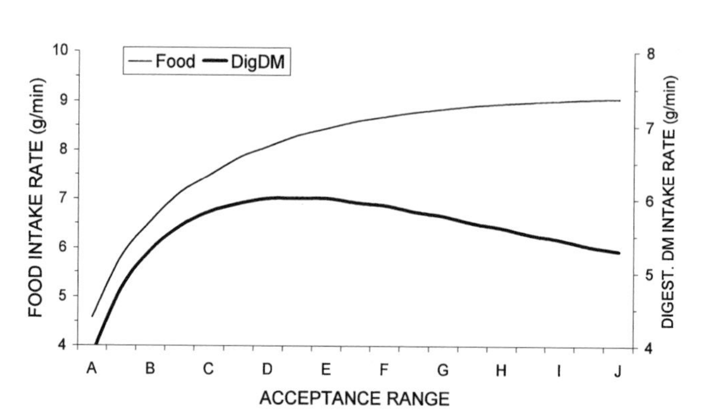

Figure 5.1. Output of the diet breadth model, showing how food and nutrient (digestible dry matter) intake rates obtained during foraging activity change with expanding dietary acceptance range. For each successive food type ranked A–J, quality c is reduced by 0.85 times while availability F is increased by 1.5 times, with handling time h held constant. (a) Basic conditions, indicating an optimal diet breadth up to and including food type D. (b) Abundance of each food type increased five-fold, the optimal diet breadth now including only food types A and B. (c) Abundance of all food types reduced to one fifth, with the optimal diet breadth spanning all food types through H.

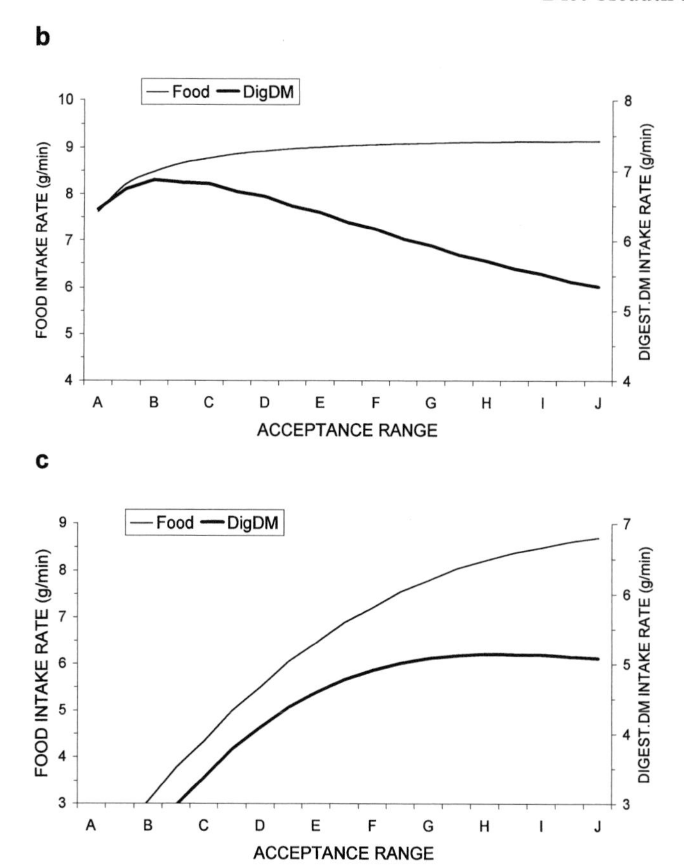

Figure 5.1. (cont.)

The intake response to widening food acceptance is gradually asymptotic if poorer-quality food types are much more abundant than better food types, as might be expected to be the usual situation. In contrast, the resultant biomass gain, G_F, peaks at some intermediate diet breadth, because the nutritional yield declines as the diet is widened to encompass poorer-quality vegetation components. The optimal dietary range (that maximizing G_F) depends on the effective food availability, or more specifically the abundance of the preferred food types (Fig. 5.1b, c).

Ingestion vs. digestion

The classical model considered only the handling time associated with food capture and ingestion, with the former being inconsequential for herbivores. However, faced with a superabundance of potential food mostly of low quality, herbivores can potentially consume food faster than they can digest it. In these circumstances, digestive processing time can become an overriding constraint on the daily food intake, and hence on the effective rate of nutritional gain. Accordingly, in modifying the classical model for large mammalian herbivores, Owen-Smith and Novellie (1982) incorporated the digestion rate limitation, as well as allowing for other restrictions on daily foraging time. Plant types were ranked simply in terms of the nutrient contents that they offered in their leaves, indexed by crude protein concentrations. Because protein concentrations are determined largely by the cell contents:cell wall ratio, and cell wall fibre governs digestion rate, this was effectively the rank order of nutritional yield relative to digestive processing time. No information on the differences in the eating rates offered by different species was available.

Verlinden and Wiley (1989) pointed out that, where digestive capacity is the overriding constraint, food types should be ranked according to their nutritional yield relative to digestive processing time, rather than handling time during ingestion. Food that can be eaten faster merely fills the gut sooner, and animals must wait for these foods to be digested before they can consume more food. However, Hirakawa (1997) noted that there may nevertheless be times when food intake rate becomes the main restriction on nutritional gains, e.g. when food quality is very high so that the digestion constraint is alleviated, or when daily foraging time is greatly restricted by adverse conditions. In these circumstances, the preference ranking of food types could appear somewhat whimsical, as animals respond to shifting constraints. This may, however, involve only those food types that offer extremely high or low eating rates.

Accordingly, maximizing the daily nutritional gain becomes a matter of balancing the short-term rate of gain obtained while foraging, relative to the potential daily foraging time, against the longer-term constraint imposed by digestive processing capacity (Fig. 5.2). As the diet is widened, food quality diminishes, causing the digestive turnover rate to lengthen. This process is governed largely by the ratio between rapidly digesting cell contents, and slowly digesting cell wall constituents, as will be discussed in Chapter 6. When both food quality and general food abundance are high, the intersection may occur to the right of the peak in nutrient intake rate, in which case digestive capacity does not influence the optimal diet (Fig. 5.2a). With reduced food abundance, the digestion constraint may come into operation, owing to expansion of the optimal diet towards lower-quality, and hence less digestible, food types (Fig. 5.2b). The optimal diet breadth is then indicated by the intersection of the ingestion and digestion constraints.

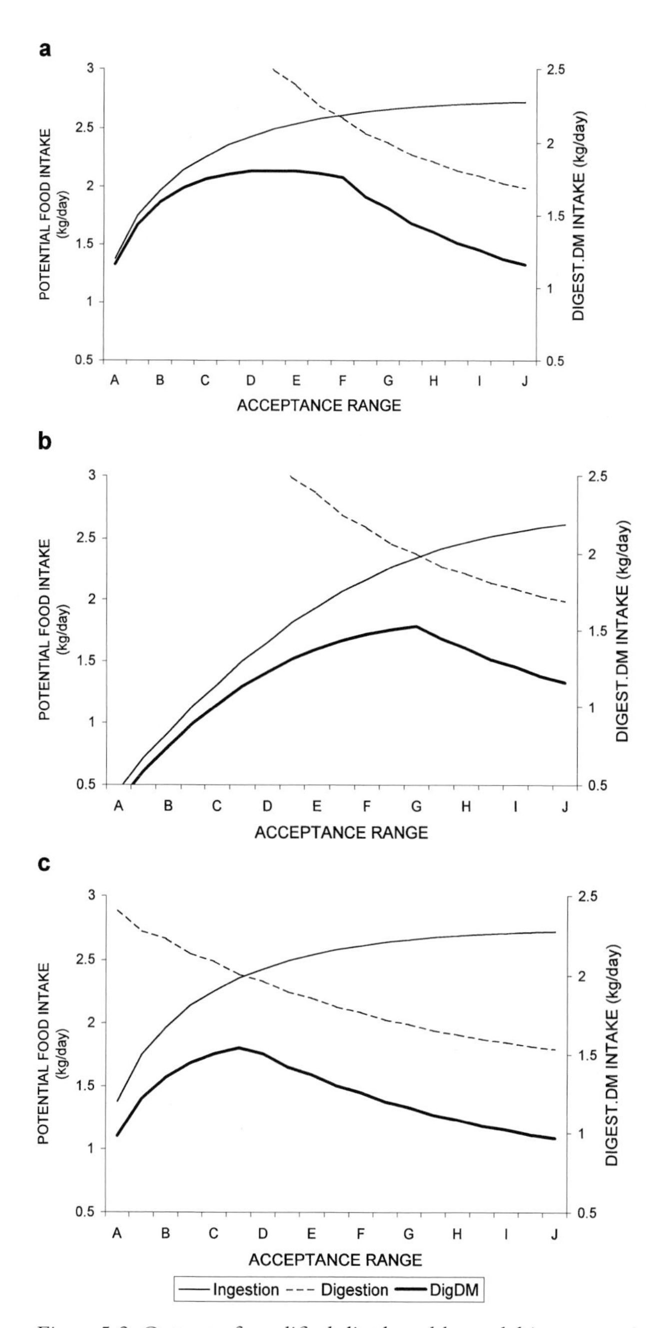

Figure 5.2. Output of modified diet breadth model incorporating a daily foraging time limitation as well as a digestive capacity constraint. Food types are ranked in quality (effectively cell content:cell wall ratio), with no allowance made for eating rate differences among food types. (a) Basic model with values for intake parameters identical to those used for 5.1a. (b) Abundance of all food types reduced to one fifth. (c) Quality of all food types reduced by one third.

A general decrease in food quality could also lead to a digestive capacity limitation, in this case associated with a narrowing of the optimal diet (Fig. 5.2c).

Linear programming (LP) presents an alternative modelling approach for balancing the constraints imposed by food intake rate, daily foraging time and digestive processing capacity (Belovsky 1984, 1986). Its main limitation is that only a few food types can be handled effectively. For the constraints to be linear, food types must be sought in alternative places, otherwise the constraint lines become broken and solutions difficult to obtain. Hence, it is more appropriate for treating selection between broad food categories, such as graminoids and dicotyledons, available in distinct habitat patches. It does not accommodate the choices that herbivores may make among the numerous plant species and parts presented within each of these categories. Moreover, it is doubtful that herbivores do select at the crude level of such categories, given the wide variation in nutritional properties and eating rates that exists among the species constituting these categories. Other problems associated with the LP approach are discussed in Chapter 6.

Palatability classes

A fundamental problem faced by the diet breadth model (DBM) is how to accommodate the multiplicity of plant species presented to large herbivores as potential food. For example, the 213 ha enclosure of savanna vegetation at Nylsvley where we studied the foraging behaviour of browsing ungulates contained 60 woody plant species, over 100 forb species, and 20 or more grass species, most of which were consumed by one or other herbivore at some time or other. Moreover, fruits were quite distinct nutritionally from foliage. Many of these plant species occurred in low abundance, and so contributed little to herbivore diets.

Our resolution of this problem was to group woody plant species into *palatability* classes, based on the observed feeding responses of kudus and other browsing ungulates to them (Owen-Smith and Cooper 1987a, 1989). During any restricted period, species could be classified either as highly acceptable food (plant-based acceptance frequency > 0.25, site-based acceptance frequency > 0.5), or generally rejected (acceptance frequency < 0.1) (see Owen-Smith and Cooper 1987a,b for the source of these cutoff limits). By definition, *palatable* species remained consistently highly acceptable throughout the year, as long as they retained leaves (Fig. 5.3a). Species that were usually rejected, except perhaps during

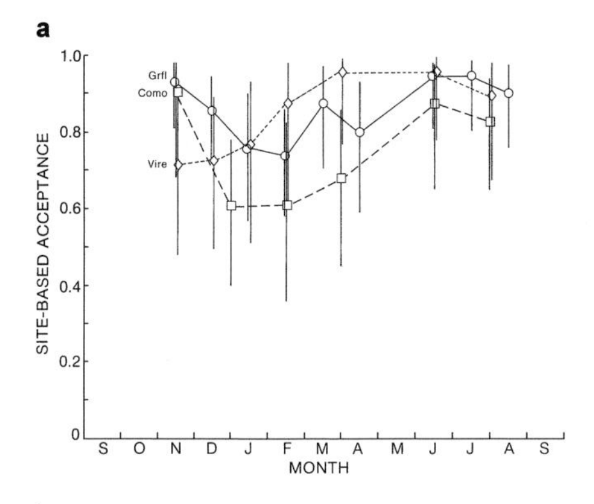

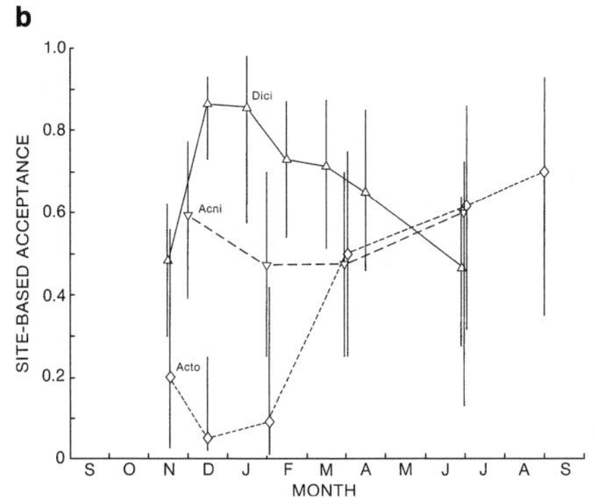

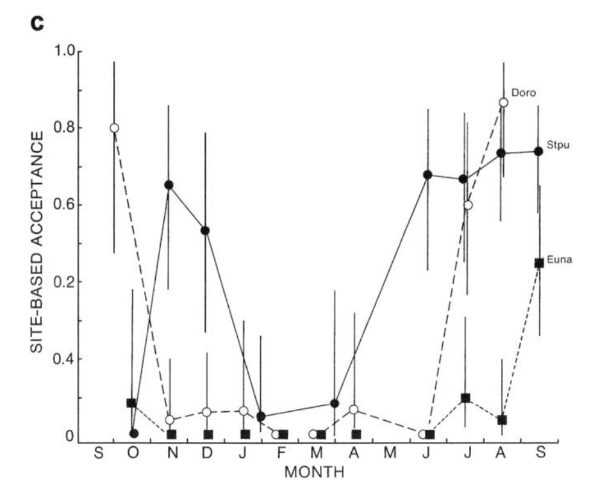

Figure 5.3. Monthly acceptance of representative woody plant species by kudus. The summer growing (wet) season extends from October to March or April, and the winter dormant (dry) season from May to September (but with some woody species showing a pre-rain leaf flush during September). Lines indicate 95% binomial confidence limits. (a) Palatable deciduous species: Grfl, *Grewia flavescens*; Como, *Combretum molle*; Vire, *Vitex rehmannii*. (b) Palatable spinescent species: Dici, *Dichrostachys cinerea*; Acni, *Acacia nilotica*; Acto, *Acacia tortilis*. (c) Unpalatable species: Stpu, *Strychnos pungens* (a relatively palatable evergreen); Doro, *Dombeya rotundifolia* (an unpalatable deciduous species); Euna, *Euclea natalensis* (an unpalatable evergreen). (From Owen-Smith and Cooper 1987a.)

restricted periods of the year, were termed *unpalatable* (Fig. 5.3c). Certain species, mostly spinescent, showed intermediate acceptances, which were ascribed to low eating rates rather than to unpalatability (Cooper and Owen-Smith 1986) (Fig. 5.3b). A further functional distinction was made between deciduous and evergreen species, on the basis of leaf retention through the dry season. This yielded five food classes for woody plants, each comprising 1–5 common species plus numerous rare species: (1) palatable deciduous unarmed, (2) palatable spinescent, (3) unpalatable deciduous, (4) relatively palatable evergreen, and (5) unpalatable evergreen.

Forbs were less clearcut in their acceptance patterns, and so were subdivided simply into either herbaceous (creepers, annuals or soft-stemmed perennials) or robust (forbs with lignified stems and dwarf shrubs) classes. Grasses, fruits and flowers formed an additional composite category, not sub-divided further by species. Through this procedure, 180+ plant species were reduced to 10 functionally distinct food classes.

The palatability classes showed corresponding differences in measures of effective nutritional quality (Cooper *et al.* 1988) (Fig. 5.4).

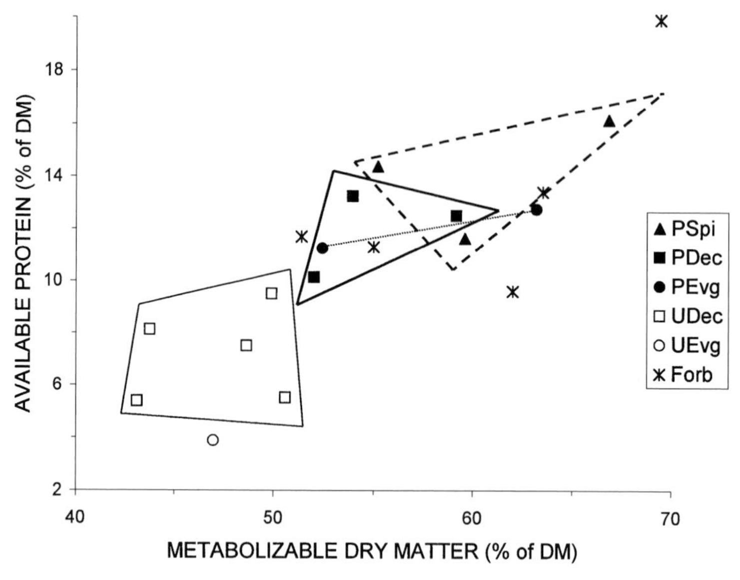

Figure 5.4. Nutritional value of woody plant species representing particular palatability categories and selected forb species, as indicated by leaf concentrations of metabolizable dry matter (cell solubles plus digestible cell wall minus unmetabolizable phenolics) and available protein (nitrogen × 6.25 minus condensed tannin × 0.5).

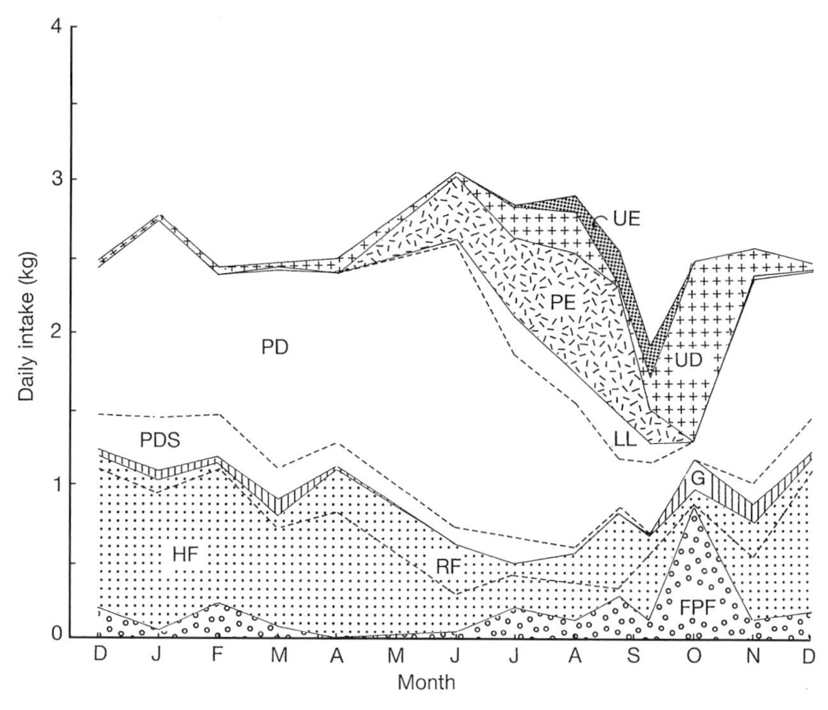

Figure 5.5. Contributions of different plant categories to the monthly dietary intake of kudus through the seasonal cycle. PD, palatable deciduous unarmed woody species; PDS, palatable deciduous spinescent woody species; PE, relatively palatable evergreen species; UD, unpalatable deciduous woody species; UE, unpalatable evergreen; LL, woody leaf litter; G, grasses; HF, herbaceous (soft-stemmed) forbs; RF, more robust (woody) forbs; FPF, fruits, pods and flowers. (From Owen-Smith and Cooper 1989.)

Spinescent woody species, plus certain forbs, yielded the highest nutrient concentrations. The main factor distinguishing unpalatable woody species was a high content of condensed tannins relative to protein, reducing protein availability as well as metabolizable energy yield. Somewhat intermediate in nutritional value were the palatable deciduous, but unarmed, woody species, together with relatively palatable evergreens and more robust forbs.

For kudus, palatable deciduous woody plants plus soft-stemmed forbs provided the bulk of the diet during the wet-season months (Fig. 5.5). Palatable evergreens become a substantial component during the course of the dry season, and certain unpalatable deciduous species that produced their new leaf flush before the rains made an important contribution during this transitional period. Fleshy fruits and leguminous pods made

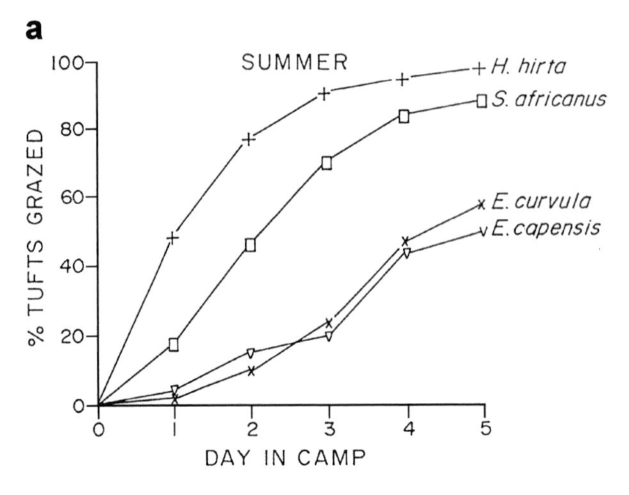

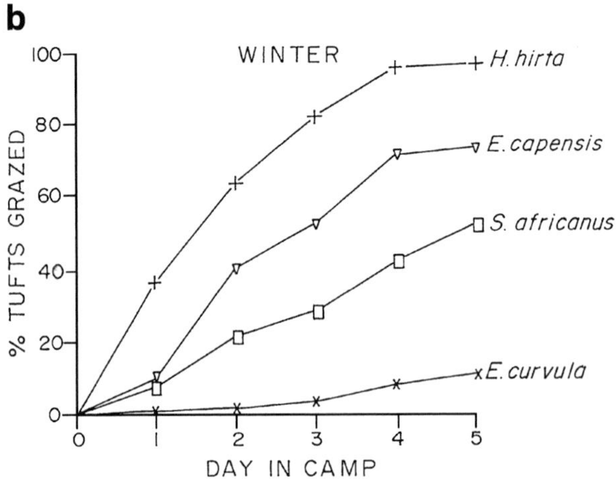

Figure 5.6. Cumulative utilization of tufts of particular grass species by cattle in relation to days of occupation in a grazing camp. (a) Summer (wet) season, comparing patterns for high (*Hyparrhenia hirta*), intermediate (*Sporobolus africanus*) and low (*Eragrostis curvula, E. capensis*) palatability species. (b) Winter (dry) season, showing patterns for the same four species. (From O'Reagain and Mentis 1989a.)

significant dietary contributions at times when they became available. Fallen leaf litter was of minor importance for kudus during the dry season, but made a more substantial contribution to the diets of impalas and goats (Owen-Smith and Cooper 1985). Only very small amounts of grass were consumed by the kudus, mainly during the early wet season when green flush became available on burns.

A palatability classification has likewise been used to group grass species, based on the propensity of domestic ungulates to consume them (Mentis 1981; Meissner *et al.* 1999). A rank order of preference is apparent from the sequence in which grass species are grazed during periods when cattle are held within restricted paddocks or camps (Danckwerts *et al.* 1983; O'Reagain and Mentis 1989a; O'Reagain and Grau 1995) (Fig. 5.6). However, for grasses, species favoured during the growing season may show decreased acceptability in the dormant season, owing to phenological changes in their leaf:stem ratio, which largely determines nutritional value. A study on eland, which consumed a diet derived largely from woody shrubs and shrublets in the study area, found that food choice depended largely upon plant growth patterns, with palatability categories being less distinct than for kudus (Watson 1997).

Evaluating the diet breadth model

In common with the classical contingency model, the modified diet breadth model predicts that the rank order of preference for different food types should correspond with the effective value of these food types to the consumer. For herbivores, the complication is whether value should be assessed based on ingestive handling time or digestive handling time. Another issue is the appropriate measure of nutritional quality: should it be assessed in terms of digestible energy yield, protein yield, or some other target nutrient? Furthermore, how should the influences of plant secondary metabolites on nutritional value be accommodated? A more stringent test is how well the observed dietary range matches that predicted from the assumed constraints. The classical model predicts, moreover, that the acceptance of a food type should be all or nothing. If a food type is worth consuming, it should always be eaten whenever it is encountered. If not, it should always be rejected. However, as soon as multiple constraints are included, this prediction falls away. The food type saturating digestive capacity should be accepted partially, up to the limit of the gut capacity. Nevertheless, this prediction is rather spurious, and never supported. There are manifold reasons why partial acceptances might arise, as outlined by McNamara and Houston (1987a) and discussed later in this chapter.

Value ranking

For kudus, we found that the acceptance frequencies for woody plant species tended to be either high (plant-based acceptance > 0.25) or low (acceptance < 0.05) during any restricted period (Owen-Smith and Cooper 1987a,b). When new species became incorporated into the diet, their acceptability changed abruptly between months (Fig. 5.3c). As the availability of the preferred deciduous species declined during the dry season, kudus widened their dietary range to include an increasing proportion of unpalatable species. The preference ranking of species was thus evident particularly from the order in which they were added to the diet during the course of the dry season. No clear distinction could be made among the acceptance frequencies shown for the palatable species that were consumed during the wet season, except for the lowered frequencies found for spinescent species offering restricted intake rates (Cooper and Owen-Smith 1986).

There was no correlation between the averaged, year-round acceptance frequencies shown by kudus for particular woody species, and measures of the ingestion rate for any target nutrient (for crude protein, $r^2 = 0.09$; for metabolizable organic matter, indexing energy yield, $r^2 = 0.002$). Palatable species could be discriminated from unpalatable species primarily through their condensed tannin content relative to protein in leaves (Cooper *et al.* 1988). Furthermore, certain of the unpalatable species become temporarily highly acceptable during the period of new leaf growth at the start of the growing season, when protein contents were elevated relative to tannin levels (Fig. 5.3c). Accordingly, much of the variation in acceptability among woody species could be explained by a measure of 'available protein', calculated as crude protein minus half the condensed tannin content (derived from the canonical discriminant axis; $r^2 = 0.472$, $n = 14$, $p = 0.01$) (Owen-Smith 1997). The correlation improved slightly when allowance was made for species differences in digestion rate, controlled by fibre content, largely because one of the evergreen species was shifted closer to the general trend. However, two *Acacia* species remained notable outliers, being only moderately acceptable despite their high protein and low fibre contents.

On account of their thorns, the *Acacia* species yielded eating rates that were much slower than their potential digestion rates. In practice, biomass gains are determined by either the digestion rate or the ingestion rate, whichever is slower. Making this adjustment, a highly significant correlation between value, assessed as effective rate of gain of 'available protein', and annual mean acceptance, emerged ($r^2 = 0.65$, $p < 0.001$), with only one outlying species (excluding this species, $r^2 = 0.82$) (Fig. 5.7). The composite index of value takes into account

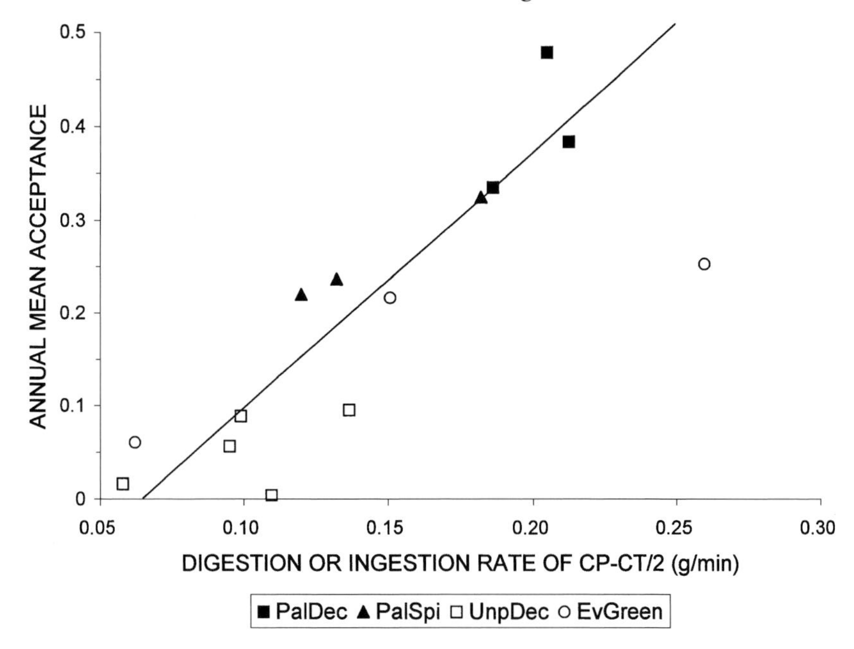

Figure 5.7. Relation between annual mean acceptance of woody plant species for kudus and a composite index of effective value, in terms of the digestion or ingestion rate (whichever is slower) for 'available protein' (crude protein (CP) minus half condensed tannin (CT) content). Symbols differentiate palatability categories. (From Owen-Smith 1994.)

both nutrient (protein) and anti-nutrient (condensed tannin) contents, as well as rate limitations on both ingestion and digestion. Presence of unidentified allelochemicals could explain the evergreen species (family Anacardiaceae) that remained a conspicuous outlier.

The strong relation between the composite index of effective value and woody species acceptances obtained in our study was due largely to the inclusion of several highly unpalatable species in the sample, because the Nylsvley study area was underlaid mostly by nutrient-deficient sandy soils. When only normally eaten species are considered, relationships with nutritional factors become more elusive, especially given the crude measures of nutritional value.

For eland studied in mixed shrubland in the Eastern Cape province of South Africa, attempts to establish a chemical basis for species preferences failed to reveal any clear pattern (Watson 1997). The acceptability of dwarf shrub species seemed to be governed mainly by fibre contents in the material consumed, which depended on leaf:stem ratios and growth phenology (Watson 1997). Likewise, the acceptance of forb species by kudus and impalas at Nylsvley seemed to depend largely on relative stemminess, underlying our distinction between 'herbaceous' and 'robust' species (Owen-Smith and Cooper 1989). No relation between forb species acceptances and any chemical factor was apparent (Cooper 1985). Presumably the forb species that were rejected contained unknown allelochemicals. A shrub known to be highly toxic because of its fluoroacetate content was not eaten by either the kudus or the impalas, at least while these animals were under observation.

For grasses, physical factors like stemminess and leaf tensile strength have a strong influence on acceptance, and leaf nutrient contents a less consistent influence, apart from the deterrent effect of aromatic compounds (O'Reagain and Schwartz 1995; Meissner *et al.* 1999). Nutritious short grasses may be neglected by large grazers such as cattle because they yield an inadequate intake rate, while being commonly favoured by medium-sized antelope (Novellie 1990; see also Owen-Smith 1991). In general, normally rejected species may be consumed when their leaves are young and nutritious, whereas favoured species become neglected when dry or stemmy. Hence, the relative preferences for grass species shown by grazing ungulates may change seasonally, depending on varying greenness, leaf:stem ratio, growth stage and prior defoliation history. Observations on migratory wildebeest in the Serengeti ecosystem of Tanzania showed that these animals selected foraging sites largely on the basis of grass greenness, and hence nutritional content, rather than height (Wilmshurst *et al.* 1999b). Nevertheless, at larger spatial scales they concentrated within regions of short to intermediate grass height and moderate greenness in both the wet and dry seasons of the year, which seemed to be the conditions yielding highest rates of energy intake.

Dietary range

To test how closely the DBM predicts the observed dietary range, and seasonal changes therein, information is needed on the abundance, chemical contents and eating rates offered by the food types potentially available for consumption. This evaluation is not quite as demanding as it may seem. Most of the food types always included in the diet need not be differentiated. Those that are rarely eaten can be ignored. Only intermediate-value foods that are candidates for dietary inclusion or exclusion need be considered.

In applying the DBM to kudus, forbs, fruits and flowers were amalgamated into a single high-quality category (Owen-Smith 1993b). For woody plants, the three most common species in the palatable deciduous class (one of them spinescent) were considered separately, with remaining species in this class amalgamated into a composite food type. Palatable evergreen species were assessed as a single food type. Among unpalatable woody plants, only the most acceptable species was considered for inclusion or exclusion. The evaluation was restricted to the *Burkea* savanna habitat, thereby excluding the *Acacia* species that gave very low eating rates. This yielded seven food types, which were ranked according to their observed acceptability to kudus. Only wet season conditions, when food abundance was high and nutritional value relatively constant, were considered.

The model calculated how the daily food intake would rise, and digestive capacity constrict, as the diet was widened to include less preferred food types (Fig. 5.8a). Nutritional yields were then calculated, both in terms of metabolizable dry matter (i.e. potential energy yield, excluding indigestible cell wall and unmetabolizable phenolics) and available protein (i.e. crude protein minus half condensed tannin content) (Fig. 5.8b). Apparently, the kudus would have become subject to a digestion constraint only if they had included the abundant but unpalatable *Burkea africana* in their diet (Fig. 5.8a). Rumen capacity was estimated from the mean mass of the rumen contents recorded during the wet season, which is somewhat less than that recorded in the dry season (Owen-Smith 1994). The observed diet breadth, which included all palatable deciduous species, appeared to be somewhat broader than that maximizing either metabolizable energy or available protein gains (Fig. 5.8b). This was because incorporating the spinescent *Dichrostachys cinerea* reduced the overall food intake rate owing to its restricted eating rate. Kudus favoured this high-quality species, despite the unfavourable intake rate that it yielded. Notably, by consuming this species the digestion constraint was alleviated to some extent. In the model, daily foraging time was fixed, so that no allowance was made for adjustments in time allocation that could be made to compensate for reduced ingestion rate.

Stochastic variability in encounter rates with different food types could largely explain the deviation between the observed diet and that predicted by the model, as well as the partial acceptances shown for particular woody species (N. Owen-Smith, unpublished). Specifically, when a preferred food type is not immediately available, it is advantageous for a herbivore to consume the next best food type. This is a consequence of the saturating

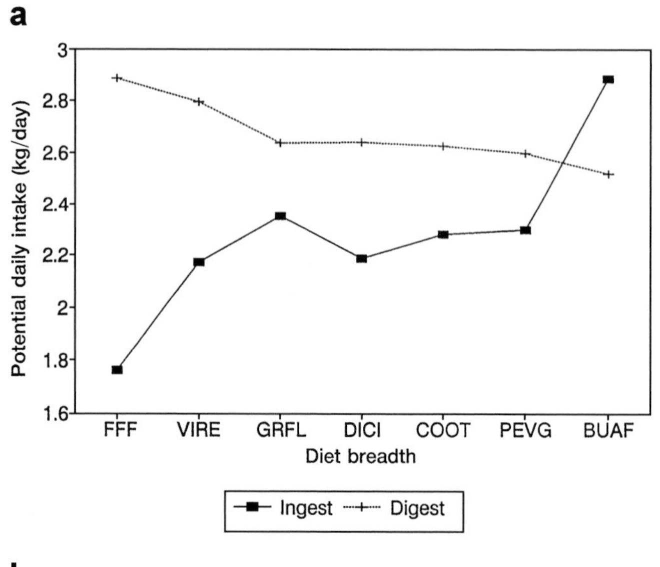

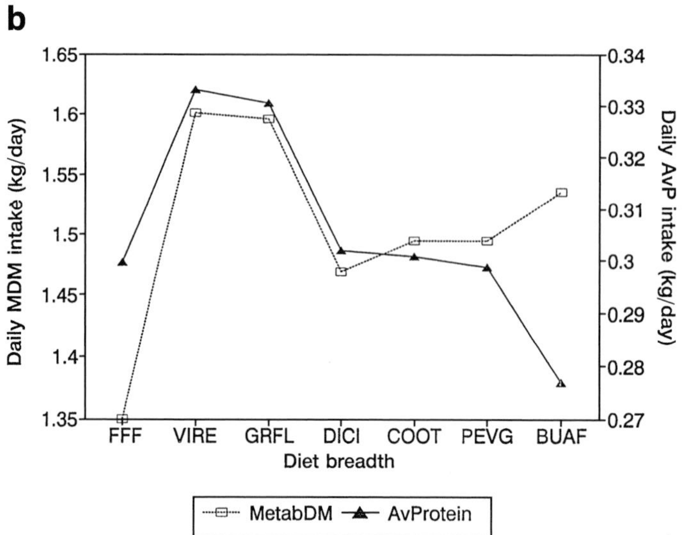

Figure 5.8. Application of the diet breadth model to kudus for late wet season conditions. FFF, forbs, fruits and flowers; VIRE, *Vitex rehmannii*; GRFL, *Grewia flavescens*; DICI, *Dichrostachys cinerea*; COOT, *Combretum zeyheri* plus other palatable deciduous species; PEVG, palatable evergreen species; BUAF, *Burkea africana*.
(a) Potential daily food intake relative to digestive processing capacity, assuming 67% of the day spent active, 60% of active time devoted to foraging, and maximum rumen fill of 1.55% (dry mass) of live mass. (b) Corresponding daily nutrient gains, expressed both as metabolizable dry matter (squares) and 'available protein' (crude protein minus half condensed tannin content) (triangles). (From Owen-Smith 1993b, with kind permission of Kluwer Academic Publishers).

form of the intake response relative to diet breadth. Losses incurred when food is sparse cannot readily be compensated when areas offering more abundant food are later encountered.

Van Wieren (1997) evaluated the applicability of the optimal diet model to Highland cattle, red deer and feral horses in the Netherlands, but without allowing for a digestion constraint. Specifically, he tested the prediction that these animals selected a diet that maximized their intake rate of digestible organic matter while foraging. He found that, although plant species offering the highest intake rate of digestible substrates predominated in the diets, a broader range of plant types was consumed than would have been optimal. Red deer diverged least and horses most from the optimal diet. He suggested that the discrepancy could have been due to unidentified nutrient or allelochemical constraints, or to the difficulty of selecting what was immediately optimal in the complex, changing environment.

The costs of acquiring the information needed to assess nutritional value are recognized to be a potential constraint on dietary optimization (Bailey *et al.* 1996). If these are substantial (e.g. numerous food types, or rapidly changing value), it may be beneficial for animals to reject only those plant species that deviate grossly in nutritional value from some target. WallisDeVries and Daleboudt (1994) suggested that the cattle they observed enhanced their daily energy intake without showing clearcut feeding preferences, simply by favouring short, nutritious patches of grass over tall patches, and by avoiding stemmy patches.

Varied diets

Besides information costs and temporal variability, other factors could cause the diets selected by large herbivores to be broader in their species content than predicted by narrowly applied principles of dietary optimization. As a striking example, the ryegrass proportion in the diets consumed by sheep did not fall below 13%, even when clover was abundantly available (Parsons *et al.* 1994). This was despite measurements showing that clover offered higher protein contents and higher eating rates, and similar digestibility and energy contents, to ryegrass, and that sheep grew faster and produced more milk when fed pure clover compared with pure grass diets. When experimentally offered a choice between ryegrass and clover, sheep selected the species opposite to that previously grazed (Newman *et al.* 1993).

Westoby (1978) suggested that large herbivores need to balance their intake of different nutrients, obtained in varying proportions from different plant types. Belovsky's (1978) LP model was based on the hypothesis that moose ate pond weeds in order to obtain sodium, which was deficient in the surrounding boreal forest. This explanation has been challenged by the observation that these moose could have obtained sodium, perhaps more readily, by visiting mineral licks (Risenhoover and Peterson 1986). Sodium is clearly a crucial mineral nutrient in short supply for most herbivores, but very few plants offer higher than average sodium contents, except in aquatic or desert environments. Phosphorus may also be a limiting nutrient for herbivores. In areas where soils are deficient in phosphorus, ungulates chew on bones to alleviate the mineral shortfall. Calcium deficiencies may cause similar responses. However, differences in P and Ca contents among plant species, relative to other nutrients, seem too small to be an important factor in herbivore diet selection (Garten 1978). For the browsing ruminants that we studied, only magnesium content was positively associated with diet selection, but only under dry season conditions when leaves were mostly old (Cooper *et al.* 1988). However, magnesium is not regarded as a limiting nutrient for herbivores, and its presence seems instead to be an indicator of the state of greenness of leaves, because of its incorporation in chlorophyll.

Nutritional requirements for protein relative to energy change with growth stage and reproductive state. Females in late pregnancy or early lactation, and young growing animals, have higher protein needs than barren females and adult males (Robbins 1993). Ungulates have the capability to alter their food preferences based on post-ingestive consequences, even for nutrients like protein that cannot be tasted directly (Provenza 1995). However, protein and energy tend to covary in their availability in foliage, being diluted in common by cell wall constituents. Balancing the consumption of leaves versus carbohydrate-rich fruits or seeds may nevertheless be an important consideration.

Secondary chemical contents may also be a factor in dietary mixing. Many dicotyledonous plant species, but relatively few grasses, contain secondary metabolites (Hagerman *et al.* 1992). The metabolic capacity of herbivores to detoxify particular allelochemicals could act as a constraint on the dietary contributions by plant species containing such compounds (Freeland and Janzen 1974). Distinctions in liver and salivary gland size (Hofmann 1989) seem to be important in limiting the proportion of dicotyledonous trees and forbs consumed by grazing ungulates (Robbins *et al.* 1995). High dietary levels of protein and energy may enhance detoxification capacity, for reasons outlined by Illius and

Jessop (1995). Kudus readily consumed several tannin-containing, and otherwise rejected species, during the new leaf phase when protein levels were elevated, although condensed tannin concentrations were no lower than in mature leaves (Cooper *et al.* 1988).

Leaf water content seemed to be an important influence on diet selection by dikdik, a dwarf antelope that is largely independent of drinking water (Manser and Brotherton 1995). Plant species avoided by dikdiks in the wet season, but favoured in the dry season, had relatively high water contents. Dikdiks could not have met their metabolic water needs in the dry season had these evergreen species not been available.

A somewhat neglected factor in diet selection is the need for ruminants to maintain the rumen chemical composition and microbiota in an appropriate state for efficient fermentation (Hungate 1975). A diet too high in nutrients relative to fibre can lead to excessive acidity in the rumen, and cause foam production and consequent bloating in cattle, although sheep seem more resistant to this effect. Mixed plant substrates may promote a diverse microbial community, facilitating seasonal dietary shifts when these are needed.

Form of the nutritional gain response

The nutritional gain response parallels the intake response in circumstances where the food resource is of uniform nutritional value, or food choice unchanging (Fig. 5.9a). Heterogeneity in resource quality, coupled with adaptive responses in diet selection, will cause the form of the nutritional gain response to deviate from that of the intake response. The form of the gain response depends moreover on the extent to which digestive capacity is limiting.

When digestive capacity is not a constraint, an adaptive widening in diet breadth could serve to maintain the daily food intake almost constant, until little food remains. Nevertheless, the corresponding rate of nutritional gain would tend to decline as poorer-quality food types become incorporated into the diet (Fig. 5.9b). If digestive capacity is limiting, both daily food intake and the resultant daily nutrient gain decline as the diet is broadened to include lower-value food types (Fig. 5.9c). In both contexts, the half-saturation level for nutritional gain is somewhat greater than that for gross food intake. Additionally, the expansion in diet breadth accentuates the 'plateau and precipice' form of both functional responses, albeit with a sloping plateau in the gain response.

For kudus, the intake response to seasonally declining food availability was humped rather than asymptotic in form (Fig. 5.10a). This was

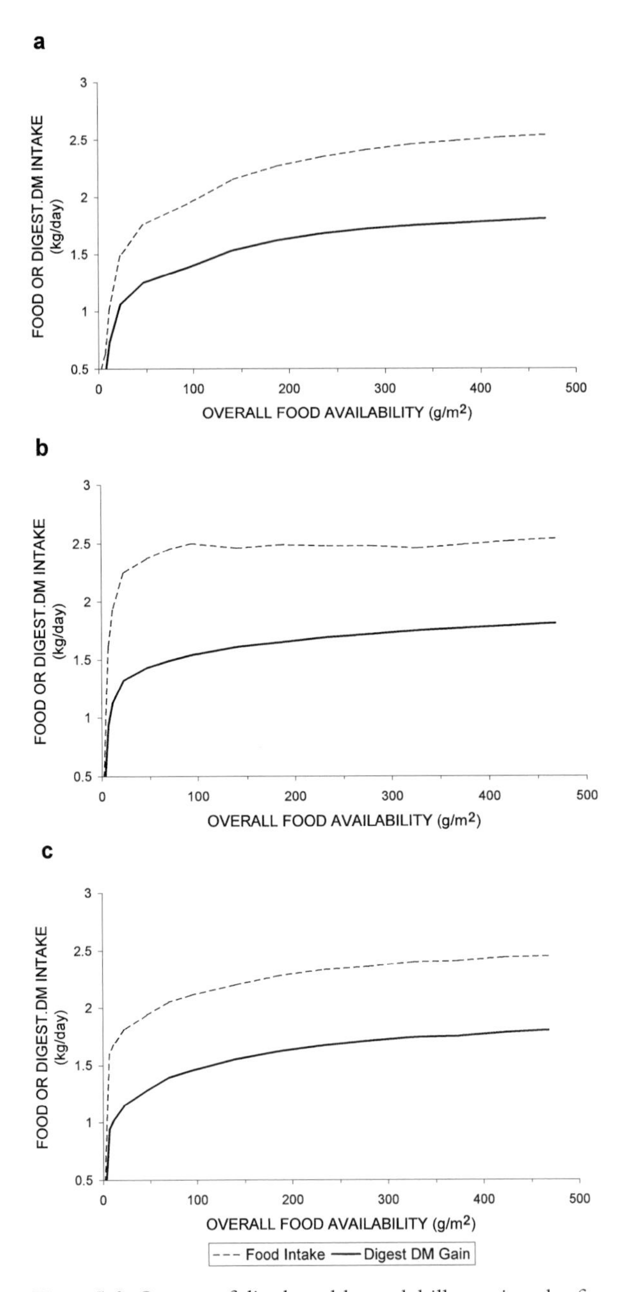

Figure 5.9. Output of diet breadth model illustrating the form of the food and nutrient intake responses to declining food abundance for different conditions (food types as in Fig. 5.1.). (a) Diet spanning food types A–E maintained unchanged as food availability declines. (b) Diet breadth expands adaptively as food availability declines, with digestive capacity not limiting. (c) Diet breadth expands adaptively under a digestive capacity constraint.

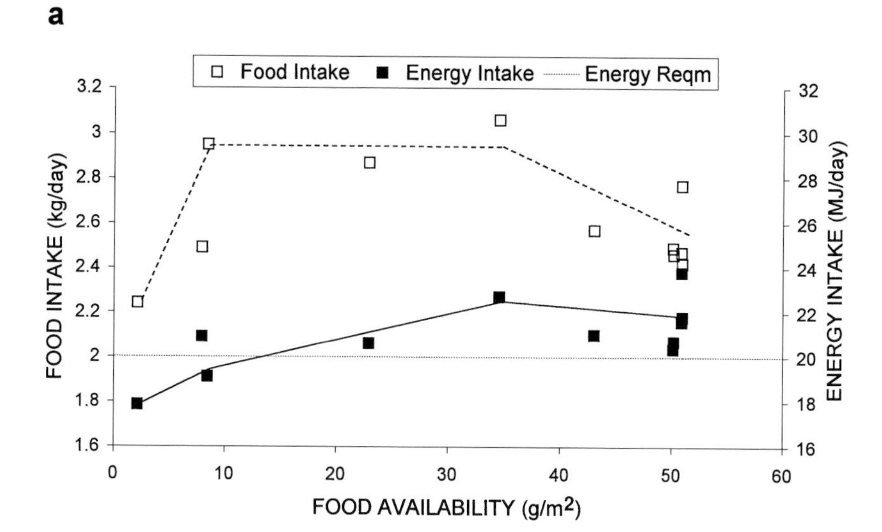

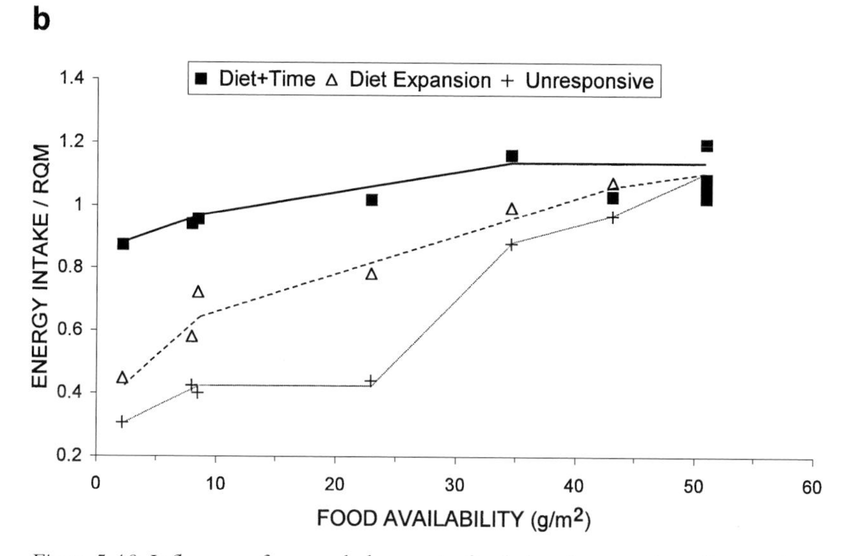

Figure 5.10. Influence of seasonal changes in food abundance and quality on the daily food intake and energy gain of kudus, averaged monthly. (a) Seasonal response in daily food intake and daily energy gain; lines indicate trend ignoring transitional months October and November, relative to energy requirement. (b) Modelled effects of adjustments in diet breadth and time allocation on daily energy intake relative to maintenance requirements. Unresponsive = dietary food types and time allocation maintained as in wet season; dietary expansion = widened diet without change in time allocation; diet + time = adaptive adjustments in both diet breadth and time allocated to feeding. (From Owen-Smith 1997.)

because animals responded to the declining nutritional intake by elevating daily foraging time, as well as the proportion of foraging time spent feeding, towards the limit set by digestive capacity (Owen-Smith 1994). Nevertheless, the daily energy gain still declined progressively owing to decreasing food quality. Various adjustments in foraging behaviour helped kudus maintain their daily energy gain at or above the maintenance requirement (excluding growth needs), until less than 10 g/m^2 of available food remained (Fig. 5.10b). Even in September, when available leaf biomass amounted to no more than 2–3 g/m^2, the daily energy gain amounted to 85% of requirements, or about 70% of the peak daily energy intake in the wet season. The commencement of the new leaf flush in September, prior to the start of the rainy season, alleviated the nutritional shortfall during this crucial period by providing small amounts of high-quality food. A half-saturation level for biomass gain of only around 1 g/m^2 of available food is indicated.

In contrast, the daily energy intake of black-tailed deer in Alaska fell below maintenance needs for over half the year (Parker *et al.* 1996, 1999). This was largely because the daily food intake of these deer dropped to barely one third of the peak summer level by the end of winter. Herbaceous vegetation became covered by snow, forcing the deer to subsist on poorly digestible stems of deciduous shrubs, plus conifer foliage. Foraging time did not increase in compensation, perhaps because of the high energy costs of activity in the snow. These deer were dependent largely upon fat reserves built up during summer for survival through winter.

Overview

The rate of biomass gain obtained by large mammalian herbivores from food consumed depends immediately on the rate of food ingestion, but over the daily cycle largely on the digestive processing capacity. Both the nutritional gain and the digestion rate are strongly influenced by food quality, controlled largely by the cell wall content in the forage consumed. The effective nutritional value is dependent additionally on contents of protein relative to digestible energy and mineral nutrients, and on the presence of various allelochemicals. Accordingly, diet selection models applied to large herbivores must balance the relative effects of food intake rates, nutrient contents, potential foraging time and digestive capacity on nutritional gains. The preference ranking of food types could change whimsically, depending on whether intake rate or digestive capacity is more limiting.

For modelling tractability, the multiplicity of plant species and parts confronting herbivores can be amalgamated into a more limited set of food types, defined by palatability classes and seasonal variation in forage retention. At least for the browsing ungulates that we studied, preference rankings among food types corresponded closely with a composite measure of effective food value, taking into account eating time relative to digestive handling and nutrient contents relative to appropriate allelochemicals. Nevertheless, animals selected a broader diet than appeared immediately optimal, which appears to be a widespread finding. Numerous factors could contribute to this pattern, including stochastic variability in encounter rates with different food types, requirements for specific nutrients, effects of allelochemicals, differences in water content, information costs, need to maintain a diverse rumen microbiota, and facilitation of dietary shifts when they become needed.

Adaptive responses to changes in food quality and quantity cause the effective nutritional gain, expressed as consumer biomass increase, to deviate from the intake response to changing food availability. In particular, nutritional intake may decline even though the daily food intake is held constant, causing the effective half-saturation level for biomass gain to be somewhat higher than that for food intake. Compensatory adjustments in foraging time may cause the intake response to seasonally declining food quantity and quality to be humped rather than asymptotic in form, and accentuate the precipitous decline in intake and corresponding gains that eventually takes hold towards very low food abundance levels.

Further reading

For conceptual development of the diet breadth model for herbivores, see Owen-Smith and Novellie (1982). The degree of success of this model was reviewed by Owen-Smith (1997). The alternative linear programming approach to herbivore diets was described by Belovsky (1986), and reviewed by Belovsky (1990). Hanley (1997) gives a thoughtful evaluation of the limitations of optimal diet models, with specific reference to deer.

6 · *Resource constraints: physiological capacities and costs*

In Chapter 5, we recognised tradeoffs between short-term constraints on the rate of nutritional gain, as governed by handling time during food ingestion, and the longer-term constraint imposed by digestive processing capacity. Additional limitations on daily food intake are imposed by foraging time, as affected by environmental conditions and other factors, and metabolic processing capacity. In this chapter, attention is focussed on how these physiological constraints limit resource gains. Optimization becomes not so much a matter of maximizing rates of gain, but of balancing multiple constraints varying over time in their effectiveness.

Standard computational techniques have been developed for identifying the best course of action, in circumstances where multiple constraints apply. These generally represent constraints as fixed upper or lower limits, which cannot be exceeded. In practice, physiological limitations are somewhat elastic, as well as changing over time. Nevertheless, costs are incurred when physiological limitations are stretched, which have to be weighed against the resultant benefits. Limitations on physiological capacity act to restrict the maximum rates of food intake and nutritional gain that can be achieved, over extended time periods. Costs may also act to increase metabolic losses, relative to nutrient gains, or reduce the effective conversion efficiency from food to consumer biomass. In terms of the **GMM** model, the difference between the relative rate of biomass gain, G, and the rate of metabolic attrition in biomass, M_P, provides an overall measure of the *rate efficiency* with which resources are converted into biomass growth potential (cf. Equations 2.18 and 2.28). This is analogous with the engineering concept of *power*, i.e. the time flux of net energy release (Brown *et al.* 1993).

In this chapter, the linear programming approach to reconciling multiple constraints on resource gains will be outlined. Restrictions on digestive processing capacity, and how these operate over different time frames, will then be examined more critically. Next, limitations on metabolic processing capacity, and related concepts, will be evaluated. Thirdly, influences of thermal tolerance on potential daily foraging time will be considered. Rather than being constant in their operation, these potential constraints change over time in their effects, over various time frames, exposing limitations and circularities in linear programming models as conventionally applied. Finally, I outline how the existence of various physiological constraints, coupled with adaptive responses by consumers, can lead to a general truncation of the intake response toward high food biomass levels.

Linear programming models

Linear programming (LP) is a computational technique to identify the optimal action when the outcome is subject to multiple potential constraints. To be tractable, the number of choices should be limited, and constraints must be linear in their operation. This outline largely follows Belovsky (1984, 1986).

Consider a herbivore confronted by just two alternative foraging areas, differing in the resource attributes they present. One area contains abundant food of relatively poor quality. If animals foraged solely in this area, daily food intake would be limited by digestive capacity. The other area offers high-quality food, but in low abundance. The daily food intake is limited by the intake rate obtained while foraging, relative to the potential daily foraging time. Assume that the areas are contiguous, so that no travel time is involved in moving between them. Under certain conditions, animals could achieve highest nutritional gains, relative to metabolic requirements, by partitioning their foraging time and hence food intake between the two areas.

For constraints to be linear, the foraging areas must be spatially discrete. The daily food intake then becomes a linear combination of the amounts of food obtained from each area. The restriction on potential daily intake is usually formulated as a time constraint. Generalizing for areas labelled $i = 1.\ldots.n$, the time taken up by foraging in all available areas must be less than the maximum daily foraging time, i.e.

$$\sum_{i=1..n} F_i/I_i \leq t_F, \qquad (6.1)$$

where F_i = mass of food consumed from area i, I_i = food intake rate ('cropping rate') obtained while foraging in area i, and t_F = maximum daily foraging time.

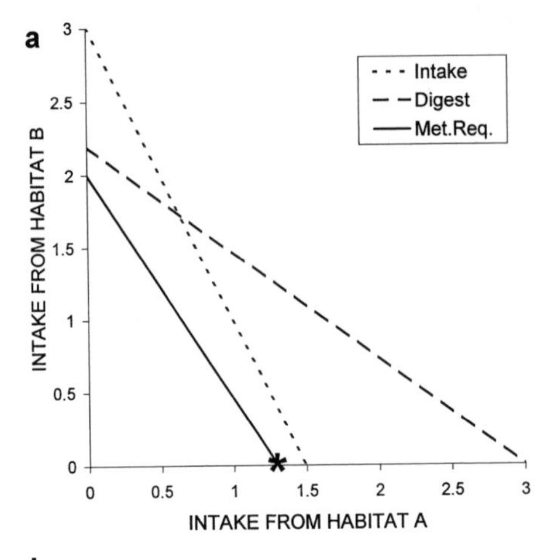
a
Intake
Digest
Met.Req.
INTAKE FROM HABITAT B
INTAKE FROM HABITAT A
0
0.5
1
1.5
2
2.5
3

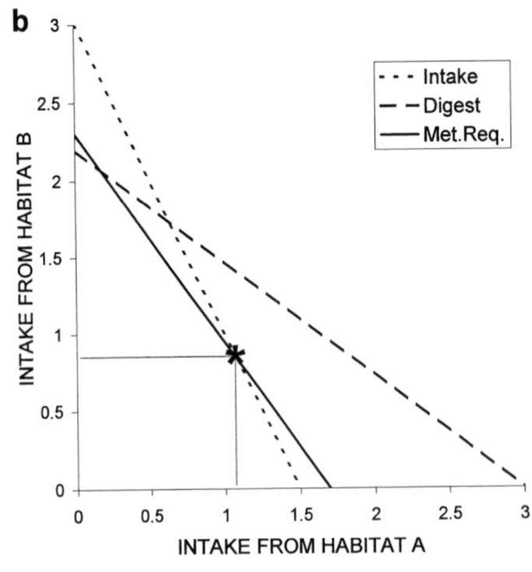
b
Intake
Digest
Met.Req.
INTAKE FROM HABITAT B
INTAKE FROM HABITAT A
0
0.5
1
1.5
2
2.5
3

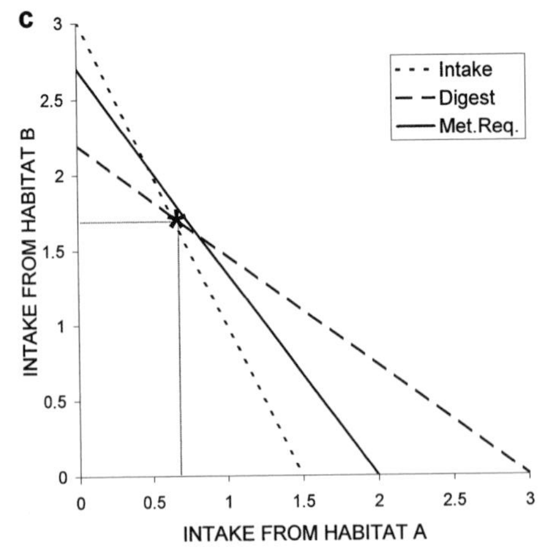
c
Intake
Digest
Met.Req.
INTAKE FROM HABITAT B
INTAKE FROM HABITAT A
0
0.5
1
1.5
2
2.5
3

Similarly, if the food types obtained from each area are digested independently, the daily digestive load is the sum of the amounts of food ingested from each. This cannot exceed the maximum digestive processing capacity, so we have:

$$\sum_{i=1..n} k_i F_i \leq d, \tag{6.2}$$

where k_i = fractional digestion rate per day of food obtained while in area i, and d = digestive processing capacity (mass per day).

The extent to which the food obtained from each area meets metabolic demands depends on the energy and material nutrients yielded. Two alternative targets can be considered: either (a) satisfying the basic maintenance requirement, within the least time, or (b) maximizing the net gain relative to this requirement. Hence, assessing overall energy needs rather than requirements for specific nutrients, we have:

$$\sum_{i=1..n} c_i F_i \leq m, \tag{6.3}$$

where c_i = conversion coefficient for food consumed in area i, and m = metabolic requirement, expressed as the mass of digestible dry matter needed to meet the metabolic energy demand per day. In the terminology of the **GMM** model, target (a) can be equated with setting $G = M_P$, while (b) requires $G \gg M_P$. The foraging target could change with the physiological state of the animal. For example, Belovsky (1986) suggested that, during the mating season, dominant male pronghorns are 'time minimizers', whereas females and non-breeding males generally forage as 'energy maximizers'.

For just two foraging areas, the optimal solution can be depicted graphically in the state space defined by the amounts of food obtained from each area (Fig. 6.1). Depending on the alignment of the constraint lines,

Figure 6.1. Linear programming model output, evaluating the optimal dietary contributions from two habitat types offering different resource features. Constraints are shown for daily food intake, digestive capacity, and metabolic requirements. Habitat type A offers food that is more nutritious and more digestible, but less available, than food in habitat type B. Asterisks indicate optimal diet. (a) Neither intake nor digestive capacity is limiting, with the diet satisfying requirements at minimal cost obtained solely from habitat A. (b) Intake is limited below requirements in habitat A, so diet is obtained partly from habitat B. (c) Intake is limiting in habitat A, while digestive capacity is constraining in habitat B, with diet minimizing the metabolic deficit below requirements obtained from habitats A and B in indicated combination.

the optimal combination of food resources from each area may be located (a) at the intersection of the maintenance requirement line with the area axis yielding lowest cost (i.e. lowest intake, unless time costs outweigh food processing costs) (Fig. 6.1a); (b) at the intersection between the requirement and an intake constraint (that effectively limiting intake) (Fig. 6.1b); or (c) at the intersection of two constraint lines, when the maintenance requirement cannot be met (Fig. 6.1c). For solution (a), the herbivore should forage solely in the most favourable habitat. For solutions (b) and (c), foraging time should be divided between the two habitat types, such as to yield the respective dietary contributions indicated in Fig. 6.1.

In the original application of LP to foraging by moose, the foraging target was identified as being to maximize daily energy intake, subject to meeting the minimum requirement for sodium (Belovsky 1978). This could be achieved by partitioning foraging between ponds containing aquatic plants, which concentrated sodium, and the surrounding boreal forest, which was deficient in sodium but more favourable in energy. Subsequently, the model was applied to various herbivore species occupying a North American grassland, assessing the relative dietary contributions of areas offering monocotyledons (predominantly graminoids) versus those presenting dicotyledons (forbs and shrubs) (Belovsky 1986). These plant types were characterized by contrasting cropping rates and digestive fill properties. However, digestive fill was calculated from the wet bulk that these plants presented, a hangover from the moose study that has been criticized as physiologically unrealistic (Hobbs 1990).

Widespread success in predicting herbivore diets has been claimed for this model (Belovsky 1990, 1994). My analysis, based on detailed data for kudus, suggested that this apparent success was largely an outcome of circularity in the way constraints were assessed (Owen-Smith 1993a). Specifically, the foraging time and digestive capacity constraints were calculated from average observed values of these measures, assuming them to be limiting, and then applied in the model to assess the outcome. This procedure will always seem to predict the observed diet, even if other constraints besides those assumed had been effective (Owen-Smith 1996b). For kudus, I found that, even though the putative ingestion and digestion constraints seemed to predict the average combination of woody versus herbaceous browse in the diet, animals did not respond to day-to-day variability in the conditions they experienced in the way expected if the constraint-balancing principle was effective. Moreover, the apparent constraint settings during the wet season when food was abundant proved somewhat elastic during the dry season when food quantity and quality become limiting (Owen-Smith 1994). More fundamentally, kudus seemed to be choosing among plant types within both the herbaceous and woody categories, rather than at the broad categorical level.

I also evaluated the LP approach more broadly with regard to the apportionment of foraging time by kudus between *Burkea* and *Acacia* savanna habitats, which offered contrasting feeding opportunities (Owen-Smith 1993a). However, the model outcome was inconclusive, since the nutritional gain from each habitat closely matched the daily energy requirement. Apparently, food availability in the more nutritious *Acacia* vegetation was depressed such that foraging returns were no better on average than those obtained from the *Burkea* vegetation (Fig. 6.2). This is the outcome expected from marginal value principles of patch use (Chapter 4). Observations showed that the kudus did spend more time foraging in the more nutritious *Acacia* habitat, relative to its proportional extent, than in the *Burkea* savanna.

A fundamental problem for LP is that the fixed upper limits representing time or digestion constraints in diagrams like those depicted in Fig. 6.1 are potentially misleading. These lines are no more than arbitrary isoclines

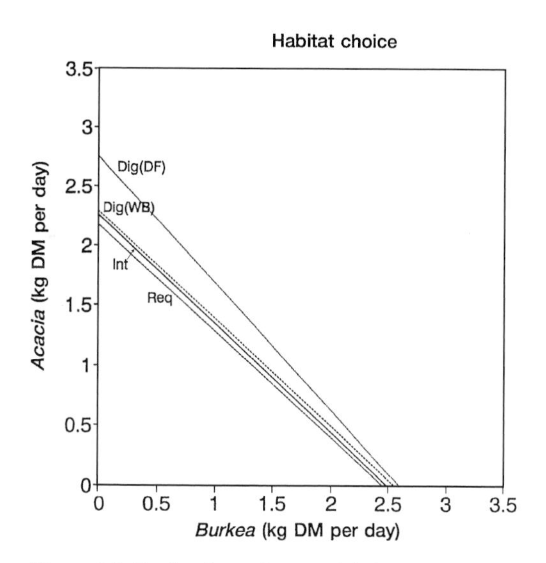

Figure 6.2. Evaluation of optimal habitat selection by kudus between the *Burkea* and *Acacia* vegetation types at Nylsvley. Int, ingest; Req, require. The digestion constraint is indicated for either (i) rate limitation on cell wall disappearance (Dig (DF)), or (ii) wet bulk limitation on rumen fill (Dig (WB)). (From Owen-Smith 1993a, with kind permission of Kluwer Academic Publishers).

in a state space of changing benefits and costs. Moreover, these costs and benefits change over time, and with circumstances. Undoubtedly factors such as digestive capacity and foraging time can limit the amount of food that a herbivore can consume per day. How do we accommodate this elasticity, as well as the changing interplay among physiological limits?

Digestive capacity

Digestion represents the breakdown of food substrates into simpler compounds, and the absorption of these products across the gut wall. Material inputs include digestive enzymes and buffering secretions. The soluble metabolites contained within plant cells (protein, soluble carbohydrates and starch) are released as soon as the cell walls enclosing them are ruptured by chewing. Degradation of cell wall constituents, primarily cellulose reinforced with lignin, occurs via the agency of bacterial or protozoan symbionts housed within particular compartments in the gut. In ruminants like cattle and sheep, microbial fermentation takes place in the rumen, a voluminous compartment located in the foregut region anterior to the true stomach. In non-ruminants like horses, the fermentation chamber includes the caecum, a blind sac at the junction of the small and large intestine, and the capacious large intestine or colon in the hindgut region.

Microbial enzymes convert cellulose into energy-yielding fatty acids. A portion of cellulose that is reinforced by lignin cannot be broken down and is passed from the gut as faeces, which include also other indigestible material together with the products of digestive secretions and gut lining. In ruminants, soluble carbohydrates and protein are initially degraded and absorbed by the microbiota, with microbial bodies later being digested by the host. Some of the digestion products absorbed into the blood are unmetabolizable, and hence are excreted in urine, together with products of tissue catabolism.

Limitations on digestive processing capacity arise from three sources: (1) the volume of relevant gut compartments, (2) the rate of digestive degradation of food material, and (3) the rate of passage from the gut of indigestible residues.

Rate limitation

Mostly simply, digestion can be represented as occurring within a well-mixed compartment receiving inputs from ingested food, enzymes and buffering secretions (Penry and Jumars 1986; Mertens 1987; Alexander

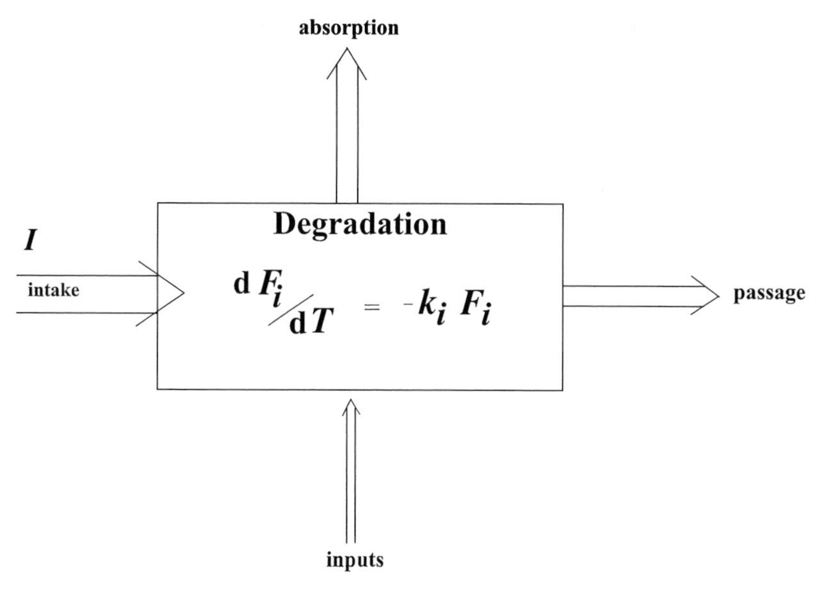

Figure 6.3. Compartment model of digestion, indicating inputs from food ingested and secretions of enzymes and buffers, and outputs via absorption or passage. (For explanation, see text.)

1991). Products are absorbed across its wall, and undigested material is passed on (Fig. 6.3). If the rate of digestion is limited only by available substrate, it can be represented by first-order kinetics, i.e. the digestible substrate disappears at a constant fractional rate:

$$\mathrm{d}F_i/\mathrm{d}T = -k_i F_i,$$

where F_i = amount of food component i, and k_i = fractional rate of disappearance of substrate i. The rate of digestion is initially rapid, but decelerates as less substrate remains undigested. Expressed on a log scale, the proportion that remains undigested decreases linearly with time (Fig. 6.4a):

$$F_i = F_{i0}e^{-kiT},$$

where F_{i0} = initial amount of food substrate i at time zero. For $k = 0.1$ per hour, digestion is 90% complete after about 24 h.

The overall rate of disappearance of food material from the fermentation chamber depends on rates of digestion or passage of the three basic components of plant tissues: (1) cell solubles, including protein, starches, sugars and oils, which are quickly and virtually completely digested;

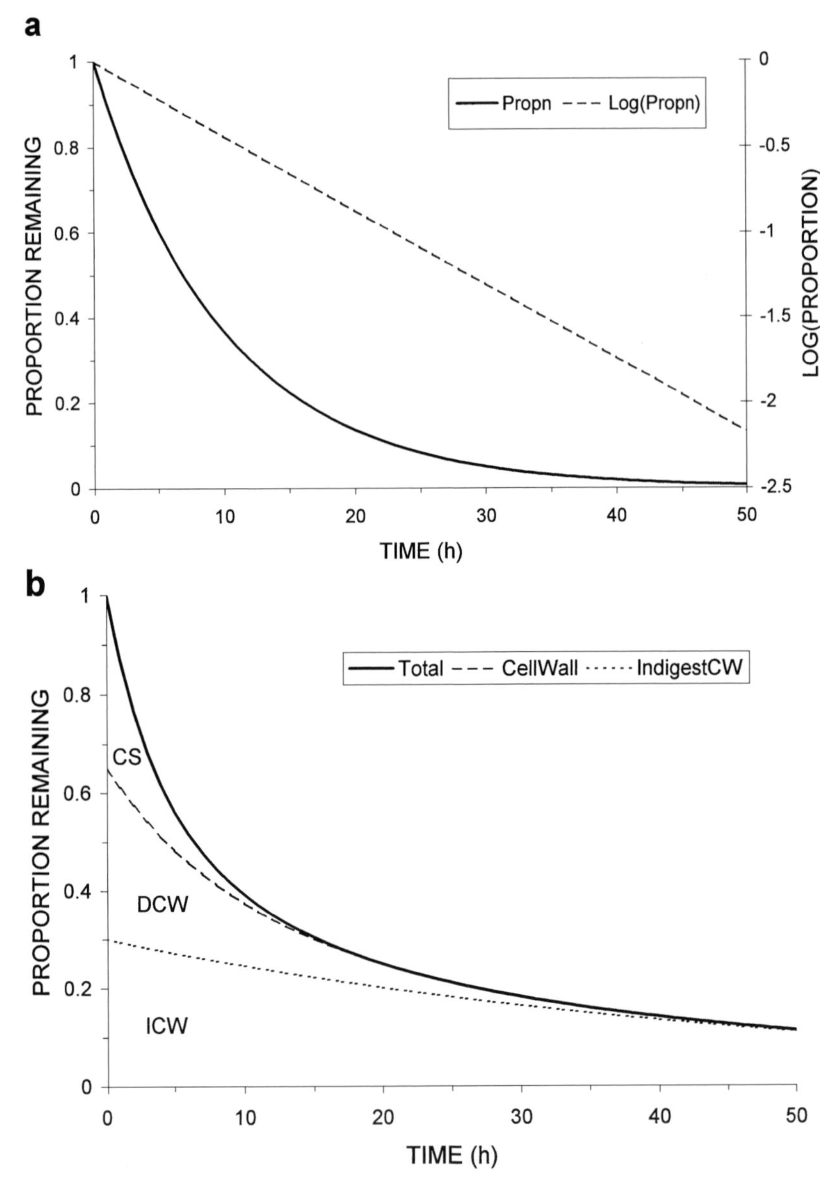

Figure 6.4. Modelled course of digestion of plant cell wall over time, assuming first-order kinetics. (a) Disappearance of digestible constituents on arithmetic and logarithmic scales, for $k = 0.10$ per hour. (b) Disappearance by digestive degradation or passage of plant material over time, assuming the following values for rate constants: k_s for cell solubles (CS) = 0.3 per hour; k_w for digestible cell wall (DCW) = 0.10 per hour; k_p for passage of indigestible cell wall (ICW) = 0.02 per hour.

(2) the digestible portion of cell wall, which degrades more gradually; and (3) indigestible cell wall constituents, together with undegradable cutins and waxes. It can be represented as the sum of three first-order processes with different rate constants (Mertens 1977):

$$dF/dT = -(k_s + k_p)F_s - (k_w + k_p)F_w - k_p F_u, \qquad (6.4)$$

where F_s = cell solubles, F_w = digestible cell wall constituents, F_u = indigestible components, k_s = fractional rate of digestion of cell solubles, k_w = fractional rate of digestion of digestible cell wall, and k_p = fractional rate of passage of both digestible and indigestible components from the compartment. For simplicity, differences in passage rates between liquid and solid residues are ignored. Because the rate of passage is the slowest process, indigestible residues persist longest and hence accumulate in the fermentation chamber (Fig. 6.4b). The cumulative proportion digested increases asymptotically towards the limit set by the ultimately digestible fraction.

In practice, there is a lag before digestion commences, because food particles must become hydrated, microbes attached and microbial enzymes secreted. Owing to secretory inputs and mastication effort, the net gain is negative before digestion gets fully under way. This lag time may be somewhat longer for the fine, fibrous leaves of grasses than for the foliage of woody and herbaceous dicots (Mertens 1977). However, browse leaves tend to be more lignified, and so show a lower ultimate digestibility of cell wall than grasses.

The optimal retention time for food substrates in the digestion chamber can be estimated by using marginal value principles (following Sibly 1981). The retention time maximizing the overall digestion rate tends to be somewhat longer for grasses, which digest more slowly due to their higher cell wall content, than for browse (Fig. 6.5a). Notably, the digestive efficiency corresponding with the optimal retention time is less than the ultimate digestibility possible. If the initial costs are taken into account, the optimal retention time is somewhat longer than if just gross yields are assessed (Fig. 6.5b). The retention time is determined basically by the size of the openings connecting the fermentation chamber to the following sections of the gut. However, passage from the rumen may be actively controlled by ruminants to some extent (Spalinger *et al.* 1993; Gross *et al.* 1996).

For non-ruminants, food substrates reach the fermentation chamber in the hindgut already well hydrated, so that the initial digestion lag is shorter than it is for a ruminant feeding on a similar food substance

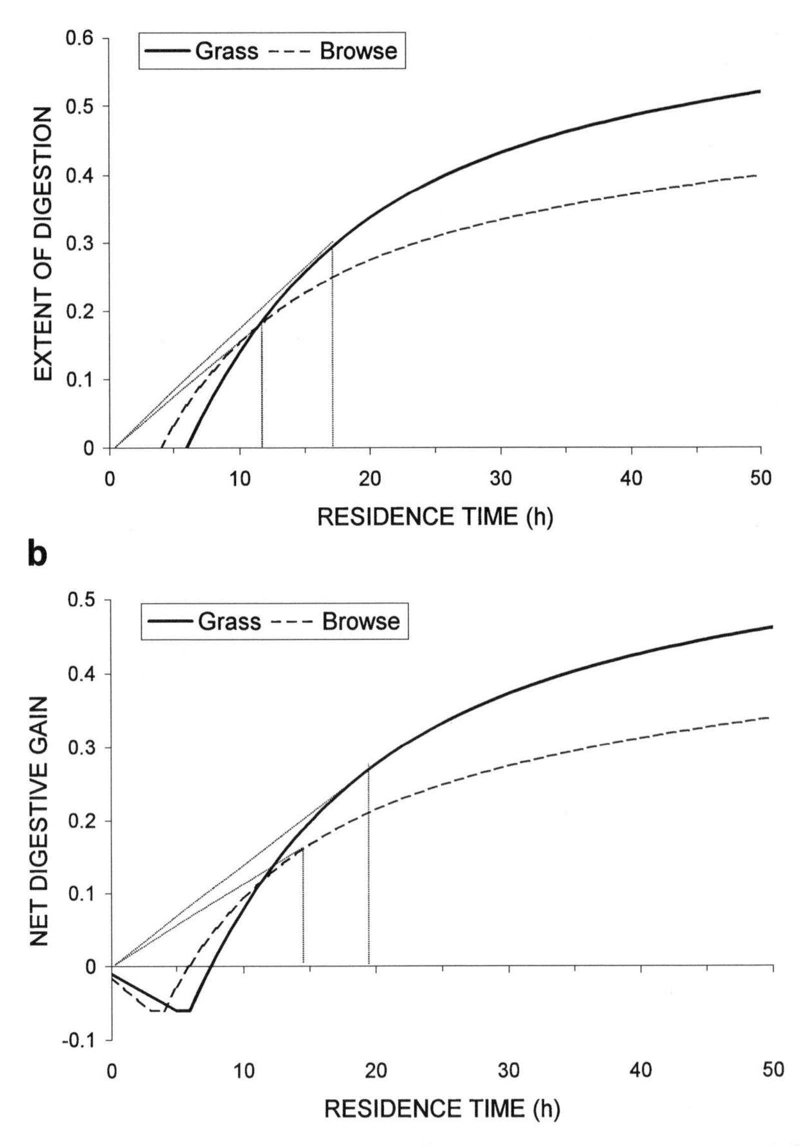

Figure 6.5. Cumulative extent of cell wall digestion, contrasting patterns typical of grasses and browse (woody or herbaceous dicotyledons). Grasses show a longer initial time lag, a slightly slower rate of digestion of digestible components, and ultimately a higher overall digestibility, than browse. The tangent from the time origin indicates the retention time maximizing the overall rate of digestion (the slope of the tangent), which is generally longer for grasses than for browse. (a) Gross extent of digestion. (b) Net digestive yield, subtracting initial costs of buffers, enzymes and masticatory effort.

(Illius and Gordon 1992). Non-ruminants also exhibit lower digestive efficiencies than ruminants, partly because they cannot re-masticate food (chew the cud). Hence Fig. 6.5, comparing grass and browse properties, could equally well represent a ruminant–non-ruminant contrast. The lower digestive efficiency of non-ruminants is compensated by a more rapid passage of material through the gut, so that the overall nutritional gain achieved may end up being very similar to that of ruminants (Bell 1971). Nevertheless, to balance the higher digestive throughput, a non-ruminant must ingest more food per day than a similar-sized ruminant.

Physical capacity

For ruminants, the digestive capacity is set largely by the volume of the rumen. The dry mass of its contents typically represents about 2.3% of the body mass of a medium-sized ungulate (Illius and Gordon 1992). The rate capacity is the product of this mass and the overall rate of exit of material, by absorption or passage. Averaging observed rates of digestion and passage for cell wall constituents (i.e. assuming equal proportions of digestible and indigestible components), we obtain a mean turnover rate of 0.07 per hour. This is equivalent to a daily turnover of 1.5 times the mean rumen contents mass, suggesting that a typical ruminant could potentially process about 3.5% of its body mass per day.

In practice, indigestible components form the bulk of the rumen contents (Fig. 6.6). Cell solubles take up space briefly following meals, but are rapidly digested and absorbed. The degree of rumen fill varies somewhat over time, and the digestion rate fluctuates correspondingly. The passage rate of the indigestible material is the primary restriction on intake rate. As a consequence, the mean turnover rate of the digesta is lowered from 1.5 times per day, as calculated above based on the food composition, to about 1.2 times per day. This lowers digestive processing capacity to less than 3% of body mass per day. Because browse foliage typically has higher cell solubles but more indigestible material in cell wall than grass leaves, the digestive contents should vary somewhat more in mass and digestion rate for browsers than for grazers (Fig. 6.6a,b). This simple model suggests that the mean digestive fill of a grazer should be about 8% higher than that of a browser, whereas the mean digestion rate should be about 8% lower. If browsers additionally have a higher passage rate of indigestible material from the rumen than grazers (as suggested by Hofmann (1973, 1989)), these differences would be somewhat greater.

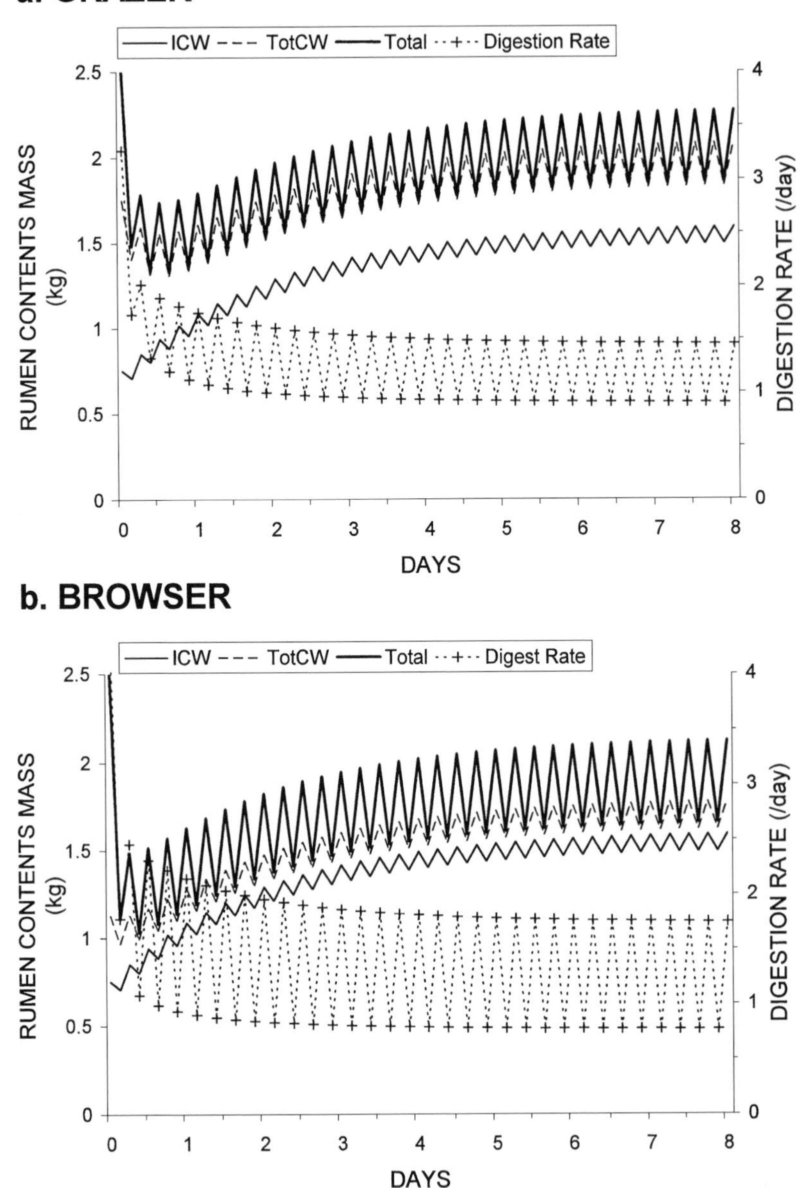

Figure 6.6. Dynamically changing substrate proportions in the rumen, and corresponding digestive turnover rate, starting with initial contents constituted by the proportion of these components in the food. The model assumes 3h meals followed by 3h rest/ruminating periods, with basic rate constants as for Fig. 6.4b. (a) Composition typical of grass: cell solubles 0.3, digestible cell wall 0.4, indigestible components 0.3. (b) Food composition typical of browse: cell solubles 0.55, digestible cell wall 0.15, indigestible components 0.3.

Illius and Gordon (1991) developed a more sophisticated model of digestive processes, taking into account the effects of rumination on particle size distribution in the fermentation chamber, and the microbial biomass component of digestive load. Their model suggested a halving in digestive processing capacity from 3% to 1.5% of body mass per day with an increase in total cell wall content (measured as neutral detergent fibre) from 30% to 80% of dry mass, for a typical medium-sized ruminant (Fig. 6.7). The simpler model presented in Fig. 6.6 generates a similar trend, while predicting a somewhat higher processing capacity, perhaps because it ignores the microbial component (amounting to 10–30% of total rumen dry mass). Empirical data presented by Hobbs (1990) indicate a somewhat lower intake capacity, and somewhat steeper decline in daily intake towards high cell wall contents, than predicted by both models (see Fig. 6.7). However, the empirical data for daily food intake show much scatter, suggesting that the observed intake was less than the maximum in many cases.

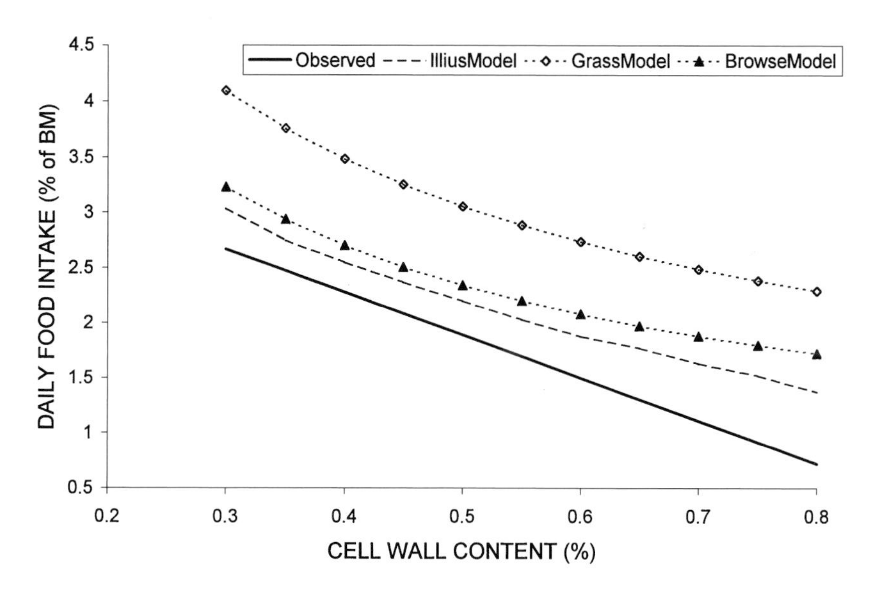

Figure 6.7. Influence of dietary cell wall content on potential daily food intake, as controlled by rumen processing capacity, comparing predictions of two models for a 100 kg ruminant against empirical observations. Observed, pattern reported by Hobbs (1990, Fig. 6). Illius Model, output from model of Illius and Gordon (1991, Fig. 3b). Grass Model, Browse Model, from the dynamic model illustrated in Fig. 6.6, using mean turnover rates for grass and browse with mean rumen contents mass assumed equal to 2.3% of body mass.

Following the analysis of Shipley *et al.* (1994), the maximum eating rate of a 100 kg ungulate is about 16 g/min. Allowing 25% of foraging time for miscellaneous activities, an animal foraging continuously for 24 h could achieve a daily food intake of 17.3 kg, i.e. about six times its digestive processing capacity. Either foraging time must be curtailed to allow digestion to catch up, or animals must adjust their intake rate somewhat below the maximum of which they are capable, or some combination of both must take place. In fact, 100 kg kudus foraged for less than 50% of the 24 h, and exhibited a mean eating rate about one third of the theoretical maximum (N. Owen-Smith, personal observations).

Hence, unless the cell wall content of the food is very low, the daily food intake will generally be constrained by digestive capacity. Nevertheless, the physical gut capacity of ruminants can be adjusted to some extent to cope with periods of high nutritional demand or highly fibrous diets. Dairy cattle can increase their effective rumen volume by as much as 40% during peak lactation (Campling 1970). Free-ranging Highland cattle in The Netherlands increased their daily food intake by nearly 40% in spring, when recovering mass lost during winter, and by over 60% during peak lactation in autumn, relative to the expected mean for the food quality (van Wieren 1992). The daily food intake of these cattle during winter averaged only half of the summer mean. The rumen contents mass of kudus in the dry season, when food quality is low, was 25% higher than that of kudus examined in the wet season (Boomker 1987, cited by Owen-Smith 1994). Similar variability in the mass of rumen contents has been documented for other ungulates (Baker and Hobbs 1987; Lechner-Doll *et al.* 1990). Impala increased the absorptive area of their rumens, by producing more surface papillae, at times of the year when they fed mainly on browse (Hofmann 1973). It is unclear what the costs are to produce the additional gut tissue.

Overall assessment

The potential daily food intake, based on maximum rates of ingestion, can vastly exceed the digestive processing capacity of large mammalian herbivores. Animals must thus respond in various ways to limit intake rates below their potential maximum. In addition, herbivores must limit the content of largely indigestible fibre in the diet, which forms most of the digestive contents mass. Because foraging takes place intermittently, animals achieve their maximum digestive fill only towards the end of a foraging spell.

Digestive capacity sets a somewhat variable limit on the daily food intake. Physical capacity may be reached only at times of the year when food quality is low, or metabolic demands especially high. Nevertheless, the gut capacity may be expanded to some degree to cope with such demands, but probably at some cost in synthesizing additional gut tissue.

Metabolic satiation

Animals eat in order to meet their physiological requirements for energy and material nutrients. Nutritional yields must support the maintenance of body tissues, thermoregulation, activity expenditures, deposition of stored reserves, growth in size and provisioning of offspring. Maintenance metabolic requirements vary interspecifically in relation to $M^{3/4}$, where $M =$ body mass. Hence small animals have higher mass-specific needs than larger species. Adult males may also have lower food needs than growing young animals, or females that are nurturing progeny.

Highest rates of metabolism are sustained during peak lactation, periods of extreme cold stress, or spells of demanding physical exercise (Peterson *et al.* 1990; Hammond and Diamond 1997). The maximum *sustained metabolic scope*, expressed relative to the resting metabolic rate, is determined largely by the relative size of the small intestine, liver, kidneys and heart. For mammals weighing over 1 kg, the highest daily energy expenditure does not exceed about four times the basal metabolic rate (Ricklefs *et al.* 1996; Hammond and Diamond 1997). Peak energy requirements for lactating dairy cattle are estimated to be around three times the maintenance level (Mertens 1987). Typical daily energy needs for non-lactating ungulates are about 1.5–2 times the resting metabolic rate, which itself is about 25% higher than the basal (fasting and inactive) rate (Moen 1973; Jiang and Hudson 1992; Parker *et al.* 1996).

When food quality is high, the potential intake of nutrients may exceed the physiological processing capacity, placing an upper limit on the daily food intake (Dinius and Baumgardt 1970; Mertens 1987) (Fig. 6.8). If more energy or nutrients are assimilated than immediately needed metabolically, the surplus must either be stored (e.g. as fat) or excreted. The potential benefits of such storage must be weighed against costs associated with processing the additional food. Besides tooth and gut wear, oxidative stress causes progressive tissue degradation, and may ultimately limit the potential lifespan (Tolkamp and Ketelaars 1992; Sohal and Weindruch 1996). Carrying stored reserves is also costly, as will be discussed in Chapter 7.

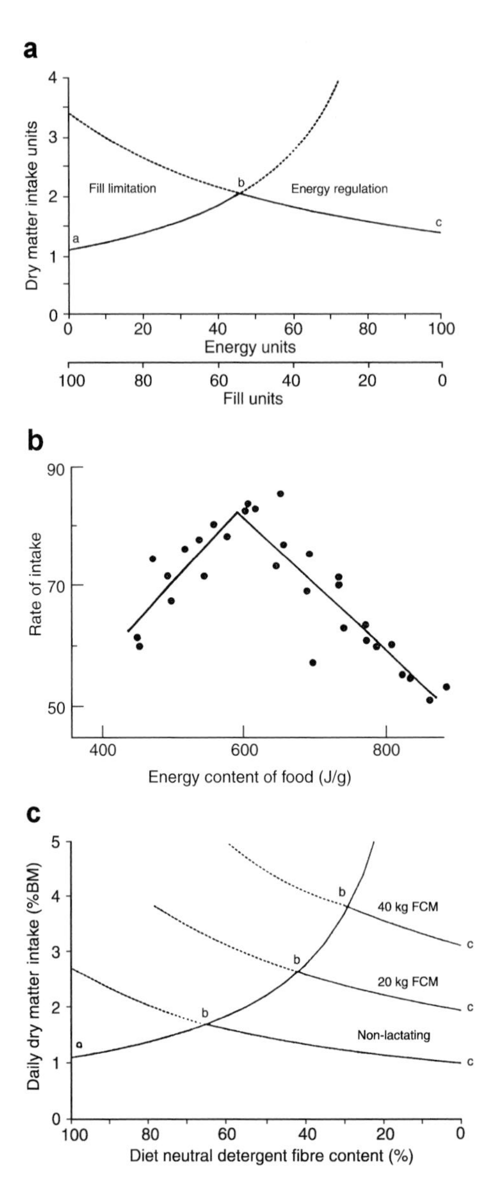

Figure 6.8. Transition from physical (gut capacity) to physiological (metabolic satiation) limitation of food intake with increasing dietary quality (i.e. decreasing cell wall content). (a) Hypothetical depiction of changing limitations with increasing digestibility ('energy units') and decreasing cell wall ('fill units'); dotted line indicates potential limit, solid line the effective limit (from Mertens 1987). (b) Observed change in 'voluntary' rate of intake of sheep with increasing digestible energy yield of forage (from Dinius and Baumgardt 1970). (c) Hypothetical effects of increased energy demands for milk production by dairy cows on daily food intake, in relation to dietary cell wall content; dotted lines indicate intake demands, solid line the effective intake limitations (from Mertens 1987).

There is widespread evidence that food intake is controlled lipostatically for energy homeostasis in the short term (Woods *et al.* 1998). However, if the upper capacity to assimilate nitrogen is reached, energy intake may become restricted (Illius and Jessop 1996). Limitations in capacity to process potentially toxic compounds can also limit the daily food intake, unless levels of digestible energy and protein in the diet are very high (Illius and Jessop 1995). In the short term, meals are terminated in expectation of the post-absorptive consequences, rather than as a response to immediate signals (Illius and Jessop 1996). Behavioural responses to palatability features of foods may lead animals to consume less than their physiological capacity (Mertens 1987).

Intake seems to be balanced against metabolic needs over periods somewhat longer than a day (WallisDeVries and Daleboudt 1994). The estimated food intake of kudus varied quite widely between days (Fig. 3.6), although averaged over three days the balance between metabolizable energy intake and needs was quite close (Owen-Smith 1997) (see also Fig. 7.1). Metabolic requirements may be reduced during periods of restricted food availability. Feral cattle in The Netherlands consumed only half as much food per day during winter as during summer (van Wieren 1992). It seemed unlikely that intake was physically limited by gut space, because these animals failed to extend their daily foraging time to compensate for the low intake rate obtained during winter, while rumination time was also lowered. Rather, the cattle seemed to be curtailing costs.

Overall assessment

An upper limit to the daily energy intake, and possibly that of other nutrients, is set by the maximum metabolic scope, amounting to about 3–4 times the basal metabolic rate. Most of the time, metabolic demands will be somewhat below this level. Energy and nutrients gained surplus to immediate requirements can potentially be stored, but at some metabolic cost. Rather than being maximized, the energy intake should be adjusted relative to immediate demands, with some day-to-day flexibility and adjustments for storage. Metabolic requirements can be reduced substantially during periods of restricted food availability, by curtailing activity costs.

Thermal tolerance

Physiologists recognize a thermoneutral zone for each animal species, encompassing the range of conditions within which activity can take place without body temperature varying beyond tolerable limits

(Schmidt-Nielsen 1975). Below the lower critical temperature, homeothermic animals must increase metabolic rate to maintain body temperature. Above the upper critical temperature, energy and perhaps water must be expended to prevent overheating. Thus animals can be active somewhat beyond the limits of the thermoneutral region, but at the cost of metabolic expenditures.

The buildup in body heat that occurs as a result of muscular activity can curtail foraging periods. For large herbivores, heat is also generated through digestive fermentation. The thermal load may be exacerbated by environmental conditions of high ambient temperature, or high solar radiation. High temperature restricts foraging activity mainly over midday, and more during summer than in winter. At the other extreme, the thermal drain associated with cold or windy conditions may lead animals to seek shelter, thereby suppressing foraging activity.

Maintenance of thermal balance may be partly responsible for the phasing of daily activity into foraging spells, separated by periods of resting or at least inactivity. Large herbivores typically forage in spells of about 2–3 h, with resting periods lasting similar or longer durations. An extended inactive period typically occurs over midday, when conditions are hottest. Inactivity may prevail at night, because of heightened predation risk rather than temperature constraints.

The gut space emptied by digestion while resting is filled during the course of the subsequent foraging spell. Gut space is physically limiting, if at all, only towards the end of foraging spells (Fig. 6.9). The target rate of intake should be governed by the potential duration of the foraging spell, as set by thermal conditions. Herbivores should tend to select for food that is more rapidly consumed during morning spells, curtailed by the heat of midday, and for more digestible food during afternoon spells, prolonged into the cool of the evening. This pattern was observed for kudus, with more woody browse being consumed during morning foraging spells and more forbs during the afternoon (Owen-Smith 1993a).

The proportion of the day spent foraging is influenced by prevailing conditions. For North American ungulates, daily active time was negatively related to mean daily temperature during summer (Belovsky and Slade 1986). However, for kudus the total time spent active was not curtailed unless the maximum daily temperature was extremely high (over 36 °C in summer and 30 °C in winter) (Owen-Smith 1994, 1998a). These conditions occurred on only about 15% of days. Thus active time was effectively unrestricted thermally on six days out of seven.

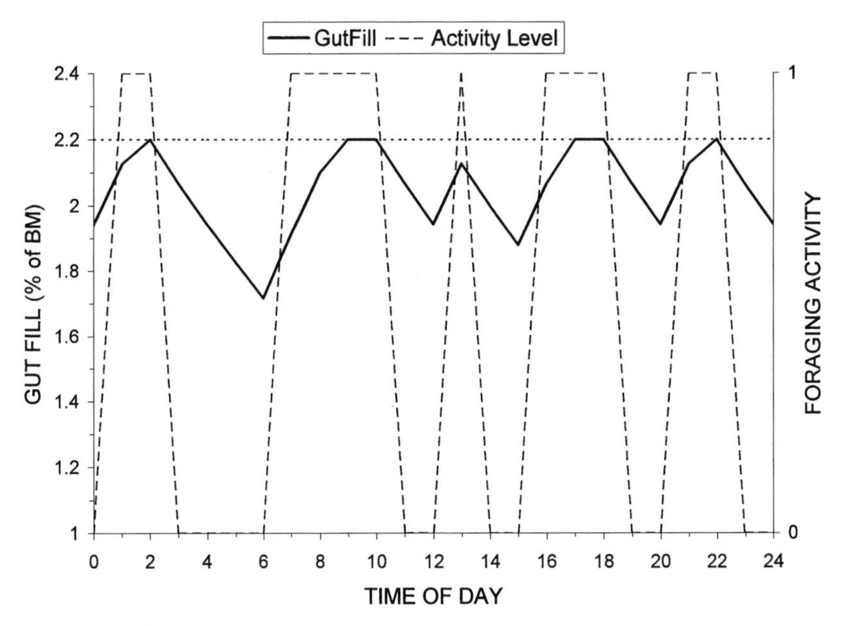

Figure 6.9. Fluctuations in rumen fill resulting from fluctuating foraging activity level over the daily cycle (assumed parameters: maximum gut fill = 2.2% of body mass, digestive turnover rate = 0.06 per hour). Average gut fill equals about 90% of the maximum fill.

North American deer remain inactive in thick cover during heavy rain, particularly when combined with cold or wind (Parker 1988). Kudus likewise seek cover in thickets during windy weather. For both kudus and deer, changes in pelage underlie seasonal differences in thermal tolerance. During snow blizzards, the foraging activity of boreal ungulates can be restricted for several successive days. Animals may need to compensate for energy deficits incurred on unfavourable days by foraging for longer when conditions improve.

Overall assessment

Thermal conditions form an intermittent restriction on foraging activity. High temperatures curtail foraging activity mainly towards midday, and low temperatures on those days when extreme cold or windy conditions occur. Thermal tolerance may be such that daily activity is restricted on only about one day in seven. Heat buildup from muscular activity forms a more fundamental limitation on the duration of foraging spells. Diet

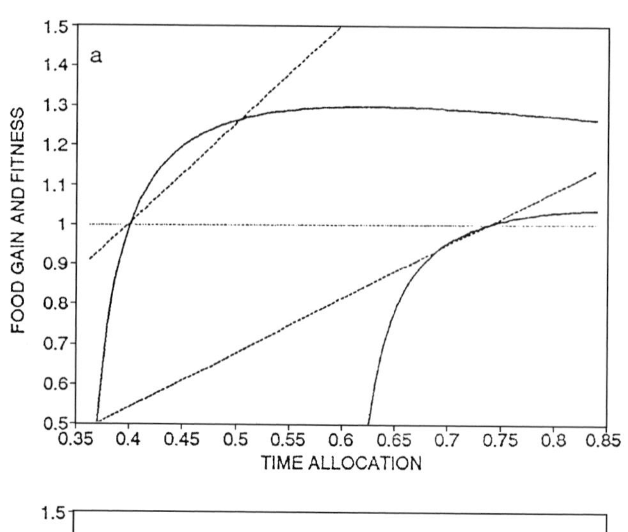
a
FOOD GAIN AND FITNESS
TIME ALLOCATION

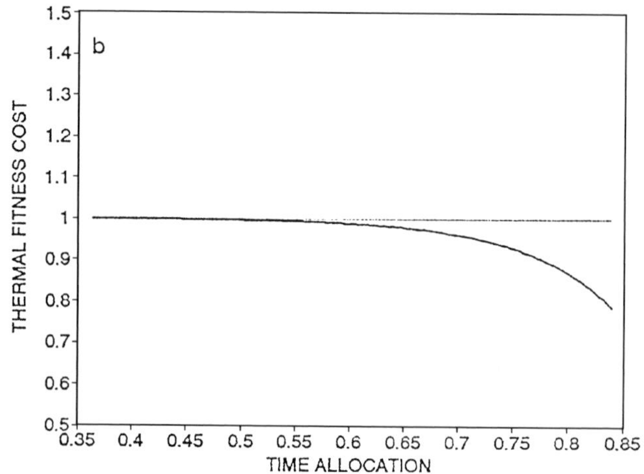
b
THERMAL FITNESS COST
TIME ALLOCATION

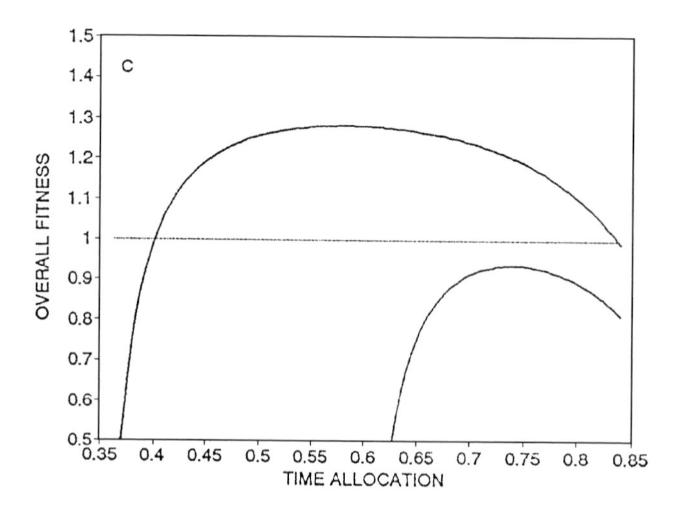
c
OVERALL FITNESS
TIME ALLOCATION

selection may be adjusted relative to the varying duration of these spells at different times of the day.

Constraints versus cost–benefit isoclines

Putative constraint settings are in practice somewhat elastic. Animals can forage beyond their thermoneutral zone, by increasing expenditures such as water losses for cooling, or shivering for heating. Digestive capacity can be expanded by increasing gut volume, at the cost of growing and maintaining the extra tissues. For ruminants, digestive turnover can be promoted by spending extra time and energy re-masticating food. The metabolic satiation level varies with prevailing requirements for activity, growth, storage and reproduction. Accordingly, the constraint lines depicted in LP models are somewhat arbitrary indicators of changing costs relative to gains. More importantly, the underlying benefit–cost relation is likely to vary non-linearly, as constraints are stretched towards the upper physical or physiological limits (Owen-Smith 1994) (Fig. 6.10). Under benign conditions, a broad plateau region may allow considerable flexibility in foraging responses, e.g. in daily foraging time or patch choice. In adverse circumstances, animals may be more narrowly restricted in their benefit–cost balance.

Using average values to assess the effective settings of the cost–benefit relations governing constraints is valid only when underlying relations are linear. Under conditions where the functional responses are strongly non-linear, and environmental conditions variable, this approach is likely to be misleading. Much of the time, foraging behaviour may be governed largely by meeting target nutrient requirements, relative to basic costs and risks of activity. Periodically, particular constraints come into operation.

Figure 6.10. Hypothetical relations between foraging time allocation, food gains and fitness. (a) Relation between food gains (dashed line), associated fitness benefits (solid line) and time allocation. Plots on the left represent conditions of high food abundance (e.g. wet season), those on the right conditions of food restriction (e.g. dry season). Food gains increase linearly with time allocation, whereas fitness shows an asymptotic relation. (b) Relation between fitness costs associated with thermoregulation and time allocation. (c) Relation between overall fitness (the product of the fitness components depicted in a and b) and time allocation. Under benign conditions (depicted on the left), fitness varies little with time allocation over a wide range. Under adverse conditions of food availability (depicted on the right), the optimal time allocation is higher despite the greater associated cost. (From Owen-Smith 1994.)

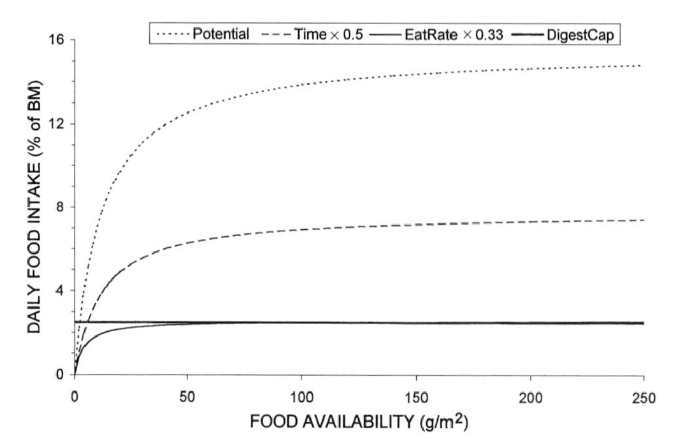
a. FORAGING vs DIGESTIVE CONSTRAINTS
Potential
Time × 0.5
EatRate × 0.33
DigestCap
16
12
8
4
0
0
50
100
150
200
250
DAILY FOOD INTAKE (% of BM)
FOOD AVAILABILITY (g/m2)

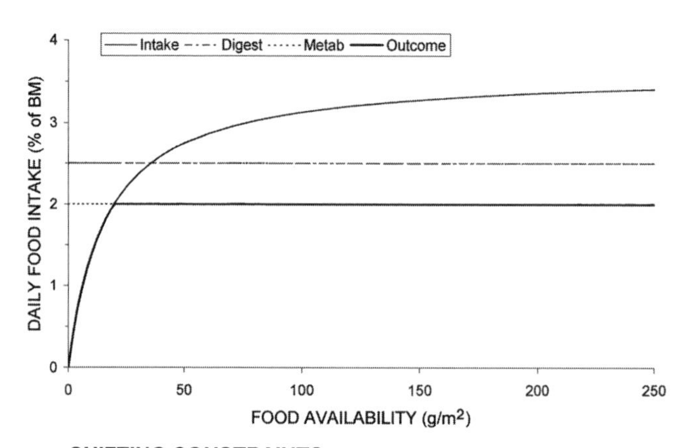
b. METABOLIC SATIATION
Intake
Digest
Metab
Outcome
4
3
2
1
0
0
50
100
150
200
250
DAILY FOOD INTAKE (% of BM)
FOOD AVAILABILITY (g/m2)

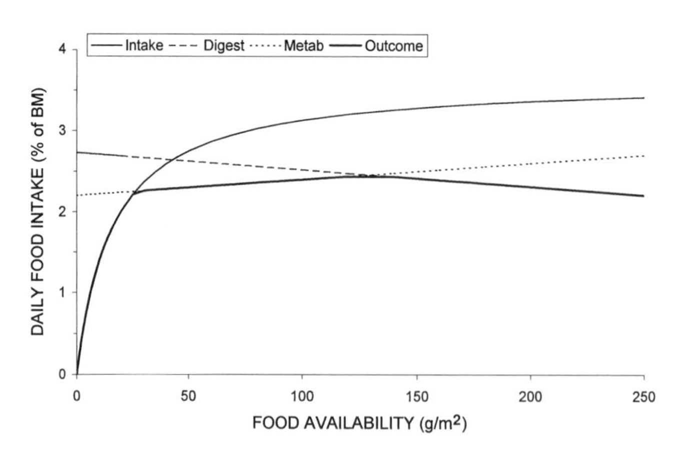
c. SHIFTING CONSTRAINTS
Intake
Digest
Metab
Outcome
4
3
2
1
0
0
50
100
150
200
250
DAILY FOOD INTAKE (% of BM)
FOOD AVAILABILITY (g/m2)

In the long term, physiological tolerance should be adjusted such that animals have the capacity to cope with extreme conditions, up to some limit dependent on how frequently these stress conditions are experienced. Thus, for much of the time animals forage unconstrained by digestive capacity, thermal tolerance or intake restrictions, and so appear slack in the amount of time spent idling (Herbers 1981). Intermittently, they must take the strain as conditions move towards their limits of tolerance.

Truncation of the intake response

As noted above, the potential daily food intake, based on maximum rates of food consumption, vastly exceeds the digestive processing capacity. Thermal restrictions, and other demands on time, restrict the maximum time that can effectively be devoted to foraging. Nevertheless, digestive physiology sets the upper limit to the amount of food that can be consumed per day over much of the range in food abundance (Fig. 6.11a).

Rather than being restricted directly by digestive capacity, animals may respond to this ceiling by adjusting their foraging behaviour. For example, bite size, and hence the food intake obtained while foraging, may be reduced, through narrowed selectivity for plant parts. To the extent that the selected parts are also lower in fibre content relative to nutrients, this also alleviates the digestive constraint to some extent. Animals could also restrict their selection of other vegetation components, such as the plant species accepted as food, although this dietary narrowing would have to be quite severe to reduce intake rate sufficiently (cf. Fig. 6.11a). Whatever the proximate mechanism, the effect is that the daily intake response to changing food availability becomes truncated well below its potential maximum.

Figure 6.11. Functional response in daily food intake to changing food abundance, as influenced by various constraints, for a hypothetical 100 kg ruminant.
(a) Comparative effect of restrictions on (i) daily foraging time, limited to 50% of 24-hour day, (ii) eating rate, reduced to one third of maximum, plus foraging time to 50% of day, and (iii) digestive processing capacity restricted to 2.5% of body mass per day, compared with potential daily intake if animals foraged continuously with eating rate (I_P) of 12 g/min. (b) Metabolic satiation restricts intake below the limit set by digestive processing capacity. (c) Situation with food quality decreasing with increasing food abundance, so that the digestion constraint has a negative slope and the metabolic constraint a positive slope. Accordingly the effective constraint shifts from intake rate at low food availability to metabolic satiation at intermediate food availability, and digestive capacity at high food availability.

At times when food quality is high, the direct limitation on intake could be metabolic satiation, with digestive capacity filled somewhat below its potential (Fig. 6.11b). Moreover, food quality could decrease with increasing food abundance, e.g. if taller grass is more fibrous. Accordingly, the plateau region would assume a negative slope towards increasing vegetation biomass if digestive capacity is limiting, but a positive slope if animals respond by eating more per day to satisfy their metabolic needs. The outcome could be an intake response showing three distinct regions: (1) a segment at low food abundance, where food availability is the main constraint on intake; (2) a segment at intermediate food abundance, where daily food intake increases to compensate for the declining nutritional value; and (3) a segment at high food abundance, where digestive processing capacity limits intake (Fig. 6.11c).

Both food abundance and food quality tend to decline simultaneously during the course of the dormant season. Accordingly, the relative inclinations of the metabolic and digestive constraints could become the opposite of that depicted in Fig. 6.11c, assessed over the seasonal cycle in vegetation biomass. The outcome remains a humped rather than asymptotic response pattern (cf. Fig. 3.6c). A gradually asymptotic, or 'Holling Type II' curve, with intake restricted solely by ingestive handling time towards high food abundance, is probably unusual for herbivores. More commonly, we can expect to find various patterns of truncation of the intake response over different time frames.

Overview

Limitations on food intake over extended time periods arise from the constraints imposed by digestive processing capacity, thermal tolerance and metabolic satiation. Herbivores may respond by restricting bite sizes below their maximum, through being more selective among plant species, or by adjusting their time allocation over the day–night cycle. Effectively, short-term gains must be balanced against the longer-term consequences. The existence of multiple potential constraints, plus adaptive responses by animals to them, may modify the form of the intake response to changing food availability. In particular a truncated, or even a humped, response pattern may be generated.

Constraints are somewhat elastic in their effective settings, since they represent the outcome of changing benefit–cost relations. Potential constraints also vary in their effectiveness over daily and annual cycles. Linear programming models based on averaged values for controlling parameters can be misleading, and vulnerable to circularity, under such conditions.

Further reading

General models of digestion processes are provided by Penry and Jumars (1986, 1987), Alexander (1991) and Penry (1993). A mechanistic digestion model for large mammalian herbivores is outlined by Illius and Gordon (1991, 1992). The variable effects of high temperatures on daily activity of kudus were assessed by Owen-Smith (1998a). A detailed analysis of the thermal tolerance of deer is presented by Parker (1988). Metabolic processing costs and limitations are discussed for herbivores by Tolkamp and Ketelaars (1992), and more generally by Ricklefs *et al.* (1996). The debate about the potential circularity of linear programming models can be followed in Owen-Smith (1993b,c, 1994, 1996b, 1997), Belovsky and Schmitz (1993) and Belovsky (1994). Laca and Demment (1996) give an overview of the operation of various constraints on foraging strategies of herbivores. A useful theoretical analysis of the influences of digestion vs. handling prey on the form of the functional (intake) response is provided by Jeschke *et al.* (2002).

7 · *Resource allocation: growth, storage and reproduction*

Surplus resources gained from the environment while foraging, in excess of basic maintenance needs, can be allocated to biomass increase in different ways. Consumers can (a) grow in individual body size, (b) add to stored body reserves, or (c) divert resources to nurture the growth of offspring. The relative advantages of these alternative investments depend on the age, size and current body condition of individual animals, and also on timing within the seasonal cycle of resource abundance. The patterns of allocation by different individuals govern how food gains become transformed into population biomass dynamics. At times resource gains may be insufficient to cover physiological losses, so that population biomass declines.

In a variable environment, tradeoffs need to be made between surplus gains at one time and deficits incurred at a later stage. Evaluating the outcome requires a dynamic approach to optimization. In particular, we must consider the consequences of allocation decisions for the future body state of individual animals over some extended period. The time frame thus expands to an annual cycle or longer, ultimately up to individual lifespans.

In this chapter, we consider first how resource gains become transformed into a biomass growth potential, taking into account losses to maintenance metabolism. The simplest decision is then considered, whether and when to allocate surplus resources to fat stores, for fully grown animals that have no growth potential. The benefits of having body reserves need to be balanced against the costs of acquiring and carrying them. Next, the tradeoffs faced by immature animals in allocating resources either to growth in size or to stored reserves are assessed. Finally, the additional option of investing surplus resources in the growth and survival of offspring is taken into account. A biomass-based perspective on life history patterns is thereby developed for seasonally varying environments.

Relative growth potential

The potential for growth in biomass to occur depends on the extent to which resource gains exceed metabolic losses. This balance is most readily assessed by using a currency of digestible (or, more strictly, metabolizable) energy, or its biomass equivalent. Protein and other nutrients may restrict the growth potential at times, but energy seems to be most generally limiting for large herbivores (Robbins 1993). Energy gained while foraging must cover energy expended during other times of the day when animals are resting or engaged in non-foraging activities. For biomass growth to occur, there must be a net energy surplus over the full daily cycle, and other nutrients contributing to the material content of this biomass must also be adequately supplied. As an example, Parker *et al.* (1996) assessed the 'foraging efficiency' of the young black-tailed deer that they studied on an Alaskan island in terms of the ratio between energy gained from the food consumed while foraging and the associated energetic cost of foraging. For these deer, this measure varied from 2.5 in summer to 0.7 during the winter months, i.e. a substantial energy surplus in the benign season changed to a deficit in the adverse season.

For cross-species comparisons, it is helpful to standardize the growth potential relative to the basal metabolic requirement. Accordingly, the *relative growth potential* (RGP) will be assessed in terms of the net difference between biomass gained from food consumed, G, and overall physiological losses in biomass for maintenance metabolism plus activity costs, M_P, divided by the basal metabolic requirement. This entails partitioning M_P between the fixed cost of basal metabolism (M_{P0}), and the variable costs of activity and other physiological processes (M_{Pa}). Hence, by modifying Equation 2.27 we obtain

$$\mathrm{RGP} = \{G - (M_{P0} + M_{Pa})\}/M_{P0}. \qquad (7.1)$$

Growth can occur if the RGP is positive, whereas biomass is lost if it is negative. Mortality losses M_Q will be ignored, although strictly the mortality risk associated with foraging activity should also be considered.

An upper limit to the growth potential is set by the 'sustained metabolic scope' for the species in question. This is the maximum metabolic rate that can be maintained over a prolonged period, as a multiple of the basal (or resting) metabolic requirement (cf. Chapter 5). For ungulate-sized mammals, estimates of this ratio are typically about 3–4 (Ricklefs *et al.* 1996; Hammond and Diamond 1997). Assuming that activity costs amount to 50% of the basal metabolic rate, the maximum RGP amounts to about twice the basal metabolic loss. In effect, mammals of this size cannot gain energy, or its biomass equivalent, faster than about twice the rate at which they would lose it on a starvation diet. For medium-sized

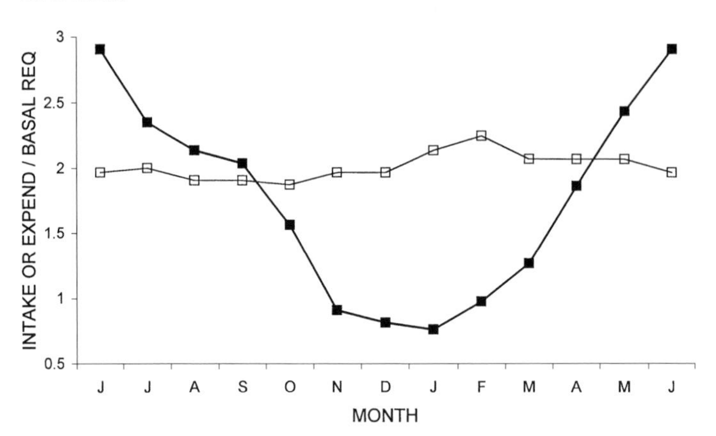

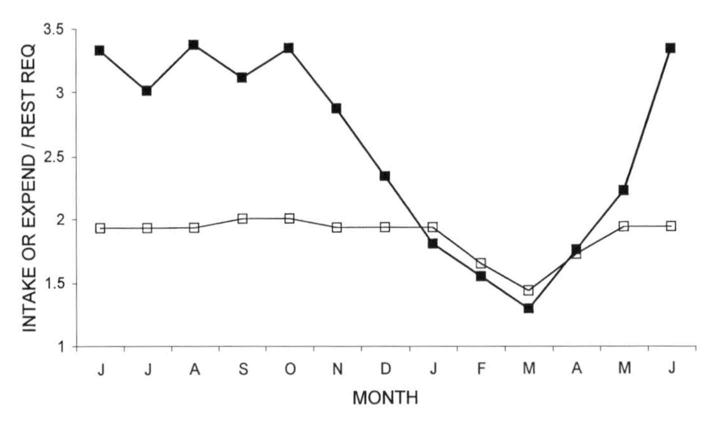

Figure 7.1. Changes in daily energy gains and expenditures over the seasonal cycle, relative to basal or resting energy requirements. (a) Young black-tailed deer in coastal Alaska (averaged over years and sexes from Parker *et al.* 1999; expenditure adjusted × 0.7 so that zero growth occurs for an intake:requirement ratio of 1.0, cf. Fig. 7.2a). (b) Yearling cattle in Colorado (adapted from Senft *et al.* 1987). (c) Adult cattle in northern Kenya, relative to maintenance requirement (from Coppock *et al.* 1986). (d) Sub-adult kudus in South Africa (modified from Owen-Smith and Cooper 1989, assuming basal metabolism equals 0.8 × resting metabolism).

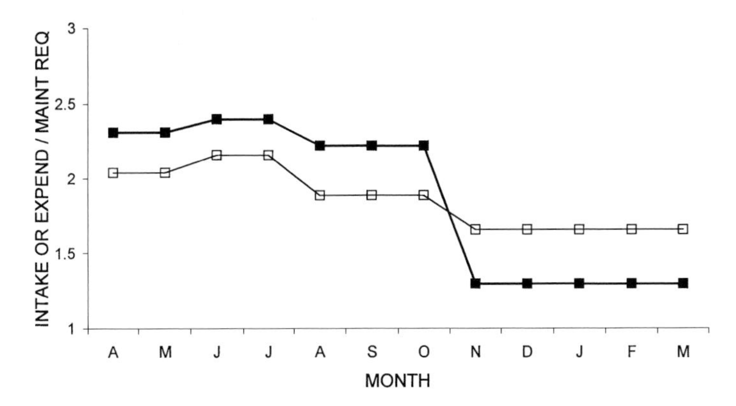

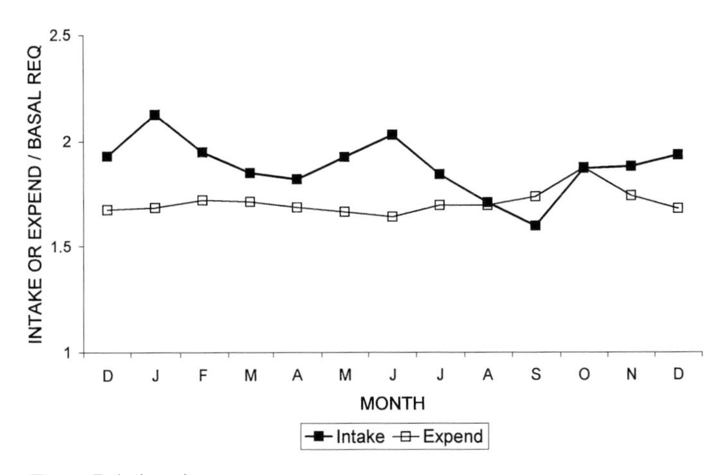

Figure 7.1. (cont.)

ungulates, the rate of mass loss due to basal metabolism amounts to about 8 g per kg of body mass per day (see Table 13.1). Hence, the maximum possible growth rate is projected to be around 15–20 g per kg of body mass per day, ignoring losses that may be associated with the conversion of nutritional gains into body growth.

For black-tailed deer in coastal Alaska, the peak energy intake in June was almost three times basal metabolism, whereas the daily energy intake during mid-winter fell somewhat below the basal requirement (modified from Parker *et al.* 1999 by personal communication) (Fig. 7.1a). Daily energy expenditure (excluding growth) amounted to about 1.5 times basal metabolism, peaking in late winter when snow was deepest. This translates into seasonal extremes in RGP of 1.5 and −0.75. The deer restricted energy deficits in winter by reducing their energy expenditure while foraging to 83% of the summer level. Protein intake remained in excess of requirements throughout the year, except in February. Daily mass gain was curvilinearly related to energy intake relative to maintenance requirements (Fig. 7.2a). The maximum growth rate amounted to around 5 g per kg of live mass per day; the most extreme rate of mass loss during winter was of similar magnitude.

Young growing cattle observed in Colorado showed a maximum RGP of 1.4 in early summer, declining to −0.1 over winter (Senft *et al.* 1987) (Fig. 7.1b). Correspondingly, these animals gained mass at a rate of up to 3.2 g per kg live mass per day in summer, while losing up to 0.4 g per kg live mass per day during winter. Seasonal variability in mass change was somewhat greater for beef breeds of cattle foraging on natural grassland in South Africa, with extremes of 3.8 and −2.0 g per kg per day being shown (Fig. 7.2b). For free-ranging cattle kept by nomadic pastoralists in arid northern Kenya, RGP averaged 0.15 over the wet season and early dry season, decreasing to −0.23 during the later dry season (Coppock *et al.* 1986) (Fig. 7.1c). These animals gained mass at a mean rate of 0.8 g per kg live mass per day over the wet and early dry season period, and lost mass at a rate of 1 g per kg live mass per day during the later portion of the dry season. However, the energy requirement estimated for these animals included the cost of lactation, which continued through the dry season.

Our studies on growing sub-adult kudus indicated a maximum monthly RGP of 0.45, and average RGP over the wet season of 0.16 (Fig. 7.1d). Energy balance became negative only for the final month of the dry season. In contrast to the deer, kudus increased their daily foraging time, and hence energy expenditure, during the adverse season. Protein intake seemed well in excess of requirements in all months. However, a RGP exceeding 1.0 could have been achieved during the wet season, had the animals expanded their foraging time and digestive capacity to the limits shown in other seasons (Fig. 7.3). Instead, foraging time and food intake rate seemed to be adjusted to match the energy and growth requirements of these animals. Intakes in excess of requirements in January (mid wet season) and June (early dry season) were perhaps related to re-gaining lost condition, and building fat reserves, respectively.

Hence, although the maximum energy gain obtained by deer over the daily cycle in north temperate environments can approach the limit set by metabolic capacity, the potential seems to be somewhat below this

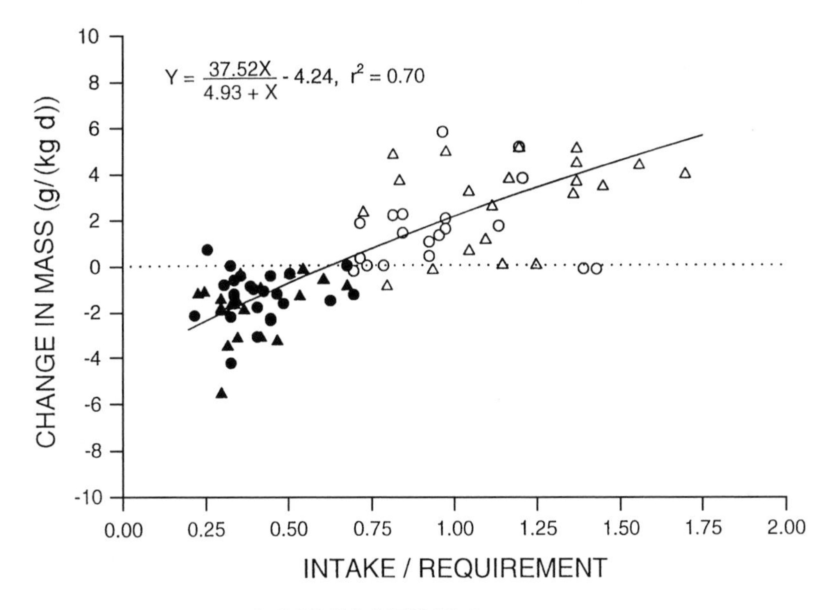

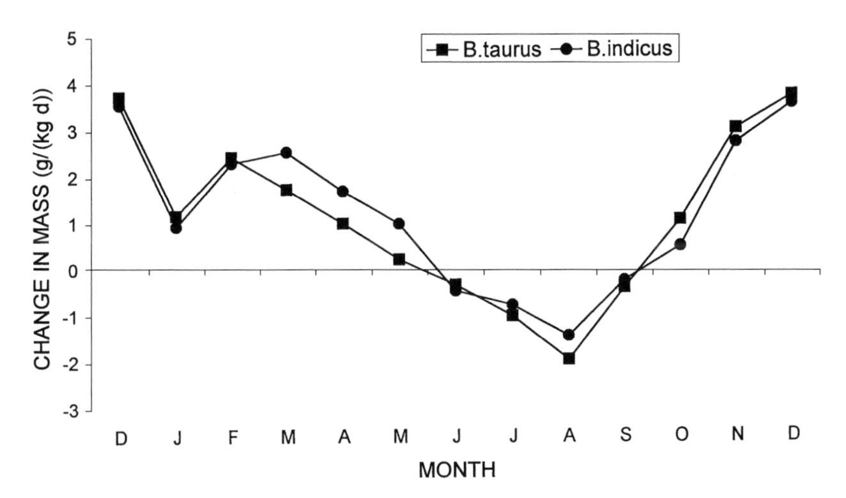

Figure 7.2. Daily increment in body mass related to food intake and month. (a) Daily mass change compared with daily energy intake relative to requirements, including growth needs, for black-tailed deer (from Parker *et al.* 1999, © The Wildlife Society). (b) Comparative daily mass change of *Bos taurus* (Shorthorn) and *B. indicus* (Afrikander) breeds of beef cattle foraging on natural grassland in South Africa. (From Joubert 1954).

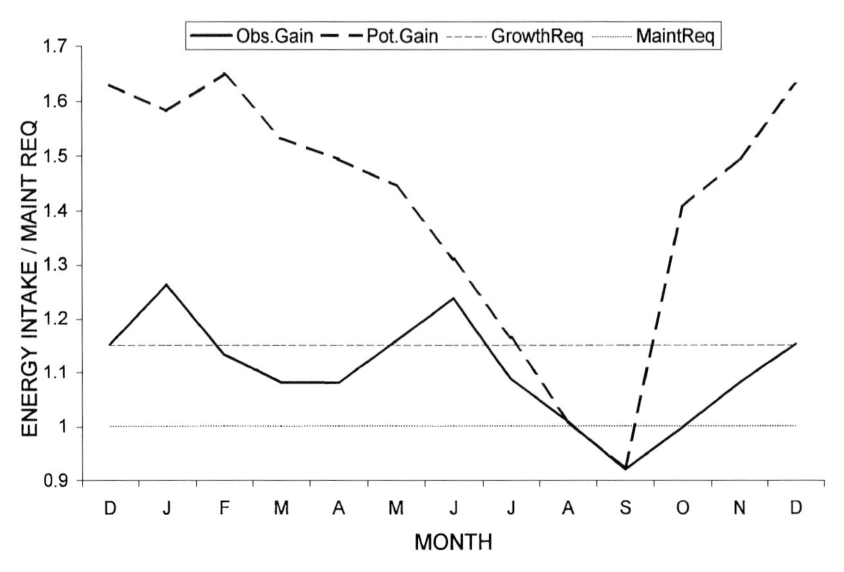

Figure 7.3. Observed versus potential daily energy gain by immature kudus, shown in relation to maintenance requirement (basal metabolism plus activity cost), and also to requirements assuming a growth increment of 1.5 g per kg live mass per day. Potential daily gain was derived by stretching time allocations to foraging to the maximum limits observed during the dry season. (From Owen-Smith and Cooper 1989.)

level elsewhere. Inefficiencies in converting the metabolic surplus into growth restrict the maximum growth potential to about 5 g per kg of body mass per day for ungulates ranging in body mass from a 60 kg deer to a 600 kg cow. In the absence of food, animals would lose mass through starvation at a somewhat greater rate than they can potentially gain it.

Storage

Storage capacity

For fully grown animals, changes in body mass are due largely to changes in body fat (Baile and Forbes 1974; Mrosovsky and Prowley 1977). Studies have shown that animals adjust their daily energy intake so as to regulate their fat reserves around a set point (Woods *et al.* 1998). However, the set point may vary seasonally and be influenced by diet composition. Upper limits to body fat may be governed by adverse consequences for heat dissipation, increased risk of predation, and perhaps other health hazards.

Among ungulates, the highest fat deposits, amounting to 15–30% of total body mass, are exhibited by reindeer living on predator-free islands in arctic latitudes (Tyler 1987). The fat reserves of deer inhabiting more temperate latitudes typically amount to between 10 and 20% of body mass (Parker *et al.* 1993). For domestic cattle, fat typically represents about 20% of body mass; for sheep, fat stores may vary between 15 and 45% of live mass. In contrast, among African ungulates fat reserves rarely exceed 5–10% of body mass (Ledger 1968; Smith 1970). The place of storage varies, with island reindeer laying fat mainly sub-cutaneously, and cattle mainly intra-muscularly. Among African antelope, fat is distributed among muscles, mesenteries and bone marrow.

Fat reserves serve as a buffer against energy deficiencies during adverse periods. One gram of fat yields 39 kJ of energy, compared with 22 kJ per gram of muscle tissue (Panaretto 1968, cited by Parker *et al.* 1993). Ungulates build up fat reserves during the vegetation growing season, and deplete this reserve during the dormant season when food availability and quality are both low. Fat may also serve as an energy reserve for periods of high demand associated with reproduction.

The mobilization of fat reserves during adverse periods entails associated losses in body protein. During winter, black-tailed deer lost fat at rates of 22–52 g/day and protein at 5–10 g/day (Parker *et al.* 1993). Reindeer lost 16% of their body mass as fat plus 9% from muscle tissue over winter (Tyler 1987). Mule deer on sub-maintenance rations depleted their fat reserves at proportionately 2.5 times the rate of their protein reserves (Torbit *et al.* 1982). The protein gain can become negative when dietary protein concentrations fall below 4–5% of food DM, owing to metabolic losses of protein in faecal material (Moen 1973). African ungulates may lose substantial body tissue as protein rather than as fat. Catabolism of muscle tissues can aid the recycling of nitrogen to the rumen contents, thereby maintaining the activity of microbes involved in generating energy substrates from food (Robbins *et al.* 1974; Duncan 1975). Chemically extractable fat includes also essential lipids, which represent about 1.5% of lean body mass.

Mammals can lose up to one third of body protein before dying of starvation (Cahill 1970; Torbit *et al.* 1985), although reindeer seem able to tolerate losses of up to 45% (Tyler 1987). A 30% protein loss results in a reduction of about 15% in lean ingesta-free body mass (Ledger 1968). Maintenance metabolism dissipates about 0.5% of body mass per day for medium-sized ungulates (assuming a conversion factor of 20 mJ per kg live mass). Movements and stress impose additional costs. Hence, an initially well-nourished ungulate could not survive for more than about a month without food.

For wild animals in natural environments, susceptibility to mortality depends on additional factors besides reaching threshold levels of tissue wastage. A decline in body condition lowers the production of antibodies to parasites and pathogens. This makes animals more vulnerable to disease, and less capable of searching for food. Weakened individuals are quickly detected and killed by predators. Cold snaps may accentuate mortality. This gives rise to 'multiplier effects', whereby small deficiencies in resources can become lethal through the mediation of pathogens and predators (Sinclair 1977; White 1983).

Dynamic storage model

Fat is stored in anticipation of dietary energy deficiencies. How much fat should be stored, and when? Should enough fat be stored to ensure survival in the worst anticipated conditions, or only for a typical year? How should the costs of building and carrying fat reserves be traded against the potentially crucial benefits of such reserves in seasonal environments?

To establish the most advantageous pattern of fat storage over the seasonal cycle, a dynamic modelling approach is needed, following Mangel and Clark (1988). The core of this approach is the relation between current body state and expected future fitness. For non-reproductive animals, fitness can be equated simply with survival chances. Body condition, i.e. the mass of muscle tissue and stored fat relative to skeletal size, represents an appropriate state variable. Changes in body condition are governed by decisions concerning resource allocation over the annual cycle, relative to the growth potential allowed by the environment at different stages. Consideration is restricted to fully grown animals, for which increase in body size is not an option.

The first need is to decide how the expectation of survival varies with body condition. Costs are associated with carrying fat, not only in terms of metabolic expenditures for transporting the additional mass, but also through thermoregulatory restrictions if the fat is sub-cutaneous, and via lessened mobility in evading predators. Predatory risks are also incurred through the extra foraging time needed to acquire surplus resources to produce the fat deposits. Survival chances are likewise reduced if lean body mass falls below some optimal mass relative to size, because animals become weakened and less able to cope with stresses, pathogens and predator attacks. Survival is likely to be affected more by losses in lean body mass than by carrying the equivalent extra mass in the form of fat. A survival (fitness) function of the appropriate form is depicted in Fig. 7.4a. How severely survival is reduced by deviations from the optimal body condition will depend on predation hazard and other environmental factors (e.g. presence of shelter).

a

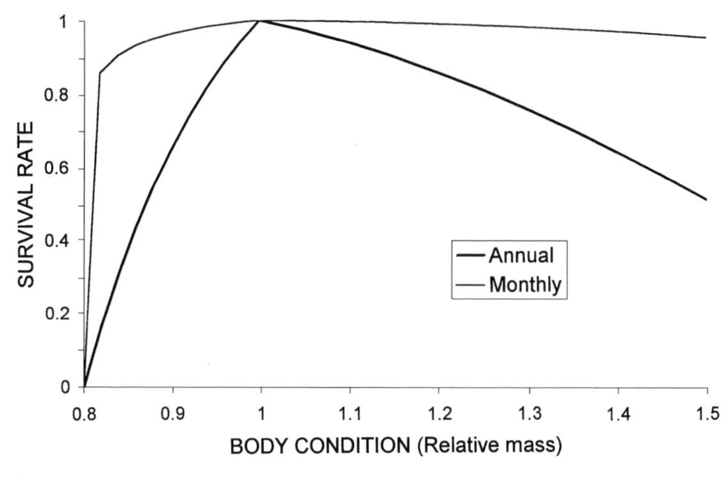

b

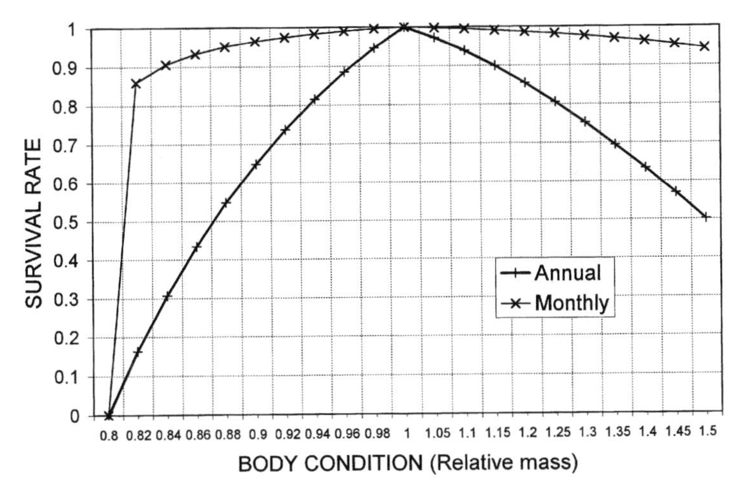

Figure 7.4. Fitness function relating monthly and projected annual survival rates to body condition, expressed as body mass relative to optimal lean mass, used in the allocation models. (a) Hypothetical variation in monthly and projected annual survival rate with departures from the optimal body mass. A body condition of 0.8, representing the loss of 20% of lean body mass, gives zero survival chance over one year in the wild. A body condition of 1.5, representing fat stores amounting to 50% of optimal lean mass, reduces the annual survival to half of that for animals in optimal condition. (b) Transformation of continuously varying condition into 20 discrete condition states for the SDP model. Each state varies by a mass increment of 0.02 for losses in body mass below the optimum mass, and by 0.05 for increases in body mass above the optimum through fat storage.

The specific procedure for stochastic dynamic programming (SDP) outlined by Mangel and Clark (1988) will be followed. For computational purposes, time is sub-divided into discrete steps. Survival over an extended period is the product of survival chances over each time step. The key to solving a dynamic optimization problem is to work backwards over time. The best state to be in at the end of the time period considered, e.g. the body condition associated with highest expected survival, is first identified. Next one evaluates how best to arrive at this state from one time step earlier, for each possible state the animal could be in at the start of this interval. This procedure is iterated back towards the start of the overall period covered. The optimal decision sequence maximizes fitness at each time step, by controlling body state appropriately. This defines the trajectory of states giving highest fitness over the entire period. The underlying principle is identical to that used in life history theory, where current reproductive output is traded against residual reproductive value (see Clark 1993).

Specifically we must determine the maximum fitness Φ that can be attained over the seasonal progression $T = 1, 2 \ldots . t_{max}$, as a result of the allocation decisions taken at each stage. The allocation of resources determines the particular body condition β reached after each time interval, and hence current fitness (expectation of survival). By definition, the function $\Phi(\beta, t, t_{max})$ is the maximum value of the probability that the animal survives from any particular time t to the terminal time t_{max}, given that at time t it is alive, and that its condition at time t is β. As a consequence of the allocation decision, the animal's condition may change from time t to $t + 1$, influencing its future fitness. The optimal sequence of decisions entails maximizing Φ at each time step.

The net energy gain obtained from food resources during the time step $t \rightarrow t + 1$ determines the maximum biomass increase (change in condition) that can occur during this period. This gain can be allocated towards either (1) increasing fat stores, (2) restoring lost lean body mass, or (3) counterbalancing metabolic attrition, so as to maintain body condition. If the potential gain falls below the metabolic maintenance level, lean body mass will decline whatever the allocation. Hence, the dynamic programming equation takes the form

$$\Phi(\beta, t, t_{max}) = \max_{i=1,2,3} \Phi(\beta - \mu + \alpha_i, t + 1, t_{max}), \qquad (7.2)$$

where β = specific body condition, α_i = allocation action (store fat, restore lean mass or maintain condition), specified as a mass change, and μ = metabolic (physiological) mass loss that occurs inevitably. The maximum allocation is constrained proximally by food availability, and ultimately by metabolic capacity. A stochastic element can be introduced by making food availability a random variable.

In solving the dynamic programming equation, a monthly time step was used. Body condition was partitioned into 20 discrete states as depicted in Fig. 7.4b, with increments in body mass more finely divided for condition states below the optimum than for states above it. An upper limit to α was set by the monthly growth potential permitted by food resources, effectively determined by energy balance. For illustration, a monthly growth potential of 0.1 per month means a maximum biomass increment of 10% of body mass from one month to the next, while a growth potential of -0.1 translates into a mass loss of 10% per month.

Four seasons were distinguished, labelled the 'gain season' (GS, i.e. summer, wet or growing season), the 'deprivation season' (DS, i.e. the winter, dry or dormant season), and two transitional seasons (labelled GD and DG, respectively) (Fig. 7.5a). The model allowed consumers to allocate one third, two thirds or all of the monthly growth potential towards fat stores. Normal, extreme, and stochastically varying patterns of monthly growth potential were considered, as depicted in Fig. 7.5a. For the stochastic model, it was assumed that normal conditions prevailed 50% of the time, high resource availability 25% of the time, and low availability 25% of the time. Based on the assumed fitness function and seasonal resource supplies, the model generated the optimal body condition for each month of the seasonal cycle.

The output suggests that, with deterministic variation in seasonal conditions, it is advantageous for consumers to delay depositing fat until near the end of the GS (Fig. 7.5b). The amount stored should be no more than needed to avoid a decline in body condition below the optimal state by the end of the DS. If the amplitude of the seasonal fluctuation in resource supply is more extreme, consumers should store more fat, and begin laying it down earlier, than under 'normal' conditions. The magnitude of the growth potential allowed by resources during the GS does not influence how much fat should be stored, but does affect how long it takes to build up the required level. If the monthly growth potential varies stochastically, it is adaptive for consumers to store somewhat greater reserves than in a predictable environment, and to retain a small reserve as a buffer against possible shortfalls during the transitional seasons.

It is not advantageous for consumers to acquire surplus energy, if all that can be done with it is to add to costly fat reserves. Hence, during the early GS fully grown but non-breeding animals should secure just enough energy to cover their maintenance activity needs, with no

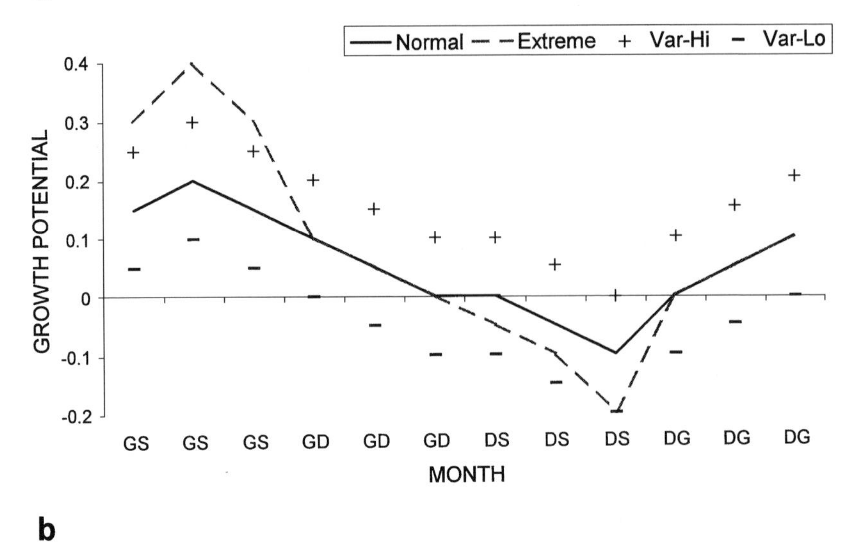

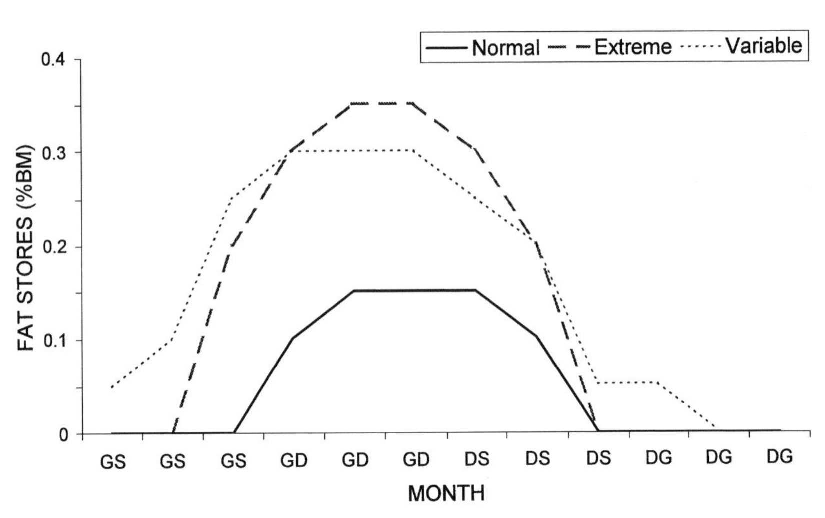

Figure 7.5. Assumed conditions and output predictions for the dynamic fat storage model. (a) Monthly variation in growth potential, as set by changing resource supplies. Two deterministic patterns are indicated, representing 'normal' and 'extreme' seasonal variability. A third pattern allows for stochastic variation in monthly resource supply around the normal level between the indicated extremes. GS, gain season; DS, deprivation season; GD and DG, transitional seasons. (b) Optimal monthly fat stores predicted by the model, for deterministic normal and extreme environments, and for stochastically varying conditions. (c) Optimal monthly change in body mass, through storage or condition loss, relative to the growth potential allowed by available resources, for the above environmental conditions.

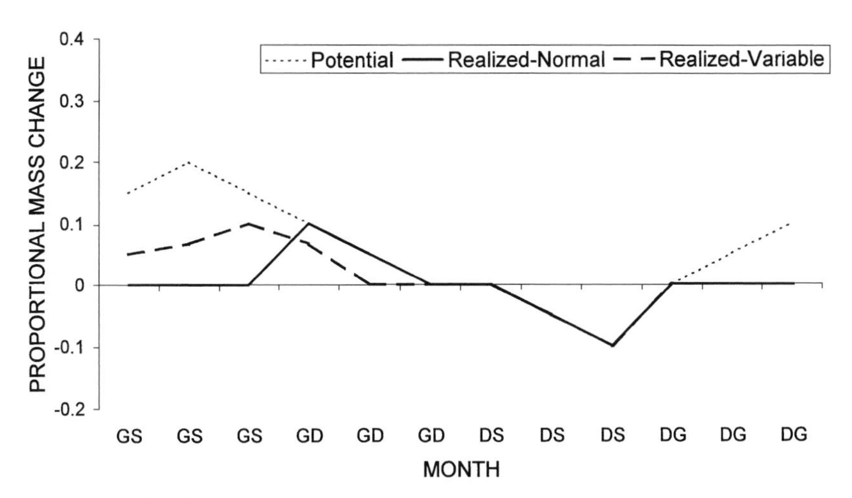

Figure 7.5. (cont.)

allocation to storage (Fig. 7.5c). In other words, maximizing daily energy gains is not optimal at this stage of the seasonal cycle. However, towards the end of the GS and through the DS, animals should forage with maximum possible intensity, so as to build fat deposits and minimize loss in condition after the resource supply becomes sub-maintenance. How the survival rate varies with body condition determines the precise tradeoff between costs of building fat reserves, versus those associated with attrition in lean body tissue. In unpredictably variable environments, it becomes advantageous to carry some fat through the GS as a buffer against the chance of losing body condition below the ideal level (Fig. 7.5c).

Observed fat deposition patterns

The predictions of this simple model are closely in accordance with the patterns of fat deposition observed for various ungulates. Fat is commonly stored towards the end of the summer or rainy season, and lost over the winter or dry season (Dunham and Murray 1982; Adamczewski *et al.* 1987; Holland 1992; Parker *et al.* 1993; Gerhart *et al.* 1996). Ungulates occupying extreme northern environments store more fat than those occupying more benign tropical environments (Ledger 1968). Highest fat reserves are observed on islands in high latitudes where predators

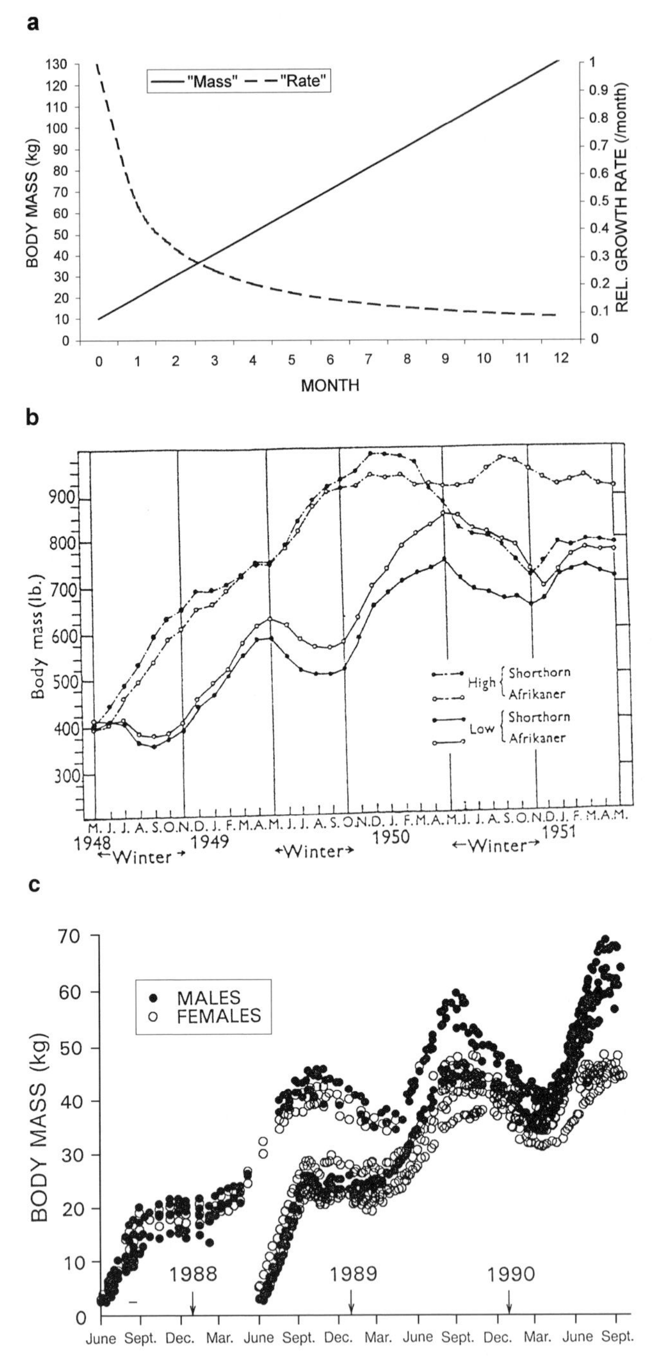

Figure 7.6. Patterns of growth in body size. (a) Projected monthly growth for a hypothetical ungulate in a constant environment, derived from formula $G_m = 10.5W^{0.75}$, where G_m = daily growth increment in grams

are lacking (Reimers *et al.* 1982; Tyler 1987; Adamczewski *et al.* 1993). However, for adult roe deer in France most fat loss takes place over summer, during the periods of highest reproductive demands for both sexes, with little loss taking place over winter (Hewison *et al.* 1996). Hence, patterns of fat storage cannot be ascribed solely to dormant season deficiencies in food availability.

Growth versus storage

Growth patterns

Animals that are not fully grown can use resource gains in excess of maintenance needs to grow either larger or fatter. Following birth, mass increase tends to be linear until adult size is neared (Robbins 1993). For ungulates, the potential growth rate relative to body size is approximately $10.5\,W^{0.75}$ g/kg per day, where W is adult body mass (modified from Robbins *et al.* 1981 and Parker *et al.* 1993). Hence, an ungulate weighing 100 kg as an adult could grow by around 10 kg per month during its early life. Although the growth increment remains roughly constant, the proportional growth rate declines with increasing size (Fig. 7.6a). A newborn animal weighing about 10% of its mother's mass can double its weight over its first month, whereas more than half-grown immatures increase in mass by around 10% per month. Smaller species show higher relative growth rates than larger species, as indicated by the dependency of growth rate on the metabolic mass equivalent.

Based on the above parameters, a medium-sized ungulate could potentially attain full adult mass within a year of its birth. However, the realized growth rate is restricted by resource availability. During the adverse season, growth is checked or may even become negative (Joubert 1954; Ringberg *et al.* 1981; Parker *et al.* 1993) (Fig. 7.6b,c). Superimposed on the body growth trend is seasonal variation in mass related to the deposition and mobilization of fat reserves. In practice, medium-sized ungulates usually take 16–24 months to reach full size, whereas cattle require about three years.

Figure 7.6. (*cont.*) per kilogram per day, and adult body mass $W = 150$ kg. (b) Comparative growth of two breeds of cattle in a seasonal grassland environment in South Africa, comparing heifers foraging on natural grassland only (low plane) with heifers provided with supplementary food during the winter dry season (high plane). The mass loss in the third year was associated with parturition and subsequent lactation (from Joubert 1954). (c) Growth from birth of two cohorts of black-tailed deer ranging freely on an island off the Alaskan coast (from Parker *et al.* 1993).

Young growing animals assign resources to growing larger at the expense of using these resources to augment body reserves. While individuals are small, relative to adult size, their survival chances are lowered, for manifold reasons: vulnerability to predation, limited reserves, high metabolic rate, inefficiency in procuring and digesting food. However, without stored reserves, animals may fail to survive periods of resource deprivation. When should immature animals allocate resources to growth, and when to storage, in seasonally varying environments?

Modelling growth–storage tradeoffs

To allow for growth in size, we introduce a second state variable Γ into the SDP model, representing the body size attained as a result of biomass gains. Accordingly, the dependence of fitness upon body size, independent of variation in fat content, must be specified. We assume that survival chances increase curvilinearly with increasing body size, relative to adult size (Fig. 7.7a). However, the mass gained also increases fitness, by bringing animals closer to the stage at which they commence reproducing (cf. reproductive value) (Begon *et al.* 1996). Hence, we will assume that effective fitness depends on the product of the expected survivorship, and the body mass gained if animals survive. The justification for this will become clearer in the next section, where reproduction is considered. Hence, the dynamic programming equation becomes

$$\Phi(\beta, \Gamma, t, t_{\max}) = \max_{i=1,2,..} \Phi(\beta - \mu + \alpha_i, \Gamma + \alpha_i, t + 1, t_{\max}) \quad (7.3)$$

where Γ = specific body size attained.

The model output suggests that juveniles should store less fat than adults or yearlings, because of tradeoffs between allocations to growth and storage (Fig. 7.7b). Accordingly, whereas adults may be able to avoid declines in lean body tissue during the DS, immature animals are susceptible to such condition losses. The precise balance in allocation depends on the relative forms of the fitness functions, as well as on environmental patterns of resource availability. The model generates the growth pattern typically shown by ungulates in seasonal environments. Increase in body size by juveniles and immatures is checked at the end of the GS, when resource surpluses become diverted towards building fat stores, followed by mass losses over the DS when stored reserves are metabolized (Fig. 7.7c).

Another assumption made was that animals commenced life as neonates at the beginning of the GS. If animals are born at other times in the seasonal cycle, their fitness, as indicated by their growth expectation, is markedly lower (Fig. 7.7d). The reduction in fitness persists through the yearling stage. Consumers that have the potential to grow should maximize the resources that they acquire at all times, subject only to the metabolic satiation limit.

Reproduction

Reproductive processes

Among mammals, growth is determinate and thus ceases when animals reach full adult size, except via changes in fat stores. Thereafter, adults contribute to population biomass increase only through the offspring they produce. The growing fetus is nourished within the uterus by resources passed from the mother's blood. The nutritional demands of the fetus increase markedly as the time of birth approaches. During parturition, the mother loses not only the mass of the neonate, but also that of the placenta and associated body fluids. Domestic cows lose about 15% of their body mass when they give birth, with the mass of the calf constituting 50–60% of this (Joubert 1954). After birth the juvenile is initially dependent completely, but later to a diminishing extent, on nutritional subsidization via maternal milk. Energy losses occur when food products are transformed into milk (Moen 1973).

Nutritional demands on the mother reach their peak during early lactation. At this time her daily energy requirement may increase almost two-fold, and requirements for protein more than two-fold, over maintenance needs (Oftedal 1984; Clutton-Brock *et al.* 1989; Chan-McLeod *et al.* 1994). For free-ranging livestock occupying a strongly seasonal environment in northern Kenya, the peak allocation of metabolizable energy to reproduction amounted to 55% of maintenance requirements for cattle, compared with 80% for sheep and goats and 33% for camels (Coppock *et al.* 1986). Even well-nourished animals may lose body mass through the high costs associated with milk production.

Only via progeny do adults have the power to generate new biomass, to offset biomass losses through mortality. Unless mass increase by offspring is taken into account, there would be no benefit to transforming parental biomass, with high survival prospects, into offspring biomass with much lower survival chances. Hence, prospective biomass gains as a result of producing offspring must be considered a component of maternal fitness. For illustration, if the mother allocates surplus resources to produce an offspring weighing 10% of parental body mass, but the offspring's survival

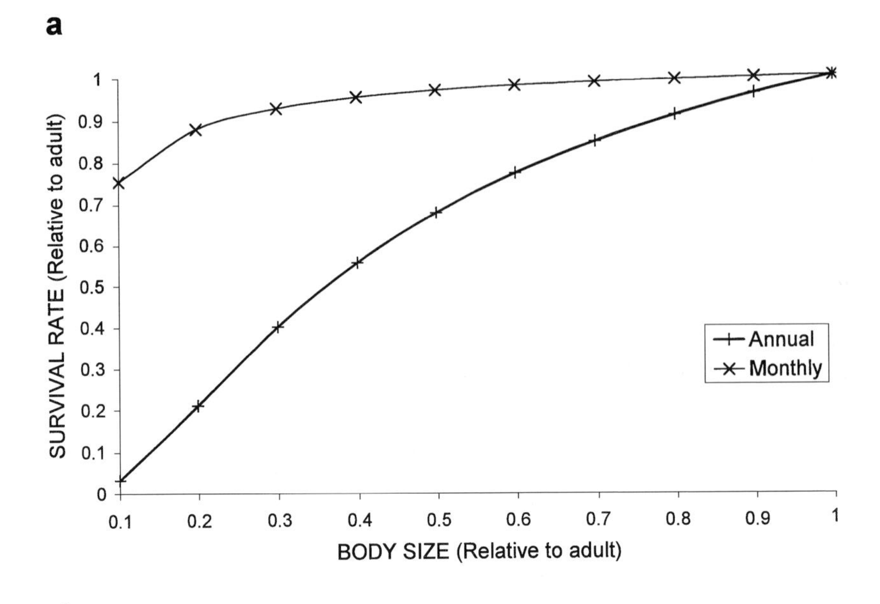

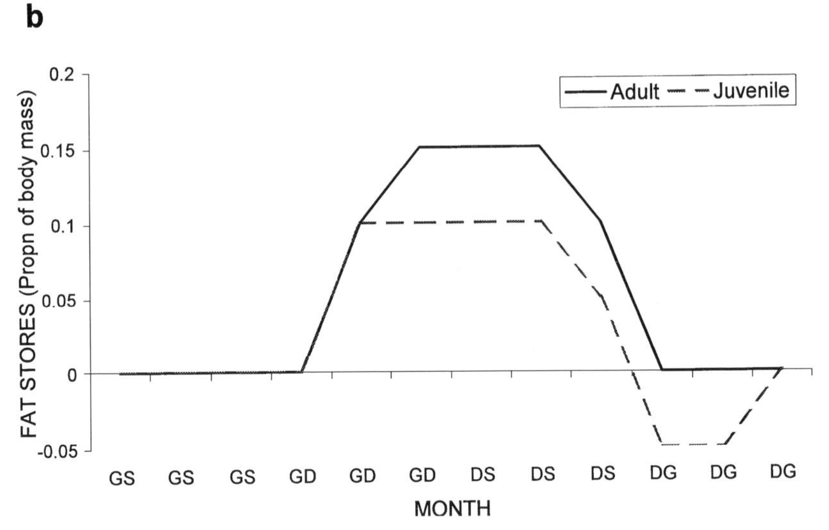

Figure 7.7. Assumed fitness costs and output predictions for the dynamic growth model. (a) Change in survival chances with increasing body size relative to adult size, based on a modified Michaelis–Menten equation: annual survival $= \{a\,X/(b + X)\}^p$, where $X =$ body size as a proportion of adult size, and a, b and p are curve-fitting parameters. Parameters were adjusted so that the annual survival rate of newborn animals is less than 5%, should they remain the same size over the whole year (equivalent to a monthly survival rate of 75%). The projected survival from birth to one year for growing animals is about 45% of the adult survival rate.

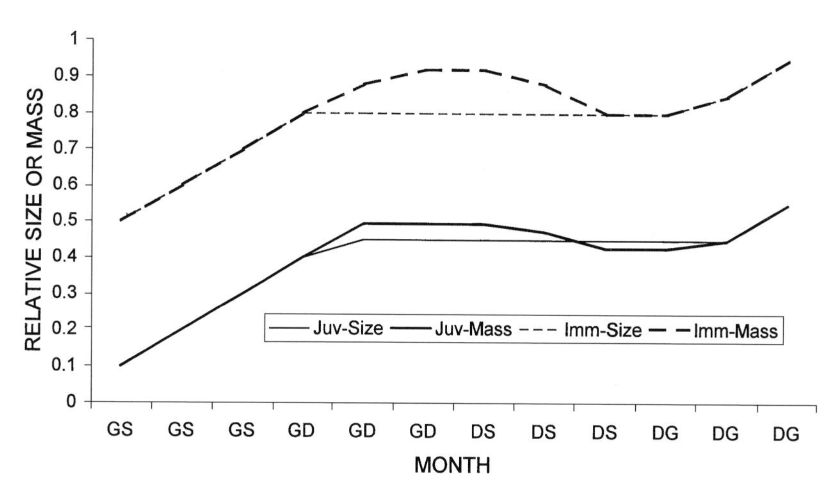

d

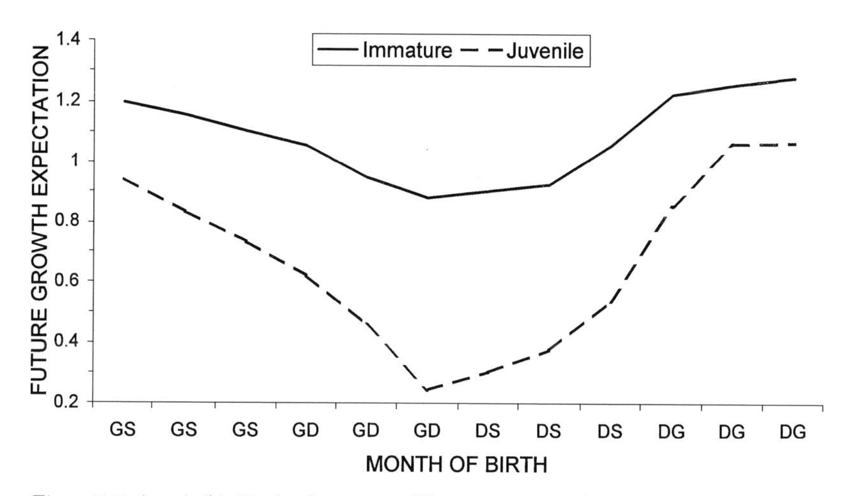

Figure 7.7. (*cont.*) (b) Optimal pattern of fat storage over the seasonal cycle, contrasting juveniles born at the start of the GS with adults. (c) Seasonal pattern of change in relative body size and relative body mass for juveniles and immatures (yearlings) predicted by the model. Relative mass includes changes in fat stores. (d) Modelled changes in fitness, as estimated from the product of growth potential and survival expectation, for animals entering juvenile (newborn) and immature (half-grown) stages in different months of the seasonal cycle.

chances are only 0.5 of those of the mother, reproduction would not be advantageous in the short term $(0.9 + (0.1 \times 0.5) = 0.95)$. However, later growth by the offspring outweighs the short-term mass loss by the parent. Fully grown adults trade their almost zero growth capacity for the high growth potential of offspring.

Life history model

The full fitness function must accordingly take into account the growth potential of both parents and progeny, as well as the survival chances of each. Hence we obtain

$$\begin{aligned}\Phi(\beta, \Gamma, t, t_{max}) \\ &= \max_{i=1,2..} \{\Phi(\beta - \mu + \alpha_i, \Gamma + \alpha_i, t + 1, t_{max}) \\ &\qquad + \Phi(\beta' - \mu' + \alpha'_i, s' + \alpha'_i, \Gamma' + \alpha'_i, t + 1, t_{max})\},\end{aligned} \tag{7.4}$$

where β', Γ', μ' and α' distinguish condition, size, physiological attrition and allocation decisions for offspring from those of parents. Allocation decisions are made by parents prenatally and by offspring postnatally. As in most population models, we restrict our focus to the female segment.

We assume reproduction to be costly. To produce an offspring of size 0.1 (relative to adult mass), the parent must invest body reserves equivalent to 0.2 of its own body mass, and repeat such an allocation once more if the full survival potential of the offspring is to be realized (otherwise offspring survival is halved, presumably through malnutrition due to inadequate milk supply). Parents can allocate either half or all of the monthly resource surplus (i.e. growth potential) to reproduction. If the magnitude of the reproductive investment exceeds the monthly growth potential, the parent must draw on stored reserves.

By evaluating the optimal pattern of resource allocation to growth, storage and reproduction from birth through to maturity, we have effectively a life history model. With reasonable assumptions, the model projects patterns in resource allocation, body condition, growth and reproductive investment that resemble those commonly observed among ungulates (Fig. 7.8). Young growing animals exhibit lower fat contents than adults, and hence incur greater losses in condition over the DS. Investing in reproduction becomes beneficial only when animals approach full size, but the body reserves available to support reproduction

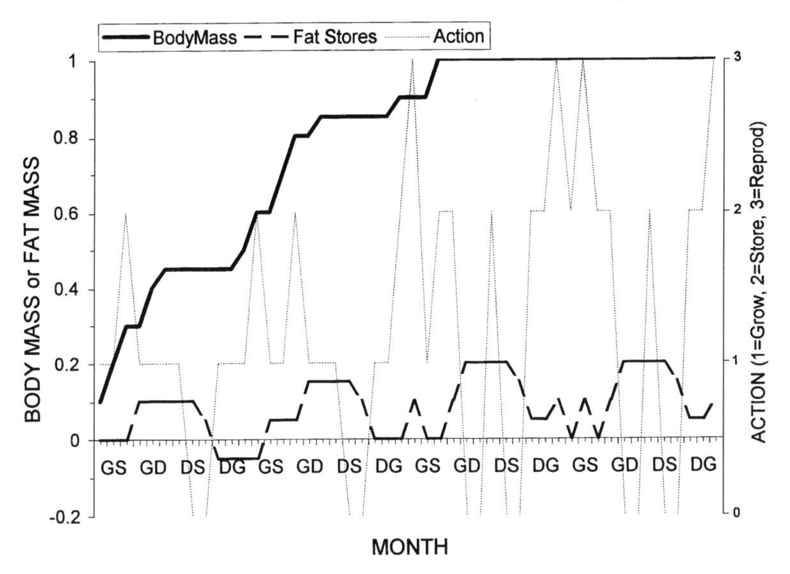

Figure 7.8. Predictions of the dynamic life history model for changes in body size and fat stores, and associated allocations to growth, storage or reproduction, from birth to maturity. Allocation actions are represented as 0 (no allocation), 1 (allocation to growth in size), 2 (allocation to storage) and 3 (allocation to reproduction). Months are assigned to seasons: GS = growing season, GD = growing – dry season transition, DS = dry season, DG = dry – growing season transition.

at the first attempt may be inadequate to ensure offspring survival. Upon attainment of maturity, it becomes advantageous for animals to retain stored reserves beyond the end of the DS, to bolster the needs of late gestation and lactation. Parturition is timed to match the seasonal peak in resource supply. The pattern of fat stores suggested by the model closely mimics that documented for young versus adult antelope over the wet–dry seasonal cycle of African savanna regions (Sinclair and Duncan 1972; Dunham and Murray 1982; Gallivan *et al.* 1995) (Fig. 7.9).

In the model, adult body size is imposed as an arbitrary species-specific constant. Presumably growth ceases when further increases in size bring little additional improvement in garnering resources, or perhaps reduce security against predation. Having reached the optimal size, animals direct resource profits towards nurturing the growth of offspring. In effect, reproduction couples high foraging power with high growth potential.

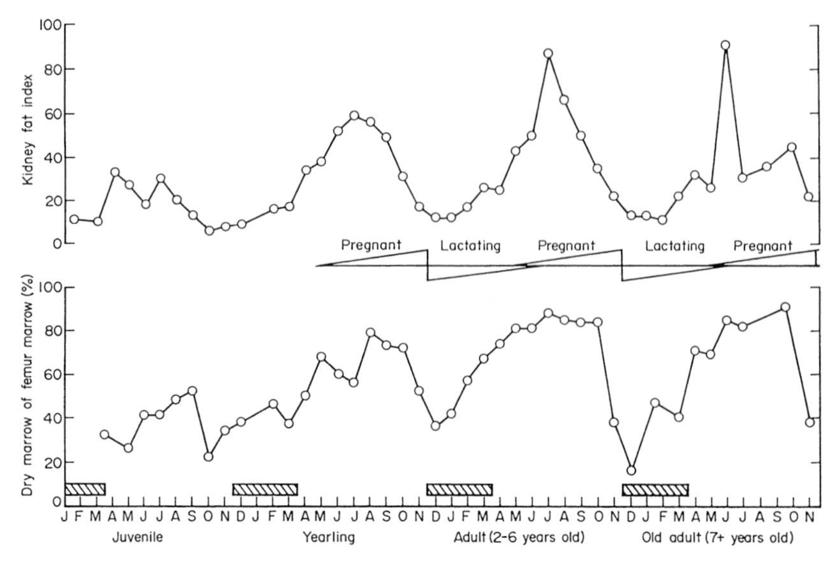

Figure 7.9. Variation with season and age of kidney fat index and bone marrow fat content for female impala in Zimbabwe (monthly means, from Dunham and Murray 1982, © Blackwell Science).

Senescence

As animals increase in age, their annual survival rate diminishes. For kudus, females older than six years showed substantially reduced survival rates compared with prime females aged 2–6 years in the same groups (Owen-Smith 1993a) (Fig. 7.10). Among impala, diminished fat stores are evident in old females (Fig. 7.9). Increased tooth wear results in substantially reduced body reserves in wild reindeer, associated with increased offspring mortality (Skogland 1988). Cumulative oxidative stress may cause body tissues to function less efficiently (Tolkamp & Ketelaars 1992; Sohal and Weindruch 1996).

The life history model could be extended to include the survival costs of cumulative senescence. There would then be no need to specify a maximum longevity. Life expectancy would emerge as a consequence of age-dependent deterioration in body state. Tissue wear is largely a consequence of food processing. Hence long-term tradeoffs arise between the benefits of maximizing current resource gains, and future survival chances. A senescence state function could be specified having as its independent variable not age, but stage of deterioration in body function, dependent upon cumulative amount of food processed.

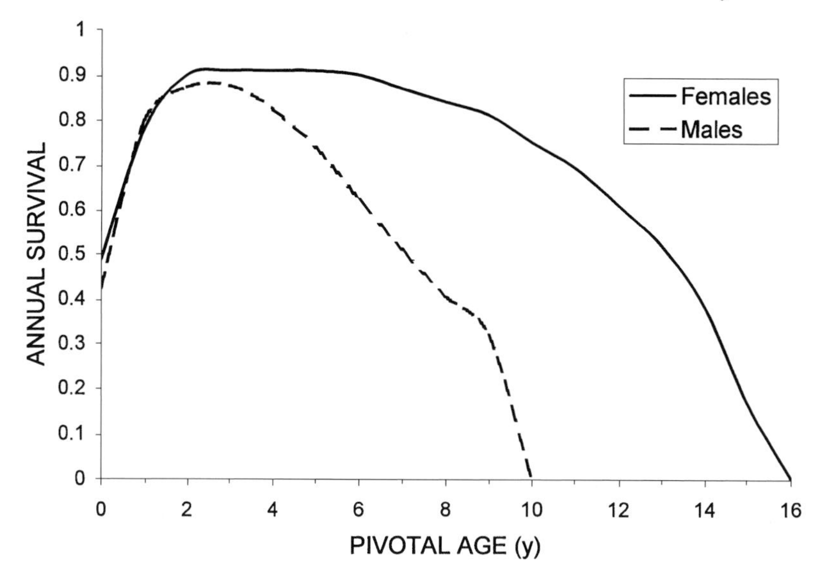

Figure 7.10. Change in annual survival rate over lifespan shown by male and female kudus in the Kruger Park. (From Owen-Smith 1993c.)

From this long-term perspective, it would not be optimal for animals to maximize their daily food intake except on rare occasions. Rather, they should minimize the amount of food and energy processed, subject to meeting target needs for material nutrients and supporting energy. This may help explain foraging patterns that seem contrary to optimality principles in the short term. For example, kudus persisted in favouring spinescent plants yielding restricted eating rates, when they could have gained more food and energy in the same time by eating lower-quality species offering higher ingestion rates (Owen-Smith 1993b). It could also account for the finding that kudus rarely foraged for the full time permitted by thermal conditions (Owen-Smith 1998a). Apparent laziness could indeed be adaptive (cf. Herbers 1981). The challenge of developing a full life history analysis, taking into account senescence, is beyond the scope of this book.

Flux- versus state-dependent mortality

This chapter emphasized how mortality depends on the body state attained as a result of resource gains accumulated over some period. In effect body condition, defined by the extent of stored reserves, provides a buffer against temporary resource shortfalls, and serves as a capital reserve to support the heightened demands of reproduction. This contrasts with the definition of the mortality function in Chapter 2 as dependent inversely upon the short-term rate of resource gain. This raises once more the issue of how to integrate across time frames.

In practice, mortality depends on resource fluxes integrated over some prolonged period. But mortality risk cannot be derived simply from the mean resource gain over the year. Biomass gains during the benign season must be traded against biomass losses during adverse times. In particular, the mortality rate is affected by how widely resource gains fluctuate seasonally. Moreover, the magnitude of resource surpluses relative to maintenance requirements during the GS matter less than the extent of resource deficits that may be incurred during the DS.

Deriving a dynamic, state-dependent model for consumers occupying complex, changing environments presents a considerable challenge. Having outlined this approach, I will back away from it in the remaining chapters of this book. Mortality losses will be related simply to resource fluxes over some restricted period (daily, weekly or monthly). This is justified because resource deficiencies not only contribute directly to a decline in body condition, with its mortality cost, but also prompt animals to invest time and incur risks in counteracting such decline, e.g. by foraging for longer. A starting outline for a model partitioning population biomass between lean body mass and stored reserves was presented by Getz and Owen-Smith (1999).

Overview

This chapter outlined how resource surpluses acquired during the benign season may be traded against anticipated deficits during the adverse season via stored body reserves. How best this is done depends on life history stage, as defined by future growth potential. Immature animals should invest primarily in growth, at the cost of vulnerability to starvation during periods of resource deficiencies. Mature animals should invest primarily in supporting the survival and growth of dependent offspring, but with reproduction timed to occur at the time of the year when resource profits above maintenance needs are greatest. Because of the costs of carrying fat reserves, such deposits should be built up late in the growing season, and amount to no more than is needed to buffer against anticipated net losses in biomass during the dormant season, allowing for stochastic variation and additional demands of reproduction. These patterns predicted by dynamic optimization accord closely with those commonly observed for large mammalian herbivores.

Short-term maximization of net energy or nutrient gains is rarely optimal in seasonally varying environments, because of the costs of carrying stored reserves. The deferred costs of metabolic processing and associated body wear need to be taken into account, as well as mortality risk entailed in acquiring resources. At times animals should merely maintain an optimal body condition, whereas at other stages they should minimize the deficits incurred. Reproduction couples the high resource acquisition power of adults with the high growth potential of young animals. From an ecological perspective, fitness can be assessed in terms of expected contributions made to future biomass increase, either through individual growth or that of offspring. Recognizing the challenges of developing a fully dynamic consumer–resource model, a simpler approach relating mortality losses to resource fluxes over some extended period will be adopted in remaining chapters.

Further reading

Dynamic optimization models of changing fat reserves over daily and seasonal cycles have been developed for birds by McNamara and Houston (1990) and Bednekoff and Houston (1994; see also Houston *et al.* 1997). Witter and Cuthill (1993) review the ecological costs of fat storage by birds. Links between resource acquisition and life history processes are reviewed by Boggs (1992). McNamara and Houston (1996) outline the dependence of biological fitness on manifold components of individual state in relation to life history stage.

8 · *Resource production: regeneration and attrition*

Vegetation constitutes a renewable resource that can be harvested through its consumption by herbivores. However, in seasonal environments the production of edible plant material does not take place continuously. Plants regenerate much of their above-ground biomass at the start of a growing season (early summer or rainy season), partly by reallocation from below-ground reserves. Later when conditions become adverse (too dry or too cold), they cease growth, and progressively shed the senescent parts. Superimposed on this seasonal cycle of growth and attrition are changes in plant populations, either through vegetative growth of new ramets, or via the recruitment of new individuals (genets) from seeds (Harper 1977).

Vegetation growth is not only phased seasonally, but also fluctuates in response to variability in weather during seasonal periods. Plants are a renewing resource for only a portion of the year, and a depleting resource for the remainder. Hence, no balanced equilibrium between production and consumption is maintained, except perhaps transiently. Because senescent parts are less nutritious than actively metabolizing leaves, herbivores must respond to seasonal changes in food quality as well as quantity. Different plant types and parts have distinctive patterns of growth and attrition in biomass and nutritional value.

In this chapter, an appropriate production function for the annual dynamics of the vegetation resource, accommodating this temporal and nutritional variability, is developed. The basis for this production, in terms of fluxes in sunlight, available moisture and mineral nutrients in soils, will not be considered explicitly. Although a full metaphysiological model of plant population dynamics is desirable, it is beyond the scope of this book. I acknowledge the important interaction between herbivores and plants via the recycling of nutrients back into the soil (DeAngelis 1992), but will not be concerned with it here.

General production function

Harvesting theory for renewable bio-resources such as fisheries is based on the principle of compensation following offtake (Walters 1986). A reduction in the biomass density of the resource tends to promote an increase in the relative growth rate of the remainder, which effectively have access to more resources per capita. This 'surplus production' can be exploited sustainably, provided harvest quotas remain below some threshold level.

The logistic equation captures this compensatory regrowth in an elegantly simple way, and has been widely invoked to model vegetation responses to consumption by herbivores (see Caughley 1976a; Caughley and Lawton 1981; Crawley 1983; Roughgarden 1997). For vegetation, mechanisms of compensation following herbivory may include a reduction in self-shading, and an increase in relative supply of soil nutrients to remaining leaves (McNaughton 1979a, 1983; Dyer *et al.* 1993). Only the above-ground component of vegetation, accessible for consumption by large herbivores and energizing overall vegetation growth via photosynthesis, need be considered. Accordingly we have

$$dV/dT = g_V V(1 - V/v_{max}), \tag{8.1}$$

where V = above-ground vegetation biomass, g_V = maximum relative growth rate of vegetation, and v_{max} = maximum biomass of vegetation that can be sustained by resources. As noted by Caughley (1976a), this formulation assumes that vegetation does not influence the supply rate of its underlying resources, i.e. radiant energy, water and soil nutrients. This may not hold for soil nutrients, where mineralization rates depend on subsidies from plants to mycorrhizas (Swift *et al.* 1979).

Rather than being constant, the potential growth rate, g_V, varies in response to the prevailing conditions. In particular, growth ceases under adverse conditions of cold or dry soils, a situation that may persist for many weeks during the winter or dry season period. During this dormant period, the vegetation standing crop undergoes progressive attrition through the shedding of leaves and other parts that become detached litter. Indeed, much of the standing material consists of dead rather than living tissues, i.e. 'necromass' rather than biomass in the strict sense. Leaf death and decay are ongoing processes even during the course of the growing season. At the start of the next growing season, foliage is regenerated from stored reserves of carbohydrates and mineral nutrients.

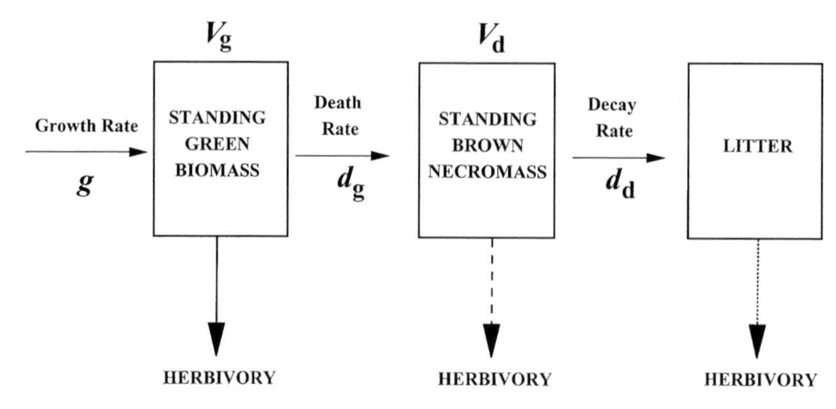

Figure 8.1. Conceptual model for the growth, death, decay and consumption of above-ground biomass of plants.

Moreover, growth through photosynthesis and the associated uptake of mineral nutrients is governed mostly by the foliage component, rather than the total plant biomass. Indeed, much of what is commonly measured as plant 'biomass' is effectively 'necromass', i.e. standing dead tissues. Over time the standing necromass decays or falls to the ground as litter.

Accordingly, the total quantity of vegetation available for consumption by herbivores can be partitioned among distinct compartments connected by fluxes of materials (Fig. 8.1). While standing necromass may hinder growth through its shading effects on active leaves, its metabolic diversion on other resources will be less than that of green leaves. Accordingly, modifying Equation 8.1 we have (in the absence of herbivory)

$$\begin{aligned} dV_g/dT &= V_g[g_V\{1 - (V_g + pV_d)/v_{max}\} - d_g]; \\ dV_d/dT &= d_g V_g - d_d V_d, \end{aligned} \qquad (8.2)$$

where V_g = green component of vegetation biomass, V_d = standing dead component, d_g = relative death rate of green biomass, d_d = decay rate of standing dead material, and p = factor down-weighting the effect of standing necromass on biomass growth. Thus the relative growth rate g_V excludes leaf mortality, and v_{max} is the combined biomass plus necromass at which the effective growth rate becomes zero.

As a simplified illustration, let us assume a stepwise seasonal cycle in which the vegetation grows continuously at its full potential rate during a six-month growing season (GS), then decays steadily over the dormant season (DS) during which the potential growth rate becomes zero.

Starting from some small nucleus at the beginning of the GS, green vegetation biomass accumulates asymptotically towards a peak, before becoming depressed by the shading effect of rising necromass (see Fig. 8.3b). The peak amount of green plus dead biomass exceeds v_{max} to some extent because of the down-weighted influence of the accumulated necromass on the realized growth rate. During the DS, green biomass declines rapidly through the death of leaves, while standing necromass does so somewhat more gradually through decay and litter fall. The accumulated necromass would inhibit growth in green biomass at the start of the next GS to some extent, unless removed, e.g. by fire. Regeneration of vegetation biomass at the start of a new annual cycle depends critically on the magnitude of the starting nucleus of green material, presumably drawn from stored reserves. Because of ongoing death of leaves, the annual production exceeds the peak standing biomass to some degree.

Year-to-year expansion or contraction in the population biomass of vegetation would presumably be dependent to some extent on the magnitude of the stored reserves built up over the seasonal cycle, which could either be accumulated internally or diverted into the seed bank. However, the potential for plant populations to expand depends also on the availability of space into which to grow. This explicitly spatial aspect to the long-term dynamics of vegetation is beyond the scope of this book, which will restrict its focus largely to within-year changes in edible plant biomass.

Herbaceous plants (grasses and forbs) differ somewhat from woody plants (trees and shrubs) in their seasonal patterns of biomass growth and decay. Accordingly, we will develop first a model for grass dynamics, and then consider how it might be modified to apply to woody vegetation.

Grass production model

Basic model

Data to parameterize a grass growth model were provided by Detling *et al.* (1979), who studied the growing season dynamics of the predominant grass species in short grass prairie in Colorado. Under conditions of plentiful water and nutrients, this grass grew at an initial relative rate of 0.5 per week to reach a peak standing biomass of about 550 g/m^2. The relative growth rate appeared to remain constant until green biomass had accumulated to about 50 g/m^2, declining thereafter presumably due to shading effects. The initial growth rate was uninfluenced by

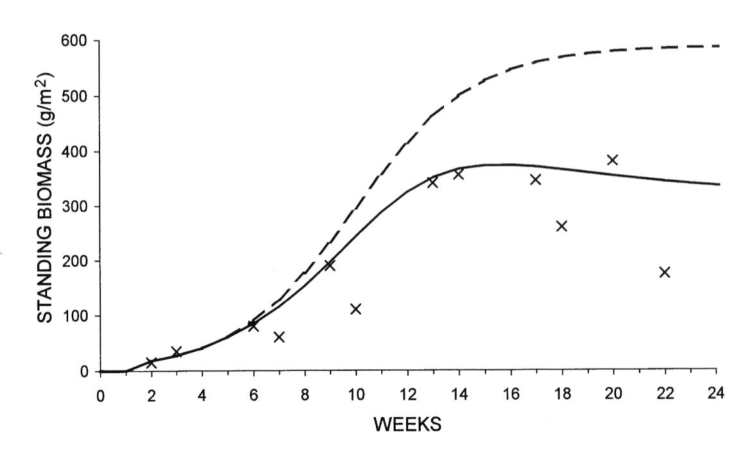

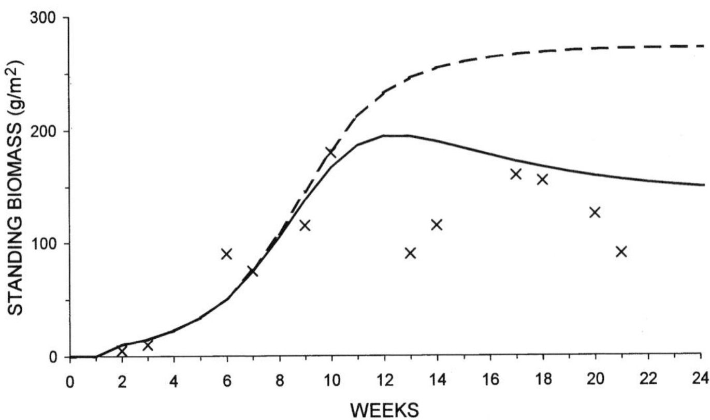

Figure 8.2. Fit between the modified logistic growth model and observed growth dynamics for North American prairie and South African savanna grasslands. (a) Prairie grassland supplied with irrigation and fertilizer ($g_V = 0.5$ per week, $v_{max} = 550$ g/m^2, threshold for density dependence = 50 g/m^2). (b) Prairie grassland irrigated without fertilization ($g_V = 0.5$ per week, $v_{max} = 250$ g/m^2, threshold = 50 g/m^2). (c) Prairie grassland supplied with fertilizer under natural rainfall regime ($g_V = 0.5$ per week, $v_{max} = 100$ g/m^2, threshold = 50 g/m^2). (d) Savanna grassland under natural conditions with erratic rainfall ($g_V = 0.5$ per week, $v_{max} = 150$ g/m^2, threshold = 50 g/m^2). (From Detling *et al.* 1979; Grunow *et al.* 1980.)

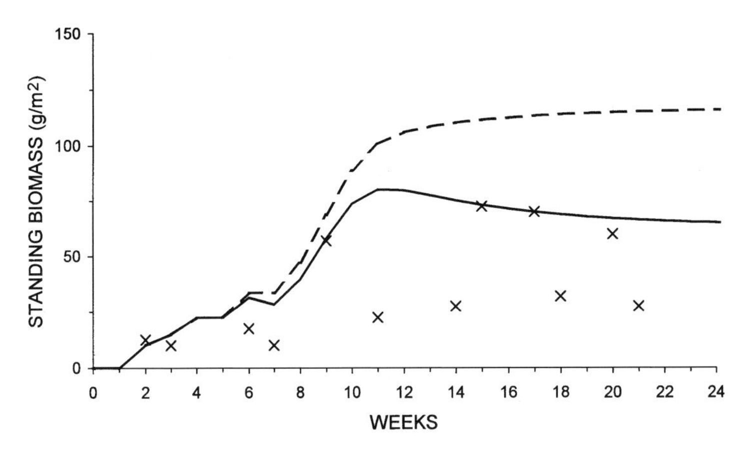

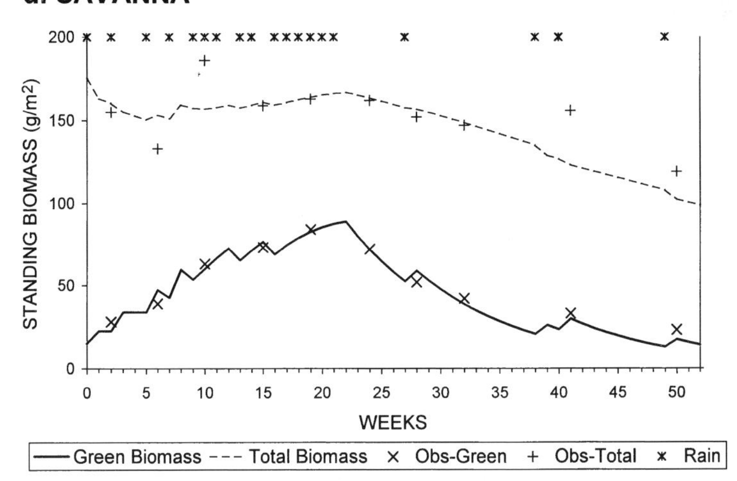

Figure 8.2. (*cont.*)

restrictions in water or nutrient supplies. The saturation biomass level was lowered in the absence of either fertilization or irrigation. With appropriate parameter values, Equation 8.2 closely mimicked the observed growth dynamics, with allowance for fluctuating rainfall and hence growth in the situation where irrigation water was not supplied (Fig. 8.2a–c).

Measurements on grass biomass dynamics in a wet–dry seasonal savanna over the full seasonal cycle were reported by Grunow *et al.* (1980), for a region in South Africa with infertile sandy soils (Fig. 8.2d). Equation 8.2 again provided a good fit to the recorded data, with appropriate allowance for cessation of grass growth in weeks without rainfall. The small amounts of green biomass measured in the late DS were probably the result of dry season rainshowers. A value for g_V of 0.5 per week likewise matched the savanna grassland data. In general, annual grass production amounts to about 0.5–1.0 gDM/m^2 per mm of annual rainfall in African savannas (Rutherford 1980), reduced somewhat by increasing woody plant density and by decreasing soil fertility.

A generic grass dynamics model incorporated the functional relations and parameter values is depicted in Fig. 8.3a. The realized growth rate declines linearly from its maximum of 0.5 per week, above a threshold grass biomass of 50 g/m^2. The factor p, down-weighting the effect of necromass on vegetation growth, was set at 0.5. The saturation biomass level of 400 g/m^2 represents a region with an annual rainfall total of 600–800 mm, towards the moister side of the middle range for tropical savanna regions. Arbitrarily, the relative death rate of green biomass during the GS was assumed to be 0.1 per week, after an initial six weeks during which no death occurred. The death rate was assumed to be 50% higher than this during the DS and during dry weeks in the GS. The death rate has an important influence on the extent to which the annual production (summed growth) exceeds the peak standing crop. The proportional decay rate of necromass was made equal to the relative death rate of green shoots during the GS, but was reduced to 20% of this value during the DS. A weekly iteration was adopted.

The initial nucleus of green biomass at the start of the GS was set arbitrarily at 20 g/m^2, i.e. 5% of the peak standing biomass. No carryover of necromass from the previous season was allowed (presumably removed by fire or grazing). In the absence of grazing, green biomass built up rapidly to reach its peak about 10 weeks after the commencement of the GS (Fig. 8.3b). Green material disappeared rapidly during the course of the DS. Standing necromass at the end of the DS amounted to about 300 g/m^2, i.e. two thirds of the peak standing crop. With allowance for erratic fluctuations in rainfall weeks during the GS, the time when peak grass biomass was reached was delayed somewhat (Fig. 8.3c). Annual grass production was reduced somewhat less than the proportional reduction in weeks receiving rainfall, suggesting that modelled vegetation growth is over-compensatory. For the selected parameter values, annual grass production amounted to about twice the peak green biomass, or 1.5 times the maximum amount of biomass plus necromass, for the situation with variable rainfall.

This model is appropriate for savanna environments where plant growth is controlled primarily by soil moisture and hence by rainfall.

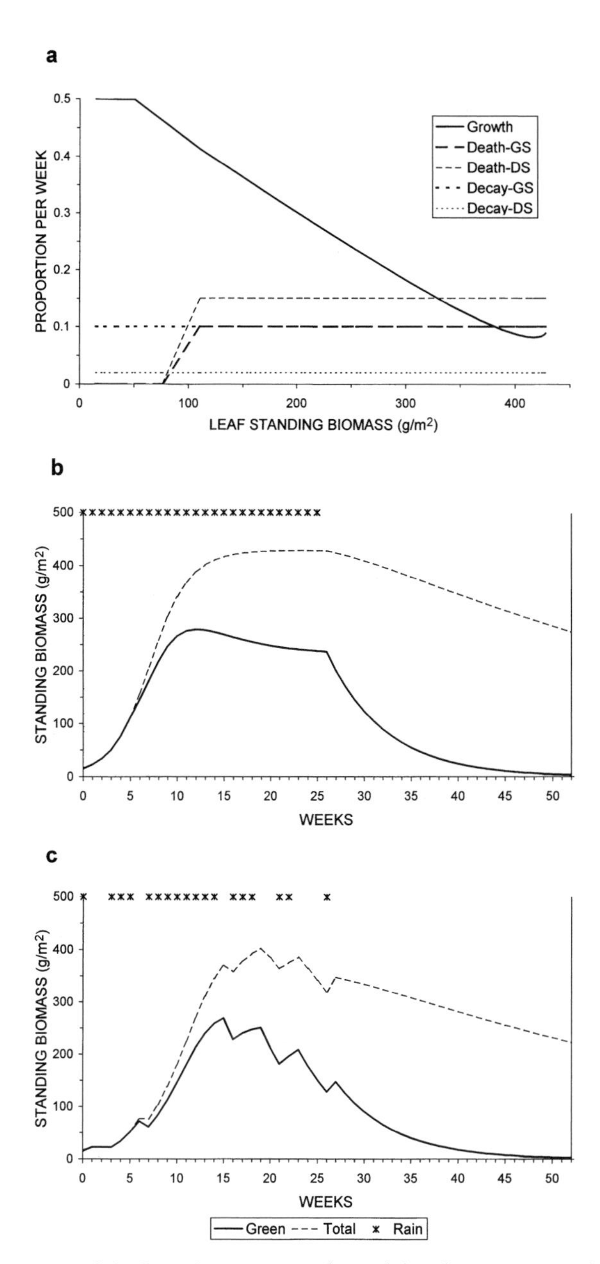

Figure 8.3. Generic grass growth model and its output under ungrazed conditions. (a) Assumed functional relations and basic parameter values incorporated into the model. (b) Model output for ungrazed grassland with $g_V = 0.5$ per week and $v_{max} = 400$ g/m^2, assuming a continuous growing season of 26 weeks followed by a dormant season of 26 weeks (derived annual grass production = 725 g/m^2). (c) Model output with same parameter values, but with rainfall controlling grass growth varying erratically during the GS as indicated (annual grass production = 590 g/m^2).

Plants simply grow when soil moisture is adequate and cease growth when soil dries out in weeks without rainfall. In temperate regions where the temperature regime is the primary influence, plant growth rate would vary in a more graded fashion over the course of the GS.

Incorporating consumption

The food intake rate of herbivores depends not on the total standing crop of vegetation, but on the fraction thereof that is available for consumption. Some vegetation biomass will be ungrazable, because it is too close to the ground, or otherwise inaccessible (cf. Noy-Meir 1975). Moreover, because necromass is nutritionally inferior, herbivores should concentrate their feeding on green biomass as far as possible. This means that the contribution of standing dead vegetation to food availability must be down-weighted relative to the green biomass component. Arbitrarily it will be assumed that the effective food availability, F, is equal to $V_g + V_d/2$, i.e. herbivores cannot avoid consuming some dead material along with green leaves. Following Equation 2.17, the relative intake rate I of herbivores is assumed to be hyperbolically saturating with increasing food availability, i.e.

$$I = i_m(F - v_u)/\{f_{1/2} + (F - v_u)\}, \tag{8.3}$$

where i_m = maximum relative intake rate, $f_{1/2}$ = food availability at which the intake rate reaches half of its maximum, and v_u = ungrazable vegetation biomass. The value of the half-saturation biomass $f_{1/2}$ was made quite small so that the form of the intake response was steeply asymptotic. Herbivore density was set in terms of the stocking measure conventionally used by cattle ranchers, with an animal unit representing one 454 kg steer. Accordingly, a unit stocking density per ha generated a maximum consumption rate of about 8 g/m^2 (DM) of herbage per week, i.e. around 2% of the peak standing crop (biomass plus necromass) of herbage.

With a stocking level of one animal unit per hectare, and constant seasonal rainfall, consumption retards the accumulation of green biomass during the GS only slightly relative to the ungrazed state (Fig. 8.4a). Annual grass production exceeds that in the ungrazed situation by about 10%, largely through the retarding effect of consumption on the buildup of necromass. Grazing reduces the amount of standing vegetation that remains at the end of the DS. The maximum sustainable stocking level is reached when just enough necromass persists to support the intake

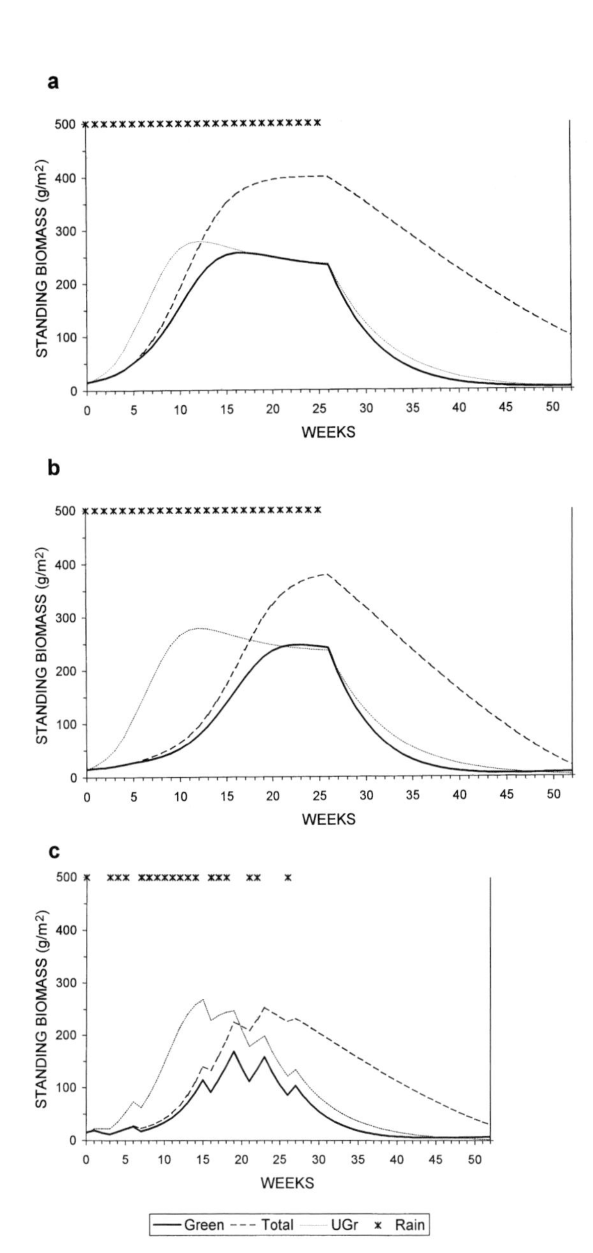

Figure 8.4. Modelled grass growth dynamics under grazing, compared with green biomass in the ungrazed (UGr) state. Grass growth parameters are as in Fig. 8.3b, with $v_u = 10$ g/m^2 and $f_{1/2} = 5$ g/m^2. (a) Moderate stocking density of 1 animal unit per hectare (annual grass production = 802 g/m^2, annual consumption = 392 g/m^2). (b) Maximum sustainable stocking density of 1.5 animal units per hectare (annual grass production = 763 g/m^2, annual consumption = 555 g/m^2). (c) Maximum sustainable stocking density of 0.85 animal units per hectare under conditions of erratic rainfall (annual grass production = 377 g/m^2, annual consumption = 290 g/m^2).

demand of the herbivores through to the end of the DS, otherwise animals starve (Fig. 8.4b). Stocking densities of this magnitude may also lead to a situation where consumption exceeds grass production at the start of the GS, leading to starvation at this stage, depending on the chosen parameter values. For the maximum stocking density, total consumption amounts to about 75% of annual grass production. If rainfall varies erratically during the GS, the maximum stocking level that can be supported is greatly reduced (Fig. 8.4c).

For these simulations, no carryover of necromass from the end of the preceding DS was allowed. Persisting dead leaves could reduce consumption pressure on the green biomass appearing at the start of the GS. This could raise the permissible stocking level, depending on the acceptability of this necromass as food. An increase in the starting amount of green biomass relocated from stored reserves would also improve the potential stocking level. However, if sufficient grass biomass does not accumulate during the GS, the herbivores run out of food before the end of the DS.

The model output emphasizes the crucial importance of early-season grass dynamics for determining whether a particular stocking level is sustainable (cf. Noy-Meir 1978). Herbivores that are highly efficient at extracting small amounts of green vegetation are most likely to suppress grass production.

Multiple grass types

Caughley (1982) suggested that, where the vegetation comprises several plant species, the effective growth rate is the harmonic mean of the growth rates of each species. However, things are not so simple when vegetation components differ in their nutritional quality, and herbivores feed selectively on them.

Plants with a high potential growth rate tend to be relatively intolerant of shading, and vice versa, because of intrinsic tradeoffs (Grime 1977). Accordingly, fast-growing grasses should cease growth at lower heights, and develop less standing biomass, than slow-growing but ultimately taller grasses. The latter will competitively dominate in mixed-species stands, maintaining a positive growth rate while the faster-growing species are being shaded out. Because they contain less structural tissue, short grasses tend to be more palatable to herbivores than taller grasses. Through their rapid re-growth, they can support higher levels of herbivory relative to their standing biomas than taller grasses, but carry less forage into the DS.

The independent biomass dynamics of two grass types differing in intrinsic growth rate and saturation biomass levels are depicted in Fig. 8.5a,

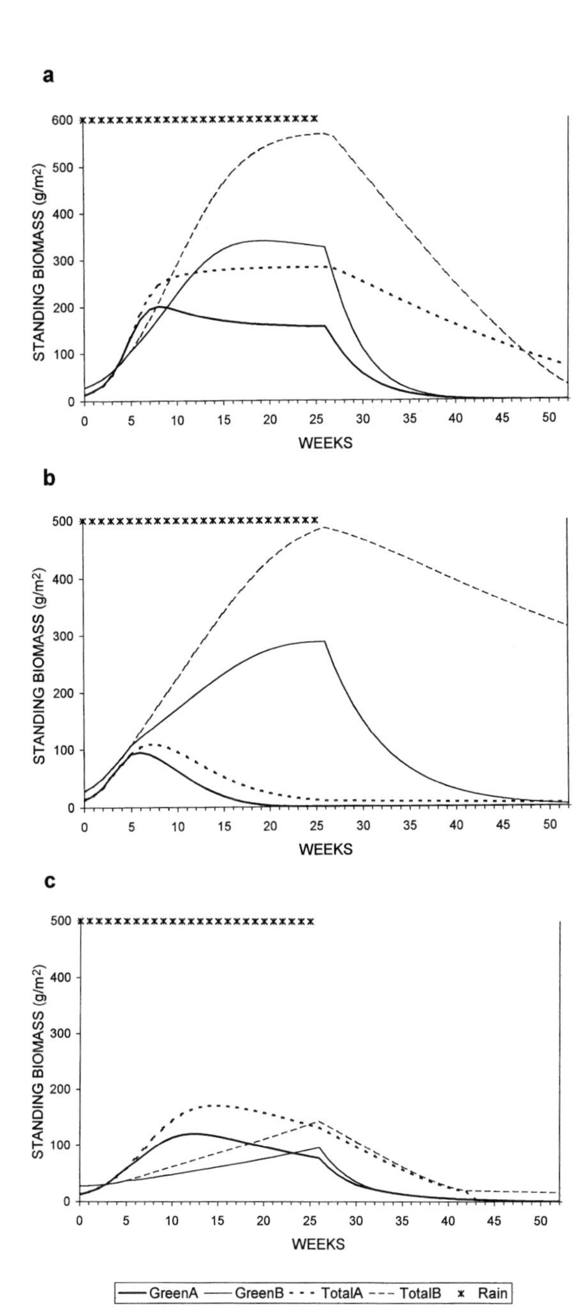

Figure 8.5. Comparative biomass dynamics of two grass types differing in relative growth rates and saturation biomass densities. For species A, $g_V = 0.65$ per week, $v_{\max} = 250$; for species B, $g_V = 0.35$ per week, $v_{\max} = 550$. (a) Independent growth dynamics under ungrazed conditions (annual production of species A = 502 g/m^2, species B = 911 g/m^2). (b) Intermingled growth dynamics under ungrazed conditions. (c) Growth dynamics when intermingled and subjected to heavy non-selective grazing (stocking density 2 animal units per hectare).

for an ungrazed situation. Parameter values have been chosen so that their mean equals that for the generic grass type represented in Fig. 8.3b. When such grass types are intermingled, so that through mutual shading the growth of each depends on the averaged standing crop of both, the growth of the fast-growing grass is soon suppressed by buildup in biomass of the initially slower-growing species (Fig. 8.5b). But if grazing pressure is sufficiently high, and evenly distributed over both grass types, the faster-growing type gains a competitive advantage through its more rapid re-growth at low biomass levels. However, insufficient herbage is retained by the fast-growing species to support the herbivores through the DS, so that the latter must move away elsewhere or starve. Thus the model illustrates how locally heavy grazing could lead to the creation of 'grazing lawns' (McNaughton 1984) where prostrate, rapidly growing grass species predominate, but support herbivores primarily during the GS. Through a prostrate habit, fast-growing grasses may gain an additional competitive advantage by having a larger amount of their biomass ungrazable to herbivores than taller species (Noy-Meir 1978).

The model suggests possible mechanisms whereby grazing pressure could promote the development of a patch mosaic of grass types, with short grasses predominating in some regions and taller grasses elsewhere. The grassland distribution may be influenced by spatial variation in soil moisture availability, with short grasses prevalent on the shallow soils of ridges and tall grasses in bottomlands. When such a spatial segregation exists, the grass types no longer directly influence each other's growth. If herbivore grazing is distributed evenly over the patch mosaic, the maximum stocking level is restricted by the capacity of the short grasses to provide forage through the DS (Fig. 8.6a). A substantially higher stocking density could be sustained if herbivores selectively concentrated their grazing in the short grass patches during the GS, then transferred mostly to the taller grass areas during the DS (Fig. 8.6b). Observations in the western Serengeti region of Tanzania documented such a pattern, with grazing ungulates shifting between ridge crests where short grasses predominate during the rainy season, and bottomlands where tall grasses prevail during the DS (Bell 1970, 1971). The seasonal migration of wildebeest and other ungulates in the broader Serengeti ecosystem is based on a landscape-scale distribution of grassland types (McNaughton 1979b; Fryxell 1991).

If grass growth fluctuates during the GS owing to erratic rainfall, herbivores can alternate between grassland types during this season, as

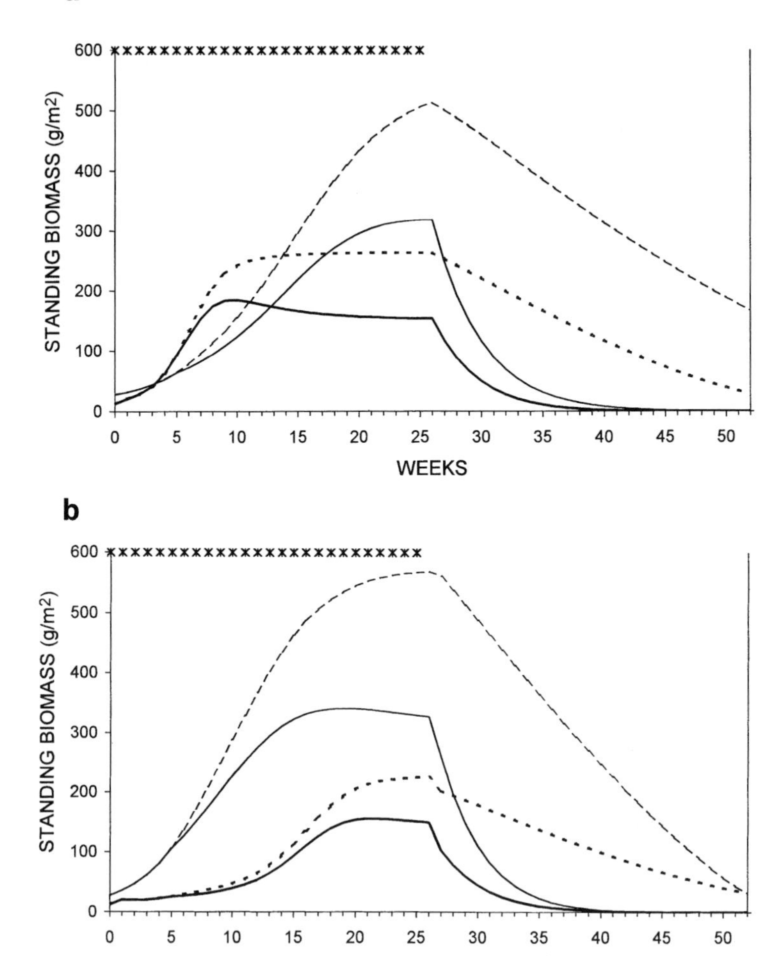

Figure 8.6. Modelled biomass dynamics of two grass types growing independently in a patch mosaic; parameter values as in Fig. 8.5. (a) Grazing distributed evenly over both grass types, with stocking density (one animal unit per hectare) at the maximum supported by species A through the dormant season (annual grass production of species A = 631 g/m^2, species B = 891 g/m^2, combined annual consumption from both species = 756 g/m^2). (b) Grazing concentrated on the fast-growing grass type during the growing season, then transferred mostly to the high-biomass type during the DS, with stocking density at the maximum supported through the dormant season (relative stocking levels, A = 2.75, B = 0.25 animal units per hectare during the GS, and A = 0.75, B = 2.25 animal units per hectare during the DS; annual grass production for species A = 745 g/m^2, species B = 922 g/m^2, combined annual consumption from both species = 1075 g/m^2).

documented for white rhinos (Owen-Smith 1988). The availability of high-biomass grass types buffers the short grasses against depletion during periods unfavourable for growth.

Above- versus below-ground biomass

Grass plants have half or more of their biomass below ground. Hence, modelling only the dynamics of above-ground biomass could be misleading. Reserves are translocated from root and stem-base tissues to generate growth above ground at the start of the GS, and perhaps also following severe defoliation by herbivores. The asymptote in above-ground biomass reached during the latter part of the GS arises partly from increasing diversion of photosynthate from shoots to support root growth, plus the replenishment of stored reserves (McNaughton *et al.* 1998). Relative allocation to roots versus shoots may be influenced by soil fertility, as suggested by the effect of fertilizer additions on asymptotic grass biomass illustrated in Fig. 8.2. Plant species seem to maintain some characteristic ratio of roots to shoots, which can be modified to only a limited extent by grazing pressure (McNaughton *et al.* 1998).

To allow for the interactive dynamics of below-ground biomass, an additional compartment can be added to the basic model depicted in Fig. 8.1a. Growth is distributed between above- and below-ground biomass, depending on the extent to which each compartment is below its saturation potential. Root tissues die slowly and hence undergo less attrition than above-ground parts during the DS. Accordingly, production is allocated primarily to shoots at the start of the GS. To support growth in total biomass, the proportional growth rate of the above-ground biomass (i.e. rate of biomass gain through photosynthesis) must be doubled.

Adding a below-ground compartment makes little difference to the modelled dynamics of above-ground grass biomass, under ungrazed conditions (Fig. 8.7a; cf. Fig. 8.3b). This is also largely true for grazed situations (Fig. 8.7b, c). However, heavy grazing tends to retard the buildup in root biomass. In particular, a high stocking density coupled with erratic rainfall can largely suppress the increase in root biomass during the GS, with the result that root biomass declines over the annual cycle (Fig. 8.7d). The reduction in root tissues could reduce the capacity of plants to regenerate above-ground biomass at the start of the following GS, and hence their ability to sustain herbivory. In this way, high stocking rates could make plants susceptible to mortality, especially when drought conditions occur.

Because of the coupling of root and shoot biomass, a decline in belowground tissues implies an overall shrinkage in the grass population. The grass species suffering this decline could become subject to displacement by other plant types: for example, by woody seedlings, leading to the range management problem of bush encroachment. However, rest periods arising from the movements of grazing ungulates elsewhere could be sufficient to allow grass populations to be maintained. In the Serengeti region, intense herbivory inhibited neither root biomass or belowground production of the grassland over both short and long time frames (McNaughton *et al.* 1998).

Browse production model

For many trees and shrubs, most of the annual production of edible browse in the form of leaves and shoots occurs within a few weeks of the start of the GS, largely via re-allocation from stored reserves (Rutherford and Panagos 1982; Rutherford 1984) (Fig. 8.8a,b). The current season's photosynthate is used largely to grow stem and root and replenish reserves. Deciduous species shed their leaves completely during the course of the DS. The fallen leaf litter is available for consumption by herbivores for a period, but becomes scattered by wind such that eventually it can no longer be harvested effectively. During the late DS, herbivores may be forced to subsist on dormant twigs and on the leaves remaining on evergreen species, which are commonly defended chemically. Trees have the potential to store somewhat larger reserves in their stems and roots than grasses and forbs.

To transform the grass growth model into a model of edible browse dynamics for woody plants, only changes in parameter values are required. Woody plants exhibit a substantially larger initial biomass drawn from reserves at the start of the GS, and slower turnover of leaves during the course of the GS (Fig. 8.8c). For deciduous species, but not for evergreens, the leaf death rate increases sharply in the DS. Leaf litter accumulates during the DS, but decays more rapidly than the standing necromass of grasses. Because trees store substantially larger reserves than grasses, they are less vulnerable to having their growth suppressed at the start of the GS by high levels of herbivory, except perhaps for early-flushing species.

Browsing ungulates cannot be sustained year-round by deciduous trees or shrubs alone, although their leaf litter is consumed while it remains

a

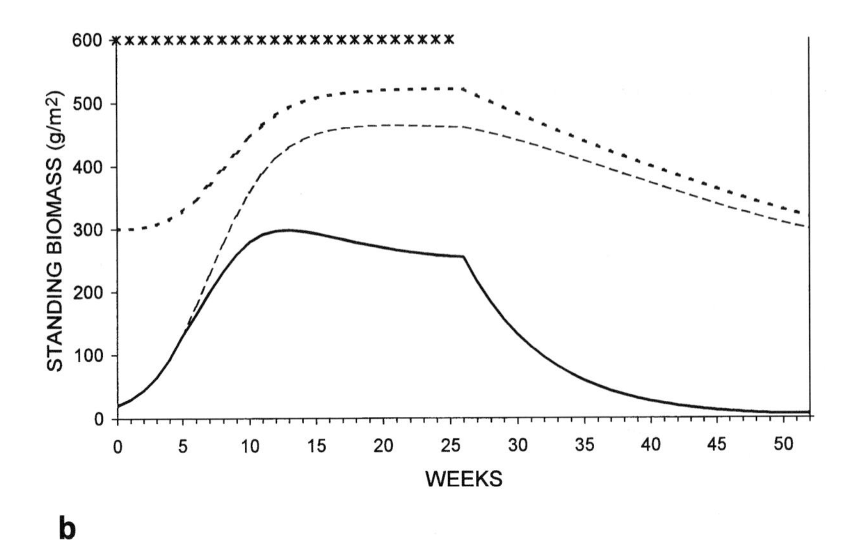

b

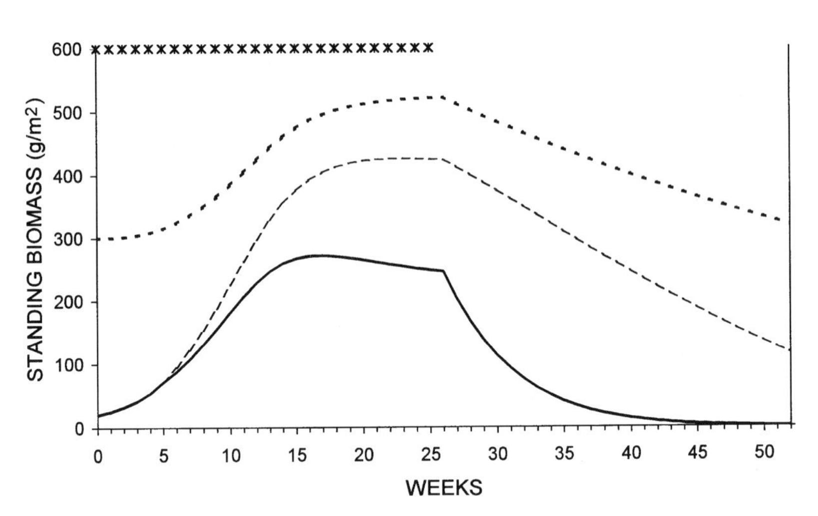

Figure 8.7. Output from grass growth model incorporating root as well as shoot biomass dynamics. (a) Ungrazed, $g_V = 1.0$ per week, above ground $v_{max} = 400$ g/m^2, below-ground $v_{max} = 600$ g/m^2 (annual biomass production above ground = 778 g/m^2). (b) Moderate stocking density of one animal unit per hectare, for same parameter values (annual biomass production above ground = 844 g/m^2, annual consumption = 382 g/m^2). (c) High stocking density of 1.65 animal units per hectare, with same parameter values (annual biomass production above ground = 813 g/m^2, annual consumption = 556 g/m^2. (d) Maximum stocking density permissible of 0.85 animal units per hectare for conditions where weekly rainfall varies erratically during the growing season (annual biomass production above-ground = 472 g/m^2, annual consumption = 301 g/m^2).

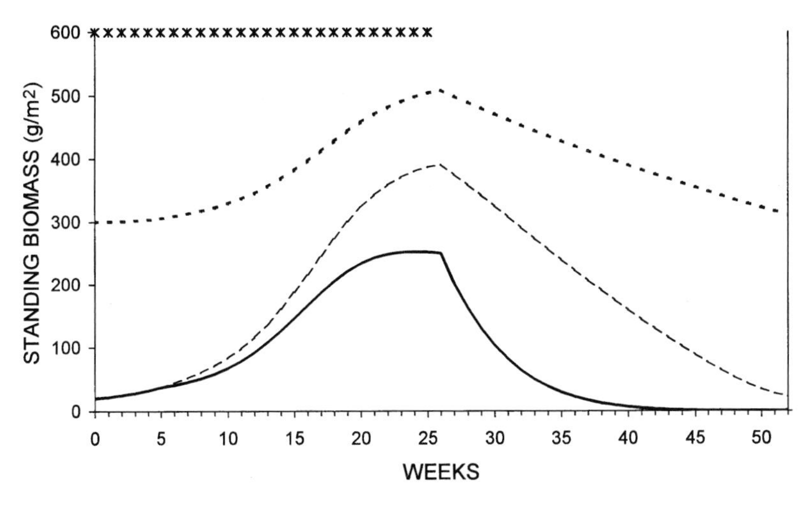

d

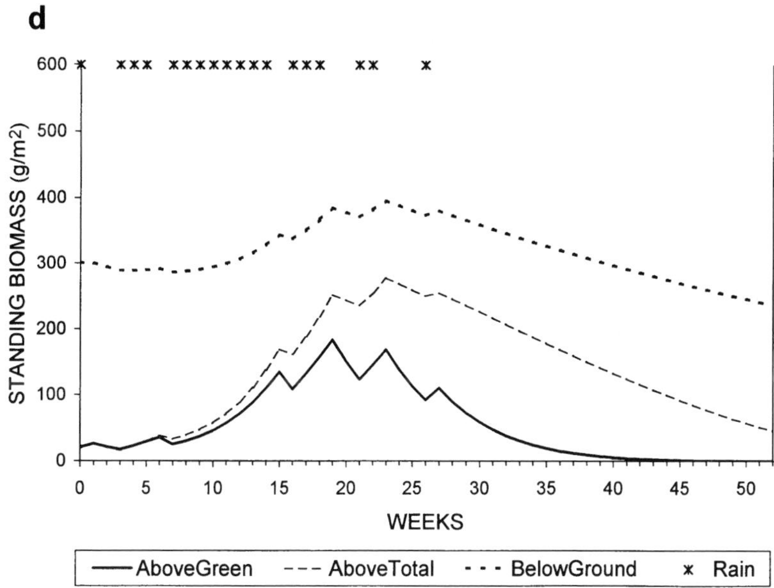

Figure 8.7. (cont.)

available (Owen-Smith and Cooper 1985). Alternative food sources are needed during the late DS, in the form of species with persistent foliage or edible dormant stems. In Chapter 11 we will consider how particular vegetation components contribute towards determining stocking densities, and ultimately habitat suitability, for browsing ungulates.

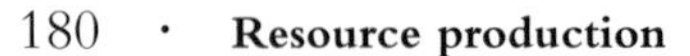

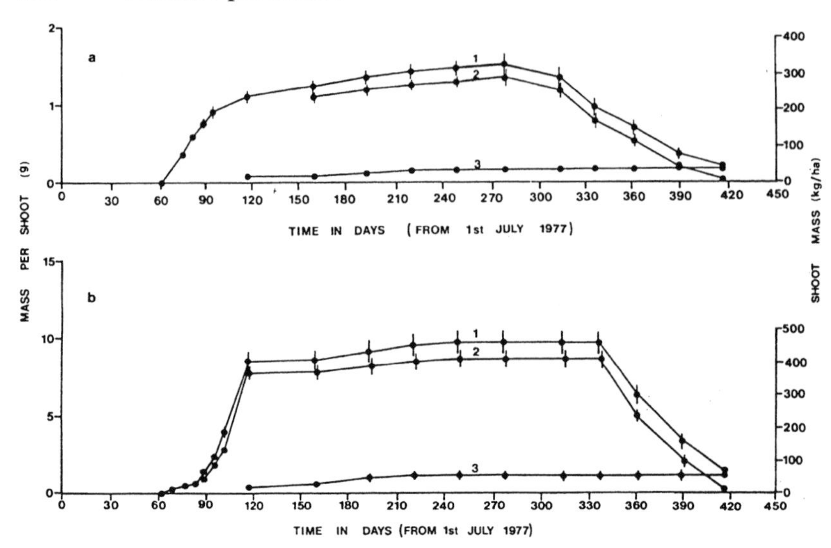

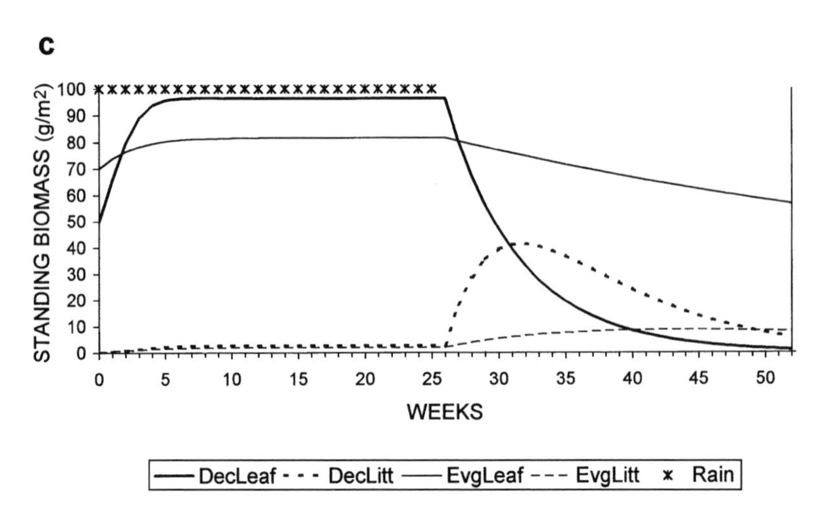

Figure 8.8. Seasonal growth dynamics of woody plants. Observed pattern of variation in (1) shoot, (2) leaf and (3) twig mass for (a) *Ochna pulchra*, the predominant shrub species, and (b) *Burkea africana*, the predominant tree species, in a South African savanna (from Rutherford and Panagos 1982). (c) Modelled biomass dynamics of green leaf and leaf litter biomass for a deciduous and an evergreen woody species, under unbrowsed conditions.

Resource-related dynamics for plants

Tilman (1980, 1982, 1988) developed a resource-based approach to vegetation dynamics which parallels some of the concepts outlined in this book. As for animal consumers, the functional relation between plant biomass accumulation and underlying resource availability could be either gradually asymptotic, or quickly saturating in form. Certain plant types capture resources rapidly when these are in abundant supply, whereas other plant types are better able to maintain growth at low resource levels (Chapin 1980). Through uptake plants reduce the pool of available resources, particularly mineral nutrients. Plants that are successful at exploiting high resource levels may flourish initially, but eventually only the plant types better able to maintain growth at low resource levels persist. However, Tilman's focus was primarily on successional replacements among vegetation components taking place over periods of several years or decades, rather than on the within-season dynamics of plant biomass.

The availability to plants of soil nutrients, as well as soil moisture, varies over the seasonal cycle in African savanna environments (Swift *et al.* 1979; Scholes and Walker 1993). Indeed, to a large extent the supply of mineral nutrients is closely coupled with that of water. The first good rains at the start of the GS promote a pulse of nutrient release from accumulated plant detritus and animal dung. The nutrient pool becomes reduced during the course of the GS as a result of rapid plant uptake. Limitations in remaining nutrient supplies could restrict the re-growth of plants following defoliation by herbivores. However, herbivores may augment available nutrients through their urine and faecal deposits, thereby facilitating plant regrowth (McNaughton *et al.* 1997). By attracting higher levels of herbivory during the GS, inherently fast-growing short grasses could secure a competitive advantage over slower-growing grasses less adept at capturing these nutrients. Plants can also actively promote the release of mineral nutrients by subsidizing symbiotic bacteria and fungi via surplus energy derived from photosynthesis.

From the perspective of seasonal biomass dynamics, it is unnecessary to distinguish growth in grass height from expansion in basal area via spreading shoots, new tillers or seedlings. If moisture and soil nutrient supplies are adequate, grass biomass can reach its maximum within a few weeks, starting from the minute biomass of seeds. Thus even if all grass tufts die as a result of severe defoliation coupled with drought, the grassland could regenerate within less than a year via the seed bank.

However, seedlings have limited resource stores and lack established roots, so that their realized growth depends strongly on the rainfall regime as well as on local nutrient availability. Usually most seedlings die without establishing.

A more elaborate model of vegetation biomass dynamics would need to accommodate the differential supplies of energy, water and mineral nutrients from disparate sources, variable resource allocations to shoots versus roots, and the spatial context for individual plant interactions. In the long term, the consequences of the underlying resource dynamics for plant population changes would need to be accommodated. Such a comprehensive model is beyond the scope of this book. The logistic model is liable to exaggerate the re-growth capability of vegetation resources following herbivory, in situations where the availability of the soil nutrients supporting plant growth changes seasonally.

Overview

In seasonally varying environments, no equilibrium between production and consumption persists for long. If plant growth exceeds the rate of consumption, vegetation biomass accumulates towards the limit set by underlying water and nutrient resources during the course of the benign season. During the adverse season, plants become a non-renewing resource for herbivores, so that animals must subsist on a diminishing capital of vegetation biomass and standing dead tissues. The capacity of vegetation to support herbivores depends on the amount of standing vegetation remaining at the termination of the growing season, and to some extent also on the initial growth rate and magnitude of the reserves translocated from below-ground at the start of the growing season. Stocking levels that appear safe when vegetation grows continually through the growing season can lead to resource depletion when growing conditions vary erratically during this period.

The generic model can be modified to represent either grasses or woody plants, merely by adjusting parameter values. Competitive interactions among grass types differing in growth rate relative to peak standing crop, and the influence thereon of grazing, can also be represented. The model demonstrates the benefits of a patch mosaic of grass types, utilized selectively in different seasons, for herbivore stocking capacity.

Incorporating below-ground biomass into the model makes little difference to the dynamics of available forage, but reveals the potential effects of consumption on root growth. Different pathways of plant biomass regeneration need not be distinguished in the short term, but can become important for time frames extending over several years. The model does not incorporate any effects of seasonal changes in soil nutrient availability on vegetation re-growth following herbivory.

Further reading

Crawley (1983) reviews models of vegetation responses to consumption by various herbivores, as well as the supporting data. Hanson *et al.* (1985) compare alternative grass growth models. Lemaire and Chapman (1996) present a thorough review of processes influencing grass growth dynamics from a grazing perspective. Thornley and Johnson (1990) provide a primary source of reference on plant growth dynamics.

9 · *Resource competition: exploitation and density dependence*

Demographic theory is permeated by the concept of density dependence, i.e. a decline in relative growth rate as population density increases towards some 'carrying capacity', where net population growth becomes zero. Density dependence arises fundamentally from intraspecific competition for limiting resources. As population density rises, each individual gets a smaller share of the resource supply. Competition may be expressed directly, through aggressive interactions, or other forms of interference with resource acquisition; or indirectly, simply via resource depression. Density dependence can also arise in other ways, e.g. through rising predation losses or parasite spread with increasing density, but these will not be the subject of this chapter.

For large herbivores, overt interference with foraging is rare. Exceptions occur in situations where high-quality food is locally concentrated, e.g. fallen fruits under a tree canopy. But generally food resources are widely dispersed, so there is little to be gained by displacing another animal from a feeding area. Competition occurs largely as a consequence of resource exploitation. As a result, the effects of competition may not be experienced immediately, but only at some later stage when less food remains as a result of the feeding impacts of conspecifics. In a seasonal environment, resource depression may be minor during the growing season when vegetation resources are renewing, but intensifies over the course of the dormant season when resource depletion is progressive.

How resource gains depend on consumer density underlies the theoretical debate whether consumption rate is a function of resource abundance, or of the ratio between resource abundance and consumer density (Berryman *et al.* 1995; Gleeson 1995; Abrams 1997; Abrams and Ginzburg 2000). For large herbivores, biomass gains over daily or shorter time frames are related directly to the prevailing food availability (see Chapters 3–6), but depend on prior herbivore densities interacting with resource production over longer periods. When functional relations are strongly non-linear, calculating the long-term gain simply from the ratio of resources to consumers could be highly misleading.

In previous chapters, we considered the adaptive choices made by individual herbivores in isolation. Now our concern is with how populations of herbivores are distributed among habitat types, taking into account the effects of competition. The classical concept of an 'ideal free distribution' (Milinski and Parker 1991) assumes conditions where the effects on resource gains (or, more generally, on fitness) are experienced immediately, either through a scramble for sparse resources or through direct interference with foraging. Neither circumstance is usual for large herbivores, for which the consequences of competition on resource availability are deferred.

This chapter considers how the annual biomass increment of herbivore populations is influenced by consumer density in circumstances where competition arises indirectly through resource depression, and resource production varies seasonally. Attention will be primarily on potential growth in herbivore biomass, with consequences for losses in biomass through mortality considered only incidentally (mortality losses will be the focus of Chapter 10). Accordingly, the appropriate context is that of domestic herbivores managed for meat production, rather than free-ranging wild herbivores. Caughley (1976b) distinguished the *optimal stocking density*, or 'economic carrying capacity', sought in agricultural contexts, from ecological carrying capacity. The former is the density at which the production of herbivore biomass per unit area (or, more strictly, the economic return therefrom) is greatest. The latter is the density at which net production is zero. In the simple logistic model, density dependence is linear, and maximum production arises at half of the density at which net production becomes zero (see Fig. 2.1). This forms our null model of departure. We will focus in particular on how the form of the *density dependent function* (DDF), relating annual change in consumer biomass to consumer density, depends on the short-term form of the biomass gain response, for different models of resource production.

Stocking density models

The stocking density concepts used by livestock ranchers are expressed in the form of a simple graphical model, following Jones and Sandland (1974) (Fig. 9.1). Growth per animal declines progressively with increasing stocking density, whereas maximum production of animal biomass

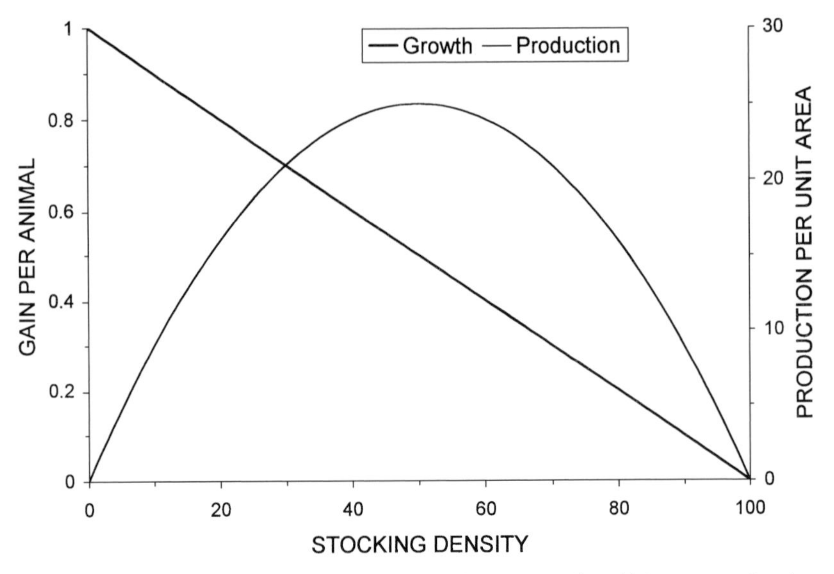

Figure 9.1. Hypothetical model relating growth per animal and biomass production per unit area to stocking density (from Jones and Sandland 1974).

per unit area is attained at about half of the density associated with zero growth. The resemblance to the logistic model is obvious. The measure of population change is merely shifted to units of biomass gain, with mortality ignored. This is appropriate for circumstances where a cohort of animals is acquired at a young age, and grown towards maturity before being dispatched to market, so that mortality losses are minimal.

Fowler (1981) suggested that, for most large mammal populations, density dependence in net population growth should be convex rather than linear in form. This means that competition becomes effective only above some threshold density, so that maximum production per unit area is attained at a population closer to ecological carrying capacity than indicated by the simple logistic model (see also McCullough 1992). However, Fowler (1987) acknowledged that, whereas birth and death rates show strong non-linearity in their responses, rates of body growth can change in a more linear fashion with increasing density.

Observations on the live mass gains of beef cattle on native pastures in Australia, extrapolated to cattle on extensive range in northern Queensland, provide some support for this model (Gillard and Monypenny 1990). However, growth per animal declined only after some threshold density had been exceeded, and thereafter density dependence in biomass gain appears gently concave in form (Fig. 9.2a). Allowing for

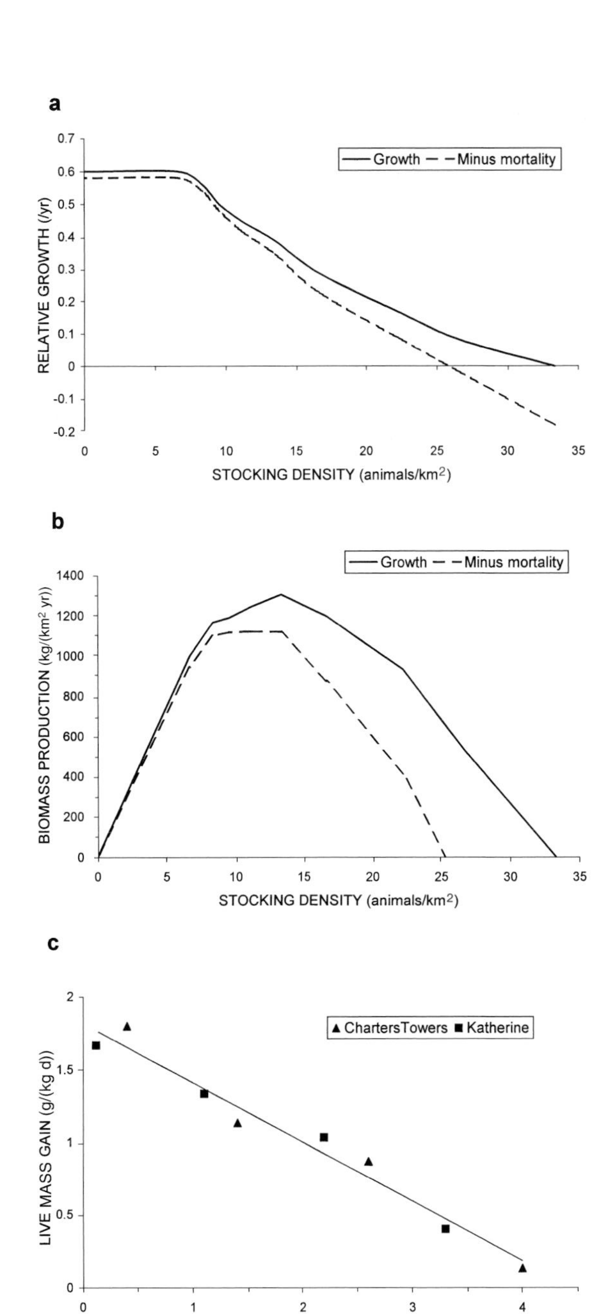

Figure 9.2. Observed relations with stocking density for animal growth and biomass production, with or without subtracting mortality losses, for cattle on natural pasture in northern Australia. (a) Relative growth of steers on extensive pastoral properties (from Table 1 of Gillard and Monypenny 1990). (b) Biomass production per unit area, derived from the relations depicted in (a). (c) Live mass gain for cattle held in paddocks in two areas of Queensland (from Ash *et al.* 1996).

mortality losses, which increased with stocking density, caused the net biomass gain to became somewhat more linearly dependent on stocking density. The density maximizing biomass production per unit area was around half of the stocking density giving zero net gain (growth minus mortality), but less than half of the density yielding zero growth per animal (Fig. 9.2b). Other Australian observations on the growth rates of cattle, obtained using animals confined experimentally within paddocks at different stocking densities, yielded a more linear decline in growth potential with increasing stocking density (Ash *et al.* 1996) (Fig. 9.2c).

Growth potential

The potential for herbivores to grow in biomass depends on the difference between biomass gained from resources consumed, and biomass losses to physiological attrition and mortality, as captured by the **GMM** model:

$$(1/H)\mathrm{d}H/\mathrm{d}T = G - M_P - M_Q, \tag{9.1}$$

where H = herbivore biomass density, T = time, G = relative (mass-specific) rate of biomass gain from food consumed, M_P = relative rate of loss in biomass through physiological attrition, and M_Q = relative rate of loss in biomass through mortality.

For simplicity, the biomass gain response to changing food availability will be modelled by using the Michaelis–Menten equation, i.e.

$$G = c\, i_{\max} F/(g_{1/2} + F), \tag{9.2}$$

where F = food availability (mass/area), c = conversion coefficient from food ingested to consumer biomass increase, $i_{\max}$ = maximum food intake rate at high food abundance, and $g_{1/2}$ = half-saturation level, i.e. the food availability at which the rate of gain reaches half of its maximum. The conversion coefficient incorporates digestibility and metabolic assimilation efficiency, as well as the transformation from plant dry mass to animal live mass. If food resources vary in quality, selective foraging leads to the most favourable components being depleted soonest, in which case c becomes a variable dependent on food availability (see Chapter 6). As a consequence, the food availability at which the rate of biomass gain is half of its maximum will differ from that at which the intake rate reaches its half-saturation level. Hence

$$I = i_{\max} F/(i_{1/2} + F), \tag{9.3}$$

where I = relative rate of food intake, and $i_{1/2}$ = half-saturation level for intake rate. The intake rate determines the impact of consumption on food resources. If food quality diminishes with decreasing food availability, $g_{1/2}$ will be somewhat greater than $i_{1/2}$.

Physiological costs vary with activity expenditures and thermoregulatory costs. We will assume that the extent of variation is small relative to that in G, so that M_P can be assumed to be effectively constant. It is helpful to normalize the biomass growth potential by expressing it relative to the metabolic maintenance requirement, following Chapter 7. From Equation 7.1, the *relative growth potential* (RGP) is given by

$$\mathrm{RGP} = (G - M_P)/M_{P0}, \tag{9.4}$$

where M_{P0} = relative biomass expenditure for basal metabolism. Growth is possible only if the dimensionless RGP is positive; if it is negative animals will lose body mass.

Mortality losses are assumed to rise with diminishing resource gains relative to metabolic requirements, following Equations 2.25 and 2.26, so that

$$M_Q = z_0 + zM_P/G, \tag{9.5}$$

where z_0 indexes background mortality from senescence and other causes, and z indexes how rapidly mortality increases as resource gains diminish.

This establishes the functional relations governing how the biomass density of consumers responds to changing resource availability. They need to be coupled to appropriate production functions for vegetation in a seasonal environment. Parameters will be set to represent a generic cow-sized herbivore weighing about 400 kg. However, their effective value will depend on the composition of the herbivore population. Fully grown adults have no growth potential, but a lower susceptibility to mortality, than immature animals. To represent a mixed-structure population, it will be assumed that the population growth potential is 50% of the biomass growth potential of immature animals, implying that half the population consists of non-growing adults. The half-saturation constant for intake, $i_{1/2}$, will be set such that the daily food intake of the herbivores does not decline much until little accessible vegetation remains, generating an intake function that is steeply asymptotic in form. Alternative values of $g_{1/2}$ relative to $i_{\max}$ will be considered, representing how strongly food quality (specifically its metabolizable dry matter yield) declines with diminishing food abundance (see Fig. 9.3a).

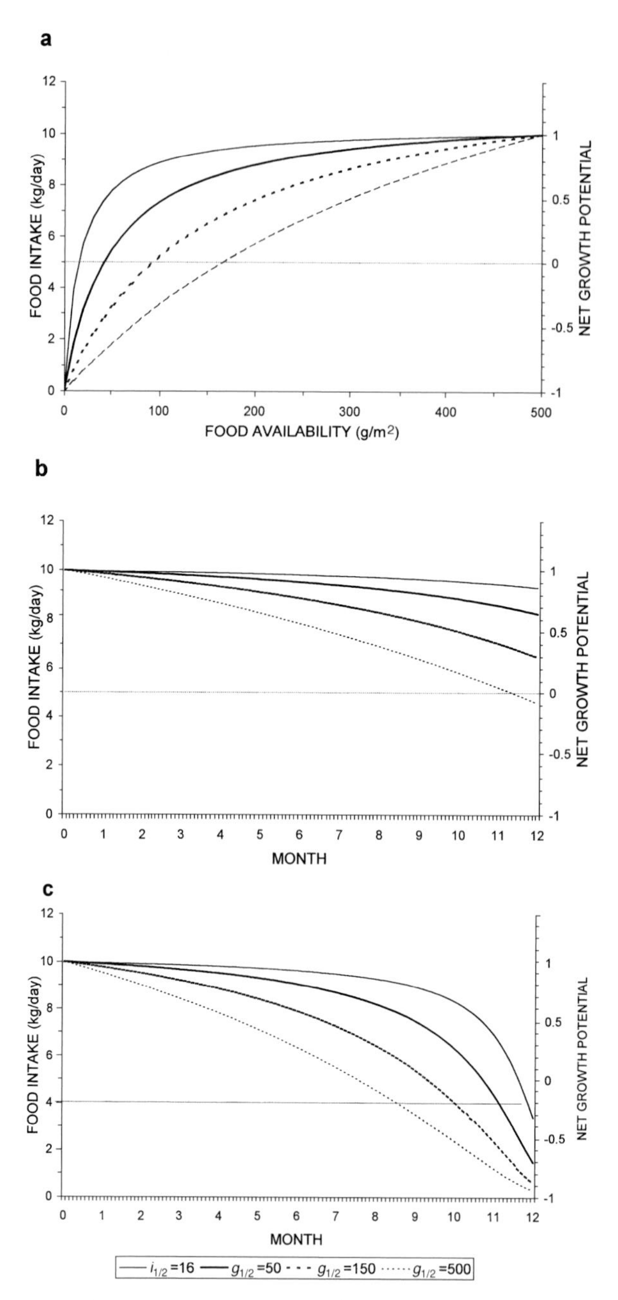

Figure 9.3. Alternative intake and nutritional gain responses (expressed as net growth potential), as controlled by the half-saturation parameter ($i_{1/2}$ or $g_{1/2}$), and their consequences for food and nutrient intake regimes over the annual cycle. (a) Hypothetical response forms. (b) Corresponding trends in intake and relative growth potential, for a moderate stocking density of 1 stock unit per hectare. (c) Corresponding trends for a high stocking density of 1.5 stock units per hectare.

Mortality parameters were set such that the inevitable deaths to senescence and accidents occurring at low population density amount to about 5% of population biomass per year, with mortality losses rising to 15–20% per year at density levels where the individual growth potential became zero. This low rise in mortality was considered to be representative of agricultural contexts, where predators are not a threat.

Alternative models of resource production

Three vegetation production models of increasing, yet manageable, complexity will be considered. They differ in the parameter values inserted into Equation 8.2, and in the seasonal oscillation in these parameters.

Production pulse model

In the simplest production model, resources are produced in a brief pulse at the start of the annual cycle. Thereafter, food availability declines progressively, owing to consumption alone. Food quality remains constant. This pattern is vaguely representative of certain tree types, which grow most of their crop of new leaves and shoots in early spring from stored reserves accumulated over the preceding growing season (Chapter 8). A three-day iteration period was used.

The output for food intake and relative growth potential over the course of the year reflects the shape of the underlying biomass gain function (Fig. 9.3b, c). If herbivore density is moderate, the growth potential remains positive year-round, unless food quality declines substantially with diminishing food abundance (i.e. $g_{1/2}$ large relative to $i_{1/2}$). A somewhat higher stocking density causes food intake to decline precipitously towards the end of the annual cycle, so that resource gains fall below the maintenance level. Accordingly, biomass growth during the early portion of the year is counterbalanced by biomass attrition later in the year. The period over which resource gains are inadequate lengthens with increasing stocking density, and with the extent of the effective reduction in food quality with diminishing food abundance.

The form of density dependence mimics that of the resource gain function. With a steeply asymptotic gain response, the DDF tends to be quite sharply convex in form, both for individual growth potential and for the net increase in population biomass allowing for mortality (Fig. 9.4a, b). Growth potential declines little, until a density is reached at which biomass gains become sub-maintenance before the end of the annual cycle. Because mortality is assumed to be hyperbolically related to resource gains,

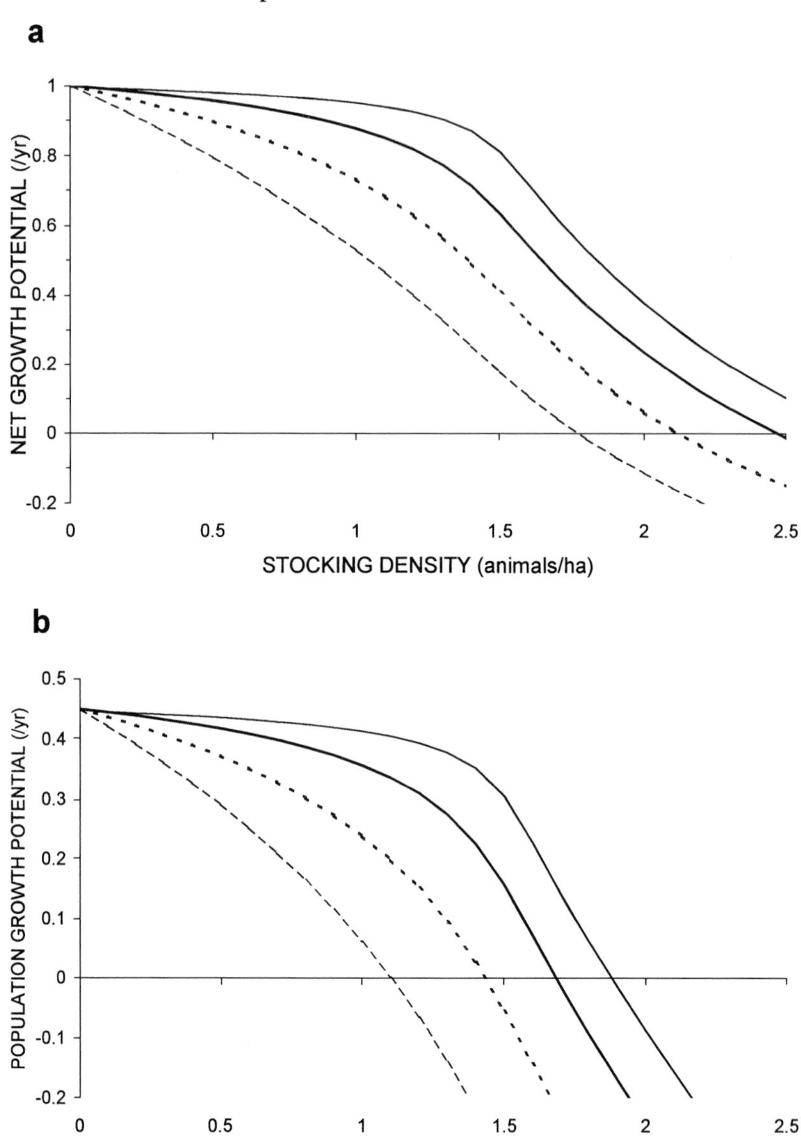

Figure 9.4. Comparative forms of density dependence in individual growth and population growth potential, subtracting mortality, corresponding with the resource gain functions depicted in Fig. 9.3a. (a) Individual growth potential. (b) Population growth potential, allowing for background mortality of 5% per year and hyperbolically increasing mortality losses with diminishing resource gains.

it influences the form of the DDF only towards high density levels. For the decline in biomass growth potential with increasing density to approach linearity, the gain response must be gradually asymptotic. Notably, the stocking density where zero growth pertains is somewhat lower for gradually asymptotic than for steeply asymptotic gain functions, for the same resource production.

Chapter 3 established why the intake response of large herbivores is generally steeply asymptotic, especially over daily or longer time frames. Accordingly, a gain function that is gradually asymptotic means that resource quality declines substantially with diminishing food abundance (or alternatively that intake rate is greatly restricted by decreasing grass height). The former situation could arise either as a consequence of plant structure (for example, because shorter grass is more stemmy), or through selective consumer depletion of higher-quality vegetation components. We need next to consider the latter mechanism more explicitly.

Extended production pulse

To allow for selective depletion, multiple food types differing in quality must be represented. For this simple model, it was assumed that food abundance remained unaffected by consumption through the six months of the growing season (GS). Whatever is eaten is immediately regenerated. The forage biomass at the start of the dormant season (DS) then becomes progressively depleted. No inherent attrition in forage biomass was allowed, nor any change in the nutritional value of food types, during the course of the DS. Specifically, four food types distributed in discrete habitat patches were represented, with poorer-quality foods producing a higher forage biomass than better-quality types. Quality was indexed by the conversion factor to consumer biomass. A small ungrazable reserve was specified.

A weekly iteration was used. At each time step, the food type yielding the highest nutritional gain was assessed. The effective gain was calculated as the product of the food intake and the biomass conversion of the chosen food type. Consumption was then concentrated solely on this food type. Herbivores accordingly switched their feeding to alternative patch types offering lower-quality food as favoured food types become depleted during the course of the DS.

The mean nutritional gain over the 26 weeks of the GS was determined from the gain estimated for week 0 prior to the start of the DS. The mean gain over the DS was calculated by averaging the changing weekly gains over the next 26 weeks. The annual gain is the average of these two values, assuming that gains during the GS offset losses during the DS without any transfer costs. The annual gain was then converted to a relative growth potential, following Equation 9.4. For biomass increase at the population level, mortality losses were subtracted.

At low stocking densities, herbivores can subsist year-round on the most nutritious food types, with a correspondingly minor reduction in their daily biomass gain over the course of the DS (Fig. 9.5a). With higher stocking densities, better-quality food types are depleted earlier during the course of the DS (Fig. 9.5b). At very high stocking densities, animals are forced to accept the poorest food type before the end of the DS, and so experience a marked decline in daily biomass gain (Fig. 9.5c).

The resultant DDF was somewhat concave in form, both for individual growth and for population increase (Fig. 9.6a). This is because even a small increase in stocking density depressed the availability of the best-quality food type, with consequent effects on biomass gains. The exact shape of the DDF depends on the differences in quality and abundance of alternative food types that are assumed in the model. The stocking density giving maximum production per unit area is somewhat lower for a population than for a cohort of growing animals not subject to mortality (Fig. 9.6b). Both the form of density dependence and the corresponding production curve derived from this model resemble those documented for free-ranging cattle (Fig. 9.2).

Seasonal production model

For this evaluation, the seasonal grass growth model developed in Chapter 8 was used. This accommodated the buildup in necromass, negatively influencing food quality, over the seasonal cycle. Situations with just a single grass type present, and with multiple grass types distributed in a discrete patch mosaic, were contrasted. Grass types differed in inherent growth rates, saturation biomass levels and nutritional quality. A simple seasonal alternation was assumed, with rainfall and hence grass growth occurring continually through the 26 weeks of the GS, then ceasing abruptly at the start of the DS. The model was run over two annual cycles, so that conditions at the start of the second GS were determined by the vegetation standing crop, including necromass, remaining at the end of the preceding DS.

Consumption by herbivores was restricted to the patch type yielding highest energy gain at each weekly iteration, as in the simpler selective

a

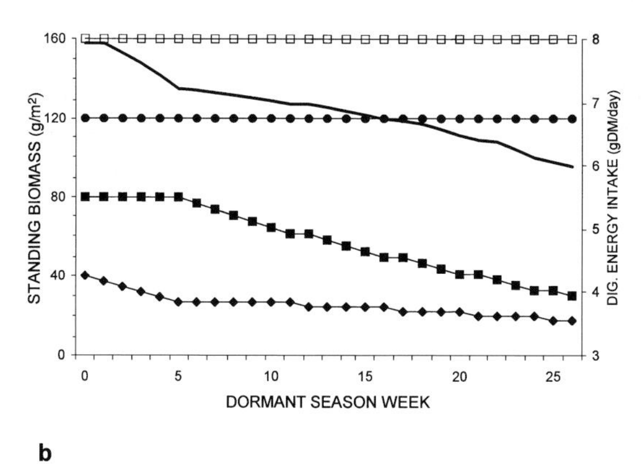

b

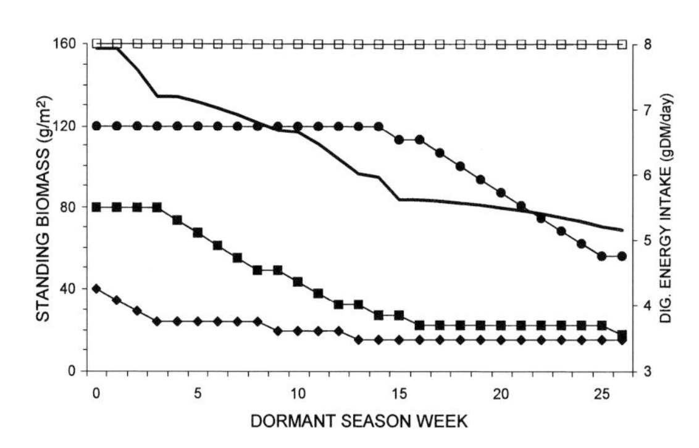

c

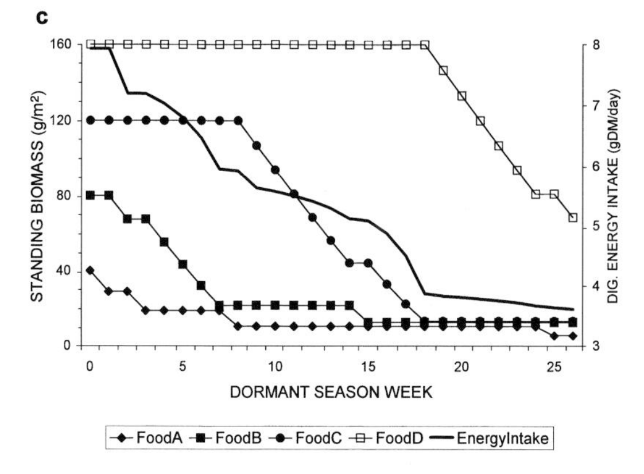

FoodA FoodB FoodC FoodD EnergyIntake

Figure 9.5. Pattern of resource depletion when four resource types are selectively consumed over the dormant season, and corresponding potential energy gain, for different stocking densities. The resource types offer relative conversion yields of 1.0, 0.8, 0.6 and 0.4 from food dry mass to animal live mass, corresponding with initial forage biomass densities of 40, 80, 120, and 160 g/m^2, i.e. mean vegetation biomass equals 100 g/m^2. (a) Low stocking density of 0.5 stock units per hectare. (b) Moderate stocking density of 1 stock unit per hectare. (c) High stocking density of 2 stock units per hectare.

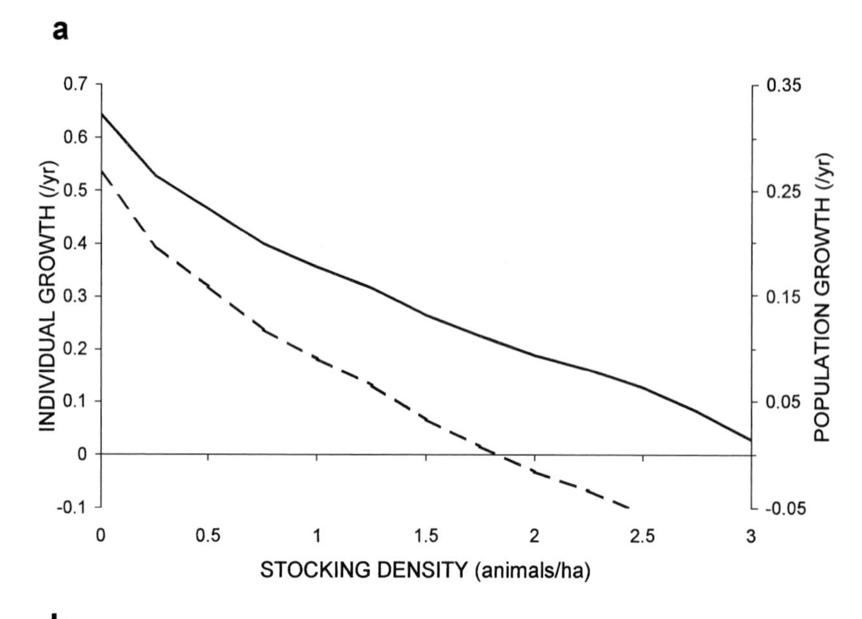

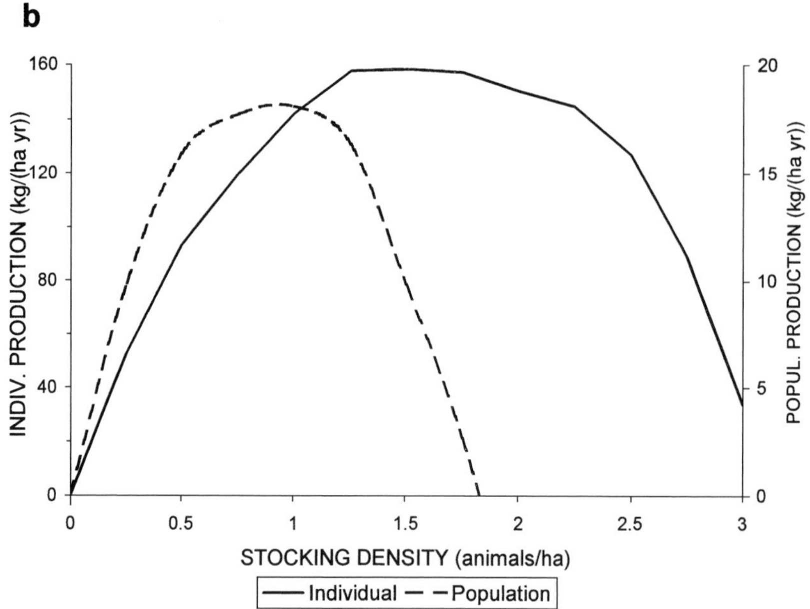

Figure 9.6. Forms of density dependence obtained from the selective depletion model. (a) Individual growth potential for growing young animals, and population growth potential, i.e. including non-growing animals as well as resource-dependent mortality. (b) Corresponding rates of biomass production per unit area, both for individual animals and for the population.

depletion model. Herbivores preferentially ingested green leaves, which offered a higher nutritional yield than dead leaves, but could not avoid consuming some necromass, as discussed in Chapter 8. Hence, food availability was represented by the combined standing crop of green foliage plus necromass, adjusted according to a selection coefficient. Unit stocking density was equivalent to one 400 kg steer per hectare.

Figure 9.7 contrasts the modelled energy gain by herbivores over the recurrent annual cycle for two contrasting grass types, presented in isolation. 'Short' grassland has a high growth rate and high nutritional value, but attains a fairly low standing crop. 'Medium–tall' grassland has a slower inherent growth, and moderate nutritional value, but produces a substantial standing biomass of forage. With only short grassland available, herbivores run out of food before the end of the DS, at high stocking densities. Medium–tall grassland gives poorer herbivore gains during the GS, but supplies forage for longer in the DS, than short grassland. However, because of the slow growth potential, grass regeneration can be suppressed at the start of the GS by high grazing pressure, unless there is sufficient carryover of necromass from the preceding DS. For the chosen parameter values, short grassland can support higher biomass gains by herbivores over the annual cycle, through being more nutritious, than can medium–tall grassland. However, animals would need to store considerable fat reserves during the GS, in order to survive periods with effectively no food that may arise towards the end of the DS in short grassland if the stocking density is high.

When both grassland types are intermingled in a patch mosaic, herbivores gain the seasonal benefits of each. By concentrating their feeding in the short-grass patches during the GS, animals obtain high gains. During the DS, they benefit from the higher grass biomass accumulated in the medium–tall patches (Fig. 9.8a). Because short grassland now occupies only half of the habitat area, the energy gains obtained by herbivores in the early part of the GS are lower than in circumstances where short grasses are prevalent everywhere. However, the improvement in growth potential during the DS more than compensates for this loss. When a third grass type is added to the model, in the form of a taller, less nutritious, slower-growing species that produces even more forage biomass, there is a substantial further gain. Nutritional shortfalls during the late DS are reduced, with only a minor cost in early GS gains. Notably, parameter values for the three grass species were chosen so that their mean equalled the values characterizing the medium–tall grass type.

Figure 9.7b shows how changing stocking density influences the annual biomass gain by herbivores, for the three-species grassland mosaic.

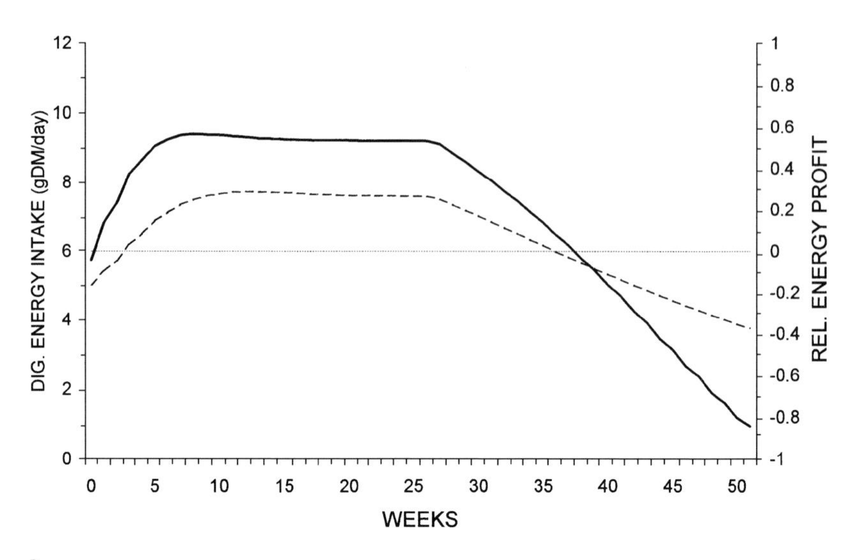

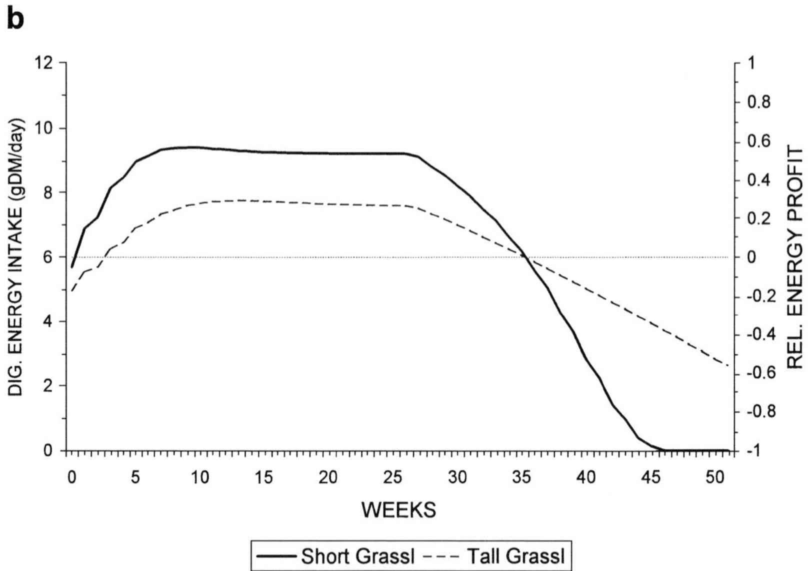

Figure 9.7. Daily biomass gain and corresponding growth potential, as controlled by the relative energy profit, over the annual cycle, obtained using the seasonal production model for alternative single grassland types. For fast-growing short grassland, $g_V = 0.75$ per week and $v_{\max} = 200$ g/m^2. For slower-growing medium–tall grassland, $g_V = 0.5$ per week, $v_{\max} = 400$ g/m^2, relative nutritional value = 0.8. (a) Moderate stocking density of 2 stock units per hectare. (b) High stocking density of 3 stock units per hectare.

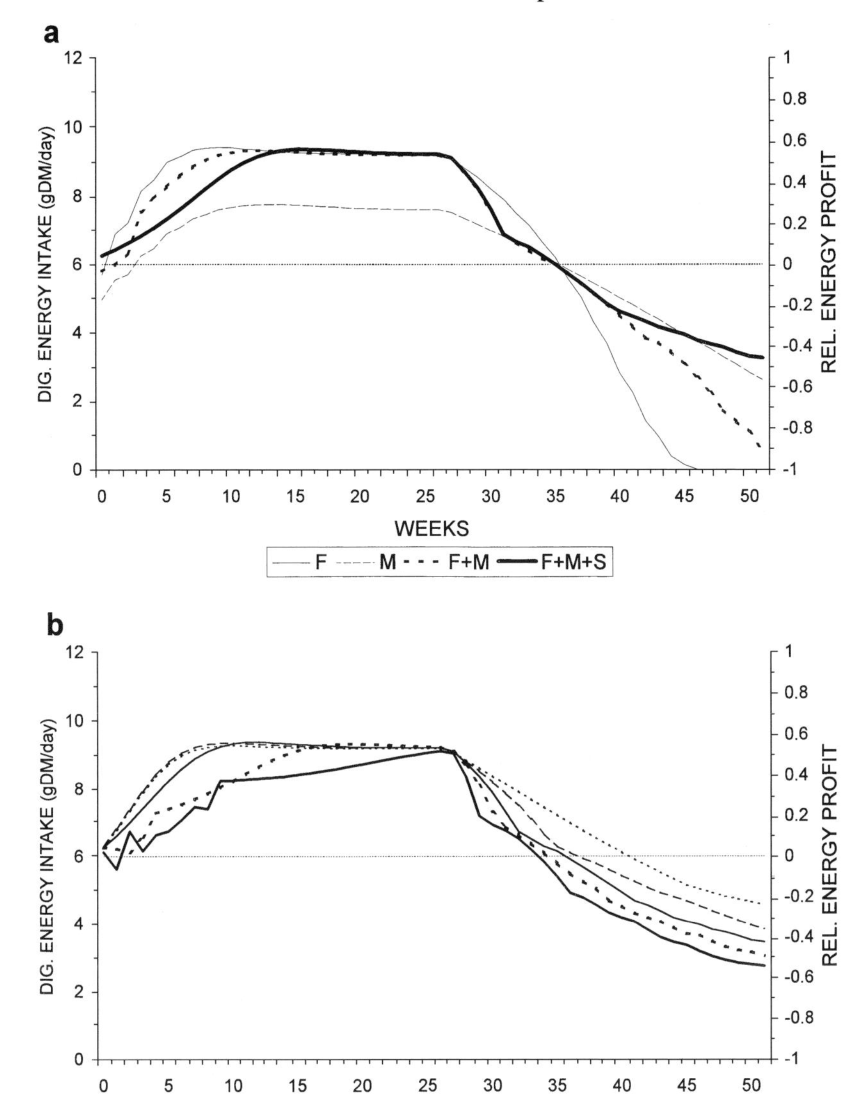

Figure 9.8. Daily biomass gains and corresponding growth potential over the annual cycle from the seasonal production model, for mixed compared with single grassland types stocked at different herbivore densities. Parameters for fast-growing (F) and medium–tall (M) types are as in Fig. 9.7. For slow-growing tall grass (S) $g_V = 0.33$ per week, $v_{\max} = 600$ g/m^2, relative nutritional value = 0.6.
(a) Comparison of two versus three grass types, relative to situations with single grassland types, for a high stocking density of 3 stock units per hectare. (b) Effects of increasing stocking density on the intake regime, for the situation with three grass types.

Even at low stocking density, herbivore gains may drop below maintenance requirements before the end of the DS, because of the poor quality of the necromass forming most of the forage standing crop in the DS. The higher the stocking density, the earlier the stage in the DS at which daily gain becomes sub-maintenance. With rising stocking densities, herbivores increasingly retard the accumulation of grass biomass early in the GS, leading to reduced gains at this stage. The presence of the poor-quality grass type prevents food from running out completely before the end of the DS, as may happen at high stocking densities when grasses are all of relatively high quality.

The form of the DDF depends on the grass types available (Fig. 9.9a). For a monospecific grassland, the DDF tends to be convex, because at low densities changes in stocking density have little effect on herbivore growth. With multiple grass types, the DDF tends towards linearity. The benefit of having poor-quality but high-biomass grass types arises mainly at high stocking densities. But without high-quality grass types, the growth potential of the herbivores during the GS is restricted.

The realized annual growth in biomass by the herbivores is the outcome of contrasting seasonal patterns of density dependence (Fig. 9.9b). Over the DS, the growth potential shows a concave relation with increasing stocking density, as was found for the selective depletion model. Over the GS, small increases in stocking density enhance nutritional gains, by reducing the accumulation of necromass. Only towards high stocking densities is herbivore growth during the GS reduced, through a retardation of the initial growth of green biomass. The presence of multiple grass types confers a substantial gain in herbivore biomass production per unit area, by allowing high stocking densities to be supported through the DS (Fig. 9.9c).

This simple model emphasizes the substantial benefits in stocking density that can arise from multiple food types with contrasting characteristics, specifically (a) a high-quality, rapidly re-growing food type enhancing nutritional gains during the GS, (b) an alternative, high-biomass food type to sustain resource gains during the early DS, and (c) a poor-quality, high-bulk food type to serve as a forage buffer during the late DS. The value of multiple resources will be enhanced in situations where rainfall, and hence vegetation growth, fluctuates over the seasonal cycle. Herbivores can transfer to the alternate grassland types at times when the preferred types become grazed down.

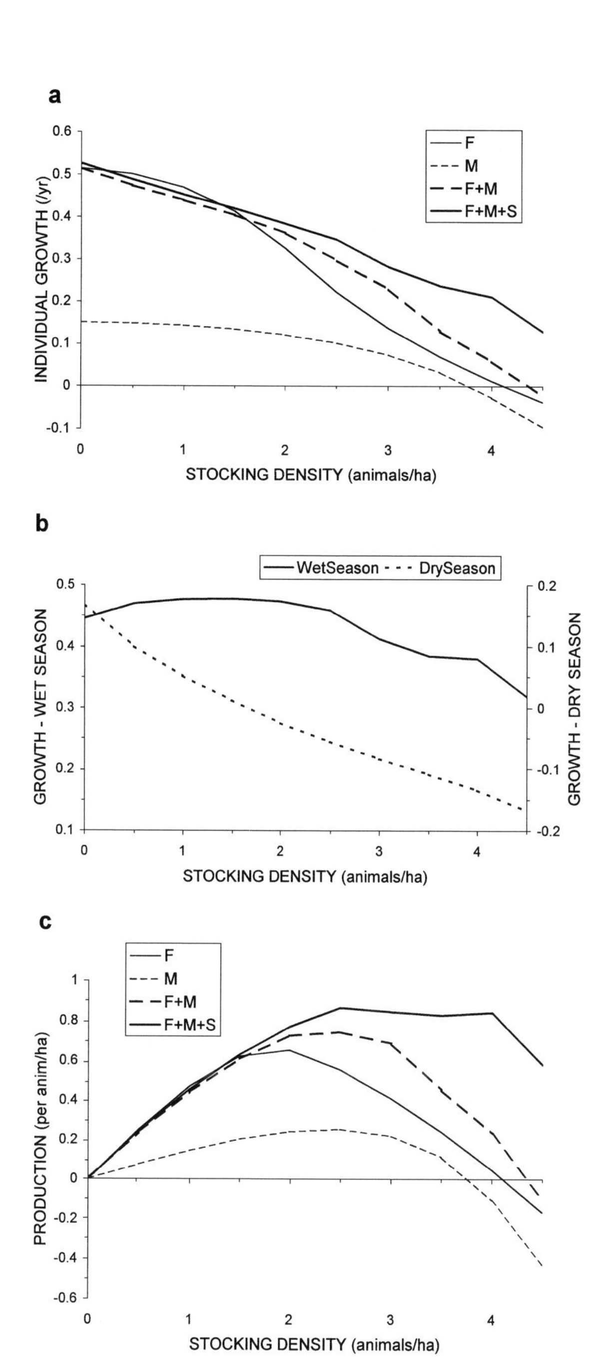

Figure 9.9. Pattern of density dependence in herbivore growth potential obtained from the seasonal production model, corresponding with the annual intake patterns depicted in Fig. 9.8. (a) Response in individual growth potential for the different grass type contexts. (b) Seasonal contrast in individual growth potential, for the situation with three grass types. (c) Herbivore biomass production per unit area for the different grass type contexts.

Competitive distributions

The classical derivation of an 'ideal free distribution' (IFD) assumes a situation where there is (a) continuous food renewal, so that no depletion occurs, (b) immediate consumption, so food does not accumulate, and (c) strong interference, so that individual gains depend on the ratio between food production and consumer abundance (Milinski and Parker 1991). Under such conditions, consumers should distribute themselves among habitat types in proportion to the relative productivity of each.

If resources accumulate, the impact of additional consumers on the resource gains of others is weakened, particularly if the gain response is decelerating rather than linear (van der Meer and Ens 1997). These are the circumstances applicable to most large mammalian herbivores. In the short term, it is advantageous for animals to aggregate in the habitat yielding highest individual gains. Congregations of individuals and herds in particular localities are indeed commonly observed for large ungulate species. However, if the local density is particularly high, some degree of short-term interference may arise. For impala, the number of bites taken per bush was less for larger herds than for small groups feeding in the same habitat (Fritz and de Garine-Wichatitsky 1996). For African buffalo, animals following in the rear of large herds obtain a reduced food intake relative to animals feeding towards the front (Prins 1996). Adult male buffalo joining these herds achieve a substantially lower resource gain than when foraging in small bachelor groups.

Over an extended period, resources in the best habitats become depleted as a result of consumption, unless the rate of re-growth is high. During the GS, herbivore concentrations within particular habitats may persist, especially if re-growth is of higher quality than available elsewhere. Movements may take place locally as particular patches are depleted. However, in African savanna regions rates of vegetation renewal are governed by rainfall, which varies erratically even during the GS. Fluctuations in resource regeneration may prompt shifts between habitats during the course of the GS, as described for zebras and for white rhinos in Chapter 4. During the DS, resources are progressively depleted. All else being equal, the proportion of time that herbivores spend in each habitat type over the DS will be determined by the relative productivity of these habitats, as reflected by the vegetation standing crop remaining at the start of the DS. It will also depend on the proportional extent of each habitat.

However, all is generally not equal, since habitat types vary in the forage quality they offer as well as in productive potential. Habitats offering more nutritious forage will be depleted to a lower level than those offering poor-quality food, with some of the latter hardly exploited at all. The extent of inherent attrition in forage availability over the DS through litter fall may also differ somewhat among habitats. As we will note later, predation risk may additionally modify habitat use. Social relations may be a further influence, with 'bachelor' males often being relegated to seemingly poorer habitat than dominant males and breeding groups of females and young (Lott 1991).

The modelled algorithm whereby herbivores transfer their consumption *en masse* between forage types in response to resource depression is the 'ideal free' equivalent for circumstances where interference is negligible, the gain function steeply saturating, and predation not an influence. Within the extent covered by each broad food type, herbivores will shift among specific foraging areas on a shorter-term rotation, in response to local resource depletion, as outlined in Chapter 4. Where social interference arises, some portion of the population may occupy inferior habitats.

Resource versus density dependence

Theoretical population ecology emphasizes density dependence, while paying scant attention to the dependence of population growth on resource supplies. This follows from the classical logistic model, which treats N (numerical population density) as a variable while regarding K (habitat capacity) as a constant. In practice, the population growth rate responds to variability in resources over multiple time scales. Indeed, the practical challenge is to identify the density dependent signal amid the environmental 'noise'. From my detailed 10-year study on kudu dynamics in the Kruger National Park, density dependence was established only for juvenile survival, and then only at a low level of statistical significance (Owen-Smith 1990). The influence of annual rainfall variability on individual survival rates was overwhelmingly dominant.

As modelled in this chapter, density dependence emerges from the interaction between resource consumption and production. The extracted form of the DDF was clear only because no variability between years in resource production, nor within years in inherent resource attrition, was allowed. When vital rates depend directly on rainfall through its influence on resource production, as is the case for kudus, density dependence can be extracted by controlling for variability in rainfall (Owen-Smith 1990, 2000). When these rates depend more on food quality, or on the seasonal distribution of rainfall, as seems to be the case for the Serengeti wildebeest population (Sinclair *et al.* 1985), allowing for changing resource availability becomes more problematic.

The debate about the merits of the resource versus resource-ratio dependence of functional responses (Abrams and Ginzburg 2000) is largely spurious. In the absence of interference, short-term gains depend directly on resource availability. However, population changes depend on resource gains integrated over the annual cycle, and thus inevitably on consumer density relative to the resource supply, but not in a linear or direct way.

Overview

For large herbivores, intraspecific competition arises primarily through resource depletion rather than via overt interference. Density dependence in the biomass growth potential emerges from the interaction between resource supplies and consumption over the seasonal cycle. The form of density dependence in relative rate of biomass growth tends to be convex when resources are fairly uniform in value, but more nearly linear when multiple resource types differing in quality are selectively exploited. Exploitation competition intensifies during the dormant season when vegetation resources are non-renewing. In the above circumstances, the 'ideal free' distribution of herbivore populations among habitat types takes the form of congregation within the best habitat in the short term, but a shifting distribution among habitats over the seasonal progression. How herbivore populations are apportioned among habitats over the annual cycle depends not only on the relative productivity of these habitats, but also on quality differences in the resources they present, and on distinctions in inherent attrition of food types through the dormant season. The density-dependent signal in population growth rate may be difficult to discern where resource production, quality and persistence vary widely from year to year.

Further reading

Fowler (1987) reviews general patterns of density dependence among large mammals, while McCullough (1992) assesses them specifically for large herbivores. Sutherland's (1996) book focuses on the ideal free distribution and its practical applications, with emphasis on birds and insects. Some insight into the debate about ratio-dependence can be obtained from Oksanen *et al.* (1992), Berryman *et al.* (1995) and Abrams and Ginzburg (2000).

10 · *Resource-dependent mortality: nutrition, predation and demography*

Losses in population biomass through mortality are the integrated outcome of many influences. Resource deficiencies can eventually deplete body reserves to critical levels of malnutrition, but only rarely can deaths be ascribed simply to starvation, at least for wild herbivores. Extreme weather heightens metabolic costs and restricts foraging activity, draining body reserves towards critical levels. Animals that are weakened through food shortfalls become more susceptible to being killed by predators. Malnourished animals are also more vulnerable to parasite and disease infestations, amplifying their susceptibility to predation (Sinclair 1977; White 1983). Food shortages cause animals to spend more time foraging, raising physiological expenditures and increasing exposure to predation and other sources of accidental death. Risk of predation affects the habitats that can be exploited securely, exacerbating food limitations (McNamara and Houston 1987b; Sinclair and Arcese 1995). Where surface water is a limitation, animals must concentrate their movements around drinking points, where predators lie in wait. Social strife over inadequate resources may lead directly to mortality, or to injuries pre-disposing animals to other sources of mortality.

Susceptibility to predation depends on the types of predator present and their hunting methods. Adaptively foraging predators should preferentially seek those prey that are malnourished and hence most easily captured and killed. Nevertheless, healthy animals may sometimes have the misfortune to become victims of predators lying in ambush.

This chapter outlines how these various factors interact to determine the form of the mortality response to changing resource availability. Furthermore, we must allow for variation in the risk of mortality among population segments distinguished by age, size or growth stage. Hence, the demographic structure of the population constitutes an additional influence on the effective loss in population biomass.

Mortality function

The proportional loss in population biomass through mortality can be partitioned among three components: (1) deaths ultimately through senescence, i.e. reaching the end of the physiological lifespan, Q_0; (2) accidental deaths through various causes, including those of otherwise healthy animals killed by predators, Q_A; (3) deaths pre-disposed by inadequate nutritional gains, whether brought about directly through starvation or through the agency of predators and pathogens, Q_G.

However, the overall mortality, M_Q, is not simply a sum of these three components. Deaths through senescence are pre-empted by mortality through other causes. Hence, the additive contribution of the latter towards overall mortality is somewhat less than the direct loss inflicted. For illustration, a population of animals with a maximum individual longevity of 10 years would incur a minimum mortality loss of 10% per year, in numerical terms, simply through individuals reaching the end of their physiological lifespan. A simple calculation shows that, if mortality through other agents eliminates 10% of the population annually, deaths through senescence diminish to 3.8% per year, so that the additive mortality is only 6.2% per year, in numerical terms. The loss in terms of biomass would be somewhat greater, because animals dying of old age are larger than the average mass.

Getz (1991, 1993) proposed that the overall mortality loss should be inversely related to resource gains. More specifically, the mortality rate should depend on nutritional gains relative to physiological expenditures (see Chapter 2), i.e. the same relative gain would be less effective at alleviating mortality at times when the costs of activity or thermoregulation are high. Hence, modifying Equation 2.27, we have

$$M_Q = z_0 + zM_P/G, \tag{10.1}$$

where M_P = overall metabolic expenditure in an appropriate currency, G = rate of resource gain, in the same currency, and z_0 and z are curve-fitting constants. Note that the difference between z and z_0 sets the mortality level applicable when the resource gain just matches metabolic requirements, while the ratio z/z_0 determines how much greater than requirements the gain must be for resource-dependent mortality to become zero.

This simple formulation does not allow for the mortality loss through senescence that would persist even when resource gains are high. Accordingly, there is a lower bound to the mortality loss, q_0, representing

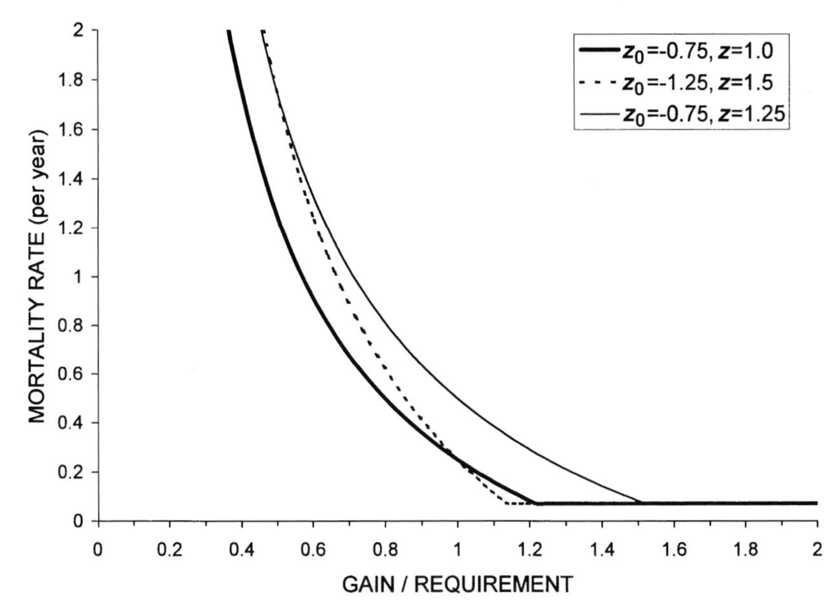

Figure 10.1. Resource-dependent mortality function, as influenced by adjustments to (i) z_0, controlling the gain level at which the mortality rate is affected, and (ii) z, controlling how steeply mortality rises with diminishing gains relative to requirements, relative to a background mortality rate q_0 of 0.07 per year. The difference between z_0 and z sets the mortality rate when resource gains just match requirements.

the minimal mortality rate. As a result, the hyperbolic *resource-dependent mortality function* (RMF) is truncated towards high rates of resource gain relative to requirements (Fig. 10.1). Other sources of mortality unrelated to resource fluxes, such as accidental predation, may elevate the background mortality rate. On the other hand, there is no upper limit to the mortality rate. As resource gains diminish relative to requirements, the mortality rate could rise from 1.0 per year to 1.0 per month or more.

However, even with no food, animals would not die instantaneously, because body reserves constitute a buffer. Illius and Gordon (1998) modelled the susceptibility to mortality of Soay sheep on a predator-free island as a direct function of the state of fat reserves, allowing for intrapopulation variability. Widening resource deficiencies during the winter period caused body reserves to be drawn down, with an increasing proportion of the population dying as a result. Nevertheless, as noted in Chapter 7, the mortality cost of fat storage may be spread more widely over the seasonal cycle, at least where predation is a factor. If this were not the case, fat stores should simply be increased to a level where mortality losses through resource inadequacies would be minimal.

Accordingly, whatever the mechanism of death, susceptibility to mortality depends on resource gains integrated over some extended period. To some degree, resource surpluses accumulated at one stage of the annual cycle can offset deficiencies at a later time. However, because of the costs of carrying fat reserves, the alleviation of mortality is only partial. For herbivores, mortality losses through malnutrition, even if mediated by predation, generally intensify towards the end of the winter or dry-season period of diminishing resources. For modelling, the time frame for assessing how the resource flux contributes to mortality will thus need to be somewhat less than a year, but certainly longer than a day. A time step of somewhere between a week and a month would seem a reasonable compromise. Nevertheless, an interval as long as a month may not adequately capture the alleviating effect of mortality on consumption, and hence on resource depletion, during the critical stage of the late dormant season. The RMF coefficients z and z_0 would need to be adjusted somewhat, depending on the time step chosen.

Density dependence

The effective resource gain governing population dynamics depends not only on the resource supply and its seasonal variation, but also on the effect of population density on resource availability, especially during the dormant season. Chapter 9 considered how the rate of biomass gain declines with increasing population density. We now consider how the mortality loss, as defined by Equation 10.1, might respond to changing population density.

For this assessment, the modified production pulse model, as outlined in Chapter 9, was used. Food resources are produced in a brief pulse at the start of the annual cycle, and persist unchanged in amount and quality through the six months of the growing season (GS). During the dormant season (DS), resources are progressively depleted solely through consumption, and are also intrinsically lower in nutritional yield relative to the GS. A progressive seasonal decline in the quality of the food consumed was represented phenomenologically by adjusting the half-saturation level for biomass gain ($g_{1/2}$ in Equation 9.2) relative to that for food intake ($i_{1/2}$, Equation 9.3), as depicted in Fig. 9.2a. The amplification of the weekly nutritional deficiency, through predation or whatever causes, was adjusted by changing the values of z_0 and z in the RMF. The background mortality, q_0, was set at 0.05 per year. The stocking density was represented as animal units made up by a 400 kg cow or equivalent herbivore, relative to an annual food production of 500 g/m^2.

The density dependence in mortality appeared gently curvilinear towards low densities, but accelerated sharply when density became high enough for food depletion to be severe (Fig. 10.2). This threshold rise in mortality was more severe if there was little seasonal decline in food quality, so that the resource gain function was steeply saturating. Elevating both z_0 and z, while keeping their difference constant, made mortality less responsive at low densities, but more sensitive to the resource depletion generated at high densities (Fig. 10.2b). Raising the value of z relative to z_0 elevated the baseline sensitivity of mortality to resource deficiencies, such that potential recruitment could be counteracted by mortality losses below the density at which mortality rose abruptly (Fig. 10.2c). Under these conditions, changes in the form of the resource gain function also made a substantial difference to the density associated with zero net recruitment, by altering the slope of the density dependence in mortality.

Overall, this exercise suggests that resource deficiencies are likely to be amplified gradually but progressively by mortality losses, under conditions where food quality deteriorates steadily over the dormant season. In contrast, if food quality varies little, so that resource gains are tied more directly to reductions in daily food intake, mortality may take effect abruptly at high densities.

Observed mortality patterns

Resource dependence

In my decade-long study of kudus in two regions of South Africa's Kruger National Park, it was possible to identify all animals individually, from birth onwards, through variations in stripe patterns and other markings (Owen-Smith 1990). This enabled me to estimate annual mortality rates separately for prime adult females (age 1.7–6.7 yr), old adult females (age > 6.7 yr), yearlings (age 0.7–1.7 yr), and juveniles (from the cow:calf ratio, thus effectively including all losses from conception to 0.7 yr). Adult males were more mobile and fluid in their social affiliations, so that survival rates could not be estimated annually for this segment.

Findings revealed how these mortality rates depended on year-to-year variability in resource production, as controlled by rainfall. The cow:calf

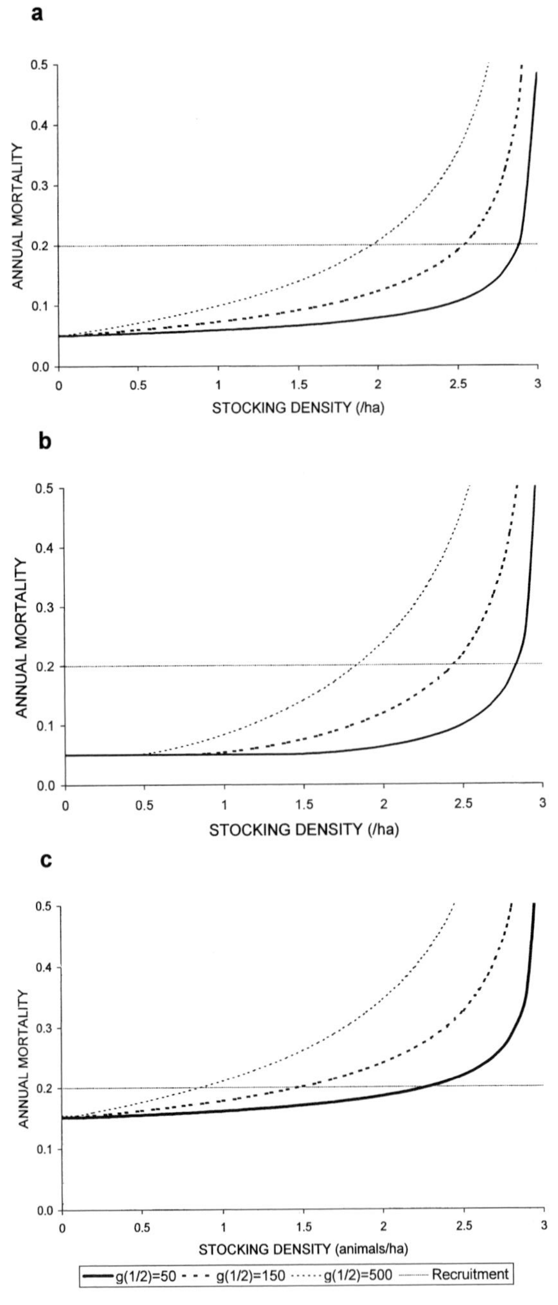

Figure 10.2. Density dependence in mortality as influenced by the form of the resource gain function, and the setting of the mortality function parameters, from the modified production pulse model. The dotted line indicates a potential recruitment rate of 0.20 per year. (a) $z_0 = -0.75$, $z = 1.0$. (b) $z_0 = -1.25$, $z = 1.5$. (c) $z_0 = -0.75$, $z = 1.25$.

ratio in the kudu population at the stage when calves were around six months old was especially sensitive to the rainfall total over the preceding seasonal cycle. However, kudu calves apparently survived better relative to rainfall during the early part of the study, when the kudu density was still fairly low, than later after the population had increased. This indicated a density-dependent influence. Accordingly, the effective food availability was indexed by the ratio between the annual rainfall total (controlling food production) and population biomass density (representing the food demand). The annual survival rate of kudu calves appeared to be linearly dependent on the logarithm of the rainfall:biomass ratio, especially if outlier points from one anomalous year were excluded, and allowance was made for the saturation in survival rates towards high food availability (Fig. 10.3). Interestingly, the data from both study populations fitted the same regression model, indicating that the 30% higher mean rainfall in one study area was counterbalanced by the correspondingly higher kudu density there.

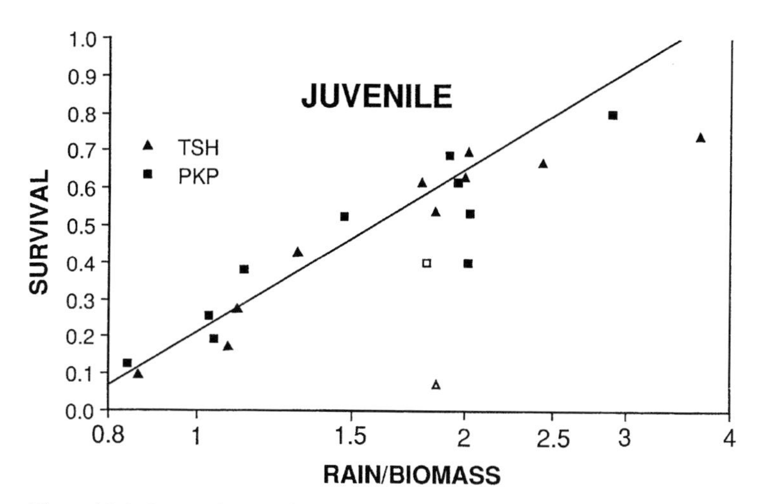

Figure 10.3. Dependence of the survival rate of juvenile kudus on rainfall relative to population density (logarithmically transformed), with annual rainfall total in mm and kudu density in kg/km^2. Symbols distinguish data from the two study areas, labelled TSH and PKP. Open symbols identify outlying points for a year in which survival was low relative to rainfall, owing to a severe cold spell. Excluding these outliers, and data from the initial year when rainfall was extremely high relative to biomass, $r^2 = 0.914$, $n = 16$, $p < 0.001$; for full data set, $r^2 = 0.847$, $n = 20$, $p < 0.001$. (From Owen-Smith 1990, © Blackwell Science).

To represent the upper and lower bounds to the annual survival rate, a model incorporating a logistic transformation of the observed survival rates was also tested, but brought no improvement in fit over the observed range in conditions. Notably, although the lower bound to annual survival is clearly zero, the upper limit to calf survival seemed to be around 0.8 per year, presumably owing to conception failures and some inevitable predation.

According to Equation 10.1, mortality rates should be inversely related to resource gains, relative to metabolic requirements. The potential resource gain may be indexed by the annual rainfall total, controlling vegetation growth, whereas the metabolic demand is dependent on population biomass density (ignoring differences in mass-specific metabolic rates as affected by body size). With these assumptions, the hyperbolic mortality function yielded a slightly improved fit to the observed data, compared with the above survival functions, for juvenile kudus, and an equally good fit for other age classes of kudus (Fig. 10.4). More importantly, this function would remain applicable over a wider range in resource availability than the linear survival relation, and is free of the assumptions about upper and lower bounds necessary for the logistic survival model. The asymptotically diminishing mortality towards high resource availability, and accelerating mortality with diminishing resources, was most clearly shown for juvenile kudus (Fig. 10.4a). The hyperbolic function also captures the expectation that even prime-aged females must at some stage exhibit rising mortality when resource availability becomes sufficiently low, although such conditions were not encountered during the study period.

The anomalously low survival rates of all age classes observed in one year of the study, despite high prior rainfall, appeared to be related to a severe cold spell that occurred during the late dry season of this year (Owen-Smith 1990, 2000). The effect of cold stress would be to elevate metabolic costs relative to resource gains, thereby leading to higher mortality than expected for the resource conditions. A notable general finding was that mortality rates were closely dependent on variability in resource production, even for a population existing at relatively low density (2–3 kudus/km^2) in the presence of numerous predators.

A hyperbolic dependency of annual mortality losses on rainfall relative to stocking density has also been documented for free-ranging cattle in northern Australia, although a parabolic curve was originally fitted to the data (Gillard and Monypenny 1990) (Fig. 10.5a). Notably, a steep rise in deaths was recorded in one year when a severe drought prevailed. A study on American elk in northern Yellowstone National Park likewise revealed that mortality rates of calves, as well as of adult males, were inversely related to the annual precipitation, especially if allowance was made for population density (Coughenour and Singer 1996) (Fig. 10.5b). Malnutrition accounted for most of the winter calf mortality, but a substantial proportion of calf losses during summer was due to predation (Singer *et al.* 1997).

For the Serengeti wildebeest population, overall mortality losses appeared to be related primarily to the amount of rain falling during the dry season, and hence to grass growth during this period, rather than to the annual rainfall total (Sinclair *et al.* 1985). A curvilinear relation between mortality rates through the dry season months, and an index of effective food availability during this period, was apparent both for calves and adults (re-calculated from Mduma *et al.* (1999)) (Fig. 10.5c–d). However, the rise in mortality towards low food availability in this season appeared somewhat gradual, and there was much scatter in observed mortality rates relative to this index, particularly for calves. This suggests additional influences, perhaps from food reserves accumulated from wet-season rainfall.

In the Kruger Park, changes in wildebeest abundance, and to a lesser extent that of zebra, appeared to be negatively rather than positively related to the preceding annual rainfall (Mills *et al.* 1995). This could be because higher rainfall had an adverse influence on food quality, by promoting higher fibre contents relative to nutrients.

Density dependence

The inverse hyperbolic dependence of mortality on rainfall relative to population density, as defined by Equation 10.1, implies that mortality should be directly and linearly related to density relative to rainfall. After controlling for resource variability in this way, a linear pattern of density dependence in mortality was shown by all kudu age classes (Fig. 10.6). Interestingly, the regression lines projected close-to-zero mortality at zero density, except for old females. Observed minimum levels of mortality were somewhat greater than zero, presumably because of some inevitable predation even under conditions of plentiful food.

Relating mortality simply to the normalized means of rainfall relative to population density implies that both factors are equally influential. However, partial regressions indicated that the influence of rainfall variability on the survival of kudu calves was twice as strong as that of density variation (Owen-Smith 1990). For other age classes of kudus, the effects of rainfall and density appeared equivalent.

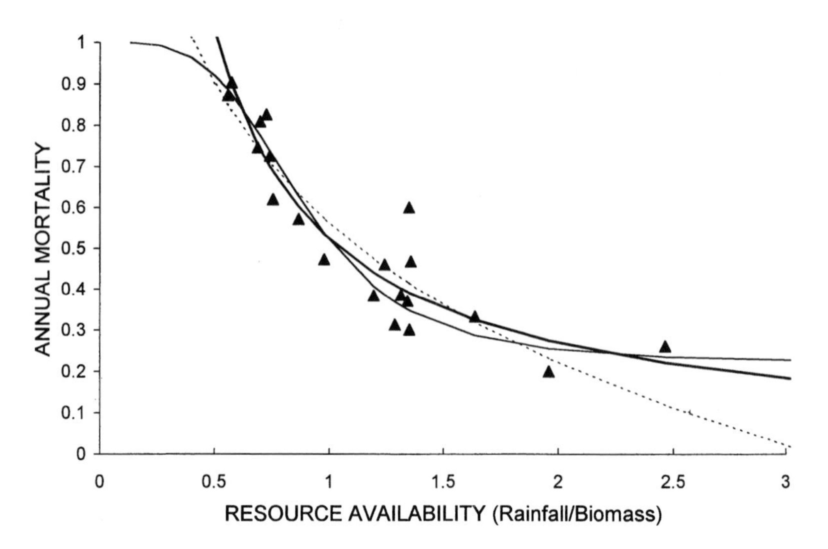

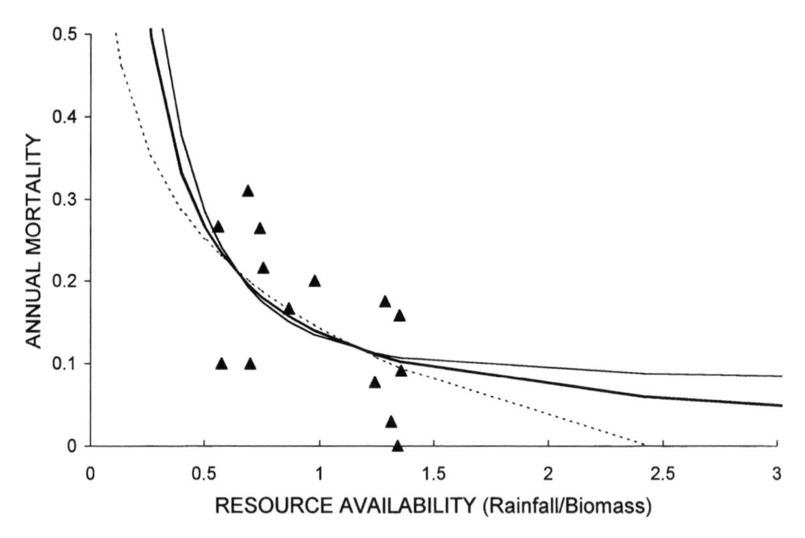

Figure 10.4. Dependence of annual mortality on resource availability indexed by normalized rainfall relative to biomass density, for different age classes of kudus. Three mortality functions are compared: (i) $Q = f\{1/(R/B)\}$; (ii) logist $(Q) = f\{1/(R/B)\}$; and (iii) $1 - Q = f\{\log(R/B)\}$. Relative goodness of fit is indicated in the caption to Fig. 10.6. (a) Juveniles. (b) Yearlings. (c) Prime adult females. (d) Old adult females. (From Owen-Smith 2000.)

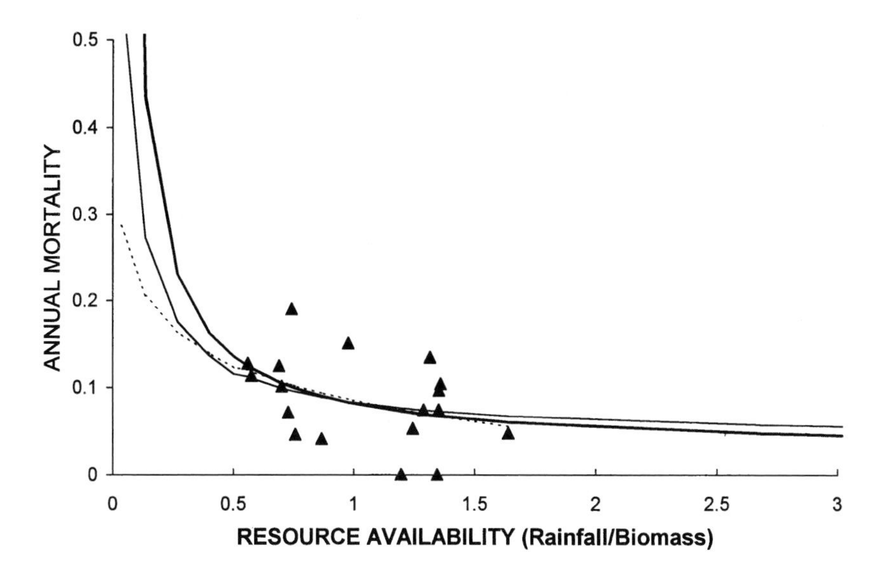

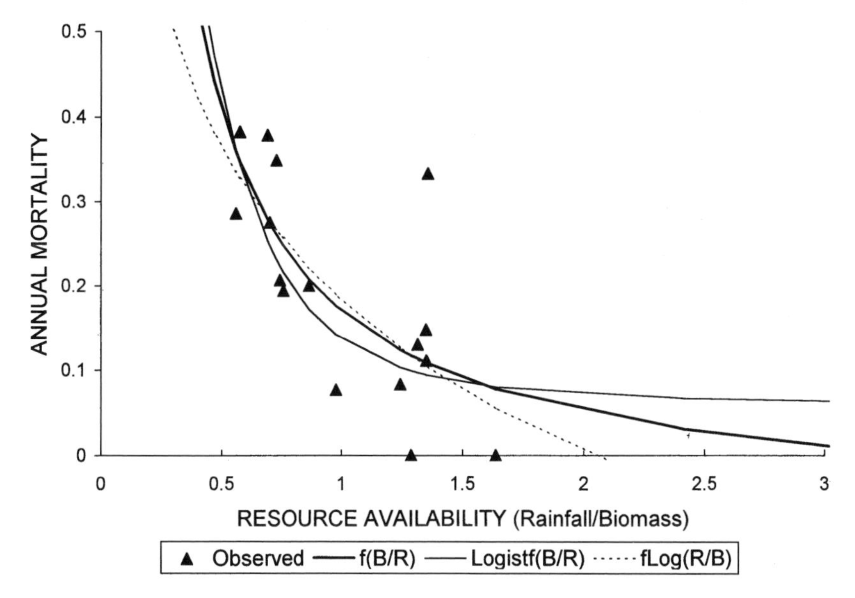

Figure 10.4. (cont.)

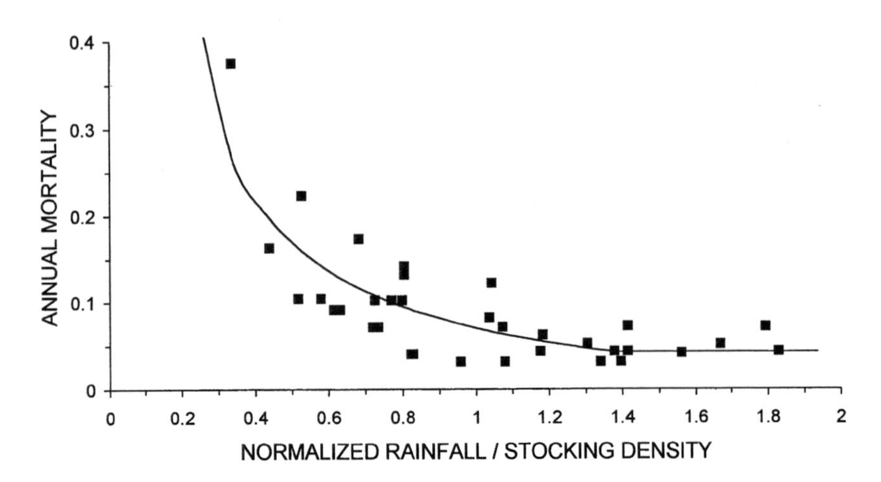

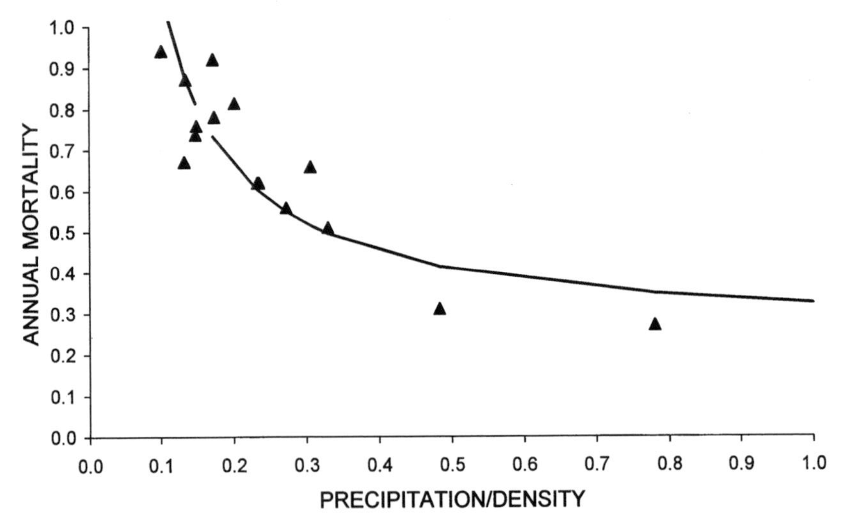

Figure 10.5. Dependence of annual mortality on rainfall relative to population density for other species. (a) Free-ranging cattle in northern Australia. (Re-drawn from Gillard and Monypenny 1990.) (b) Juvenile elk in Yellowstone National Park. (Calculated from Coughenour and Singer 1996.) (c) Juvenile wildebeest and (d) adult wildebeest in the Serengeti (re-calculated from Mduma *et al.* 1999).

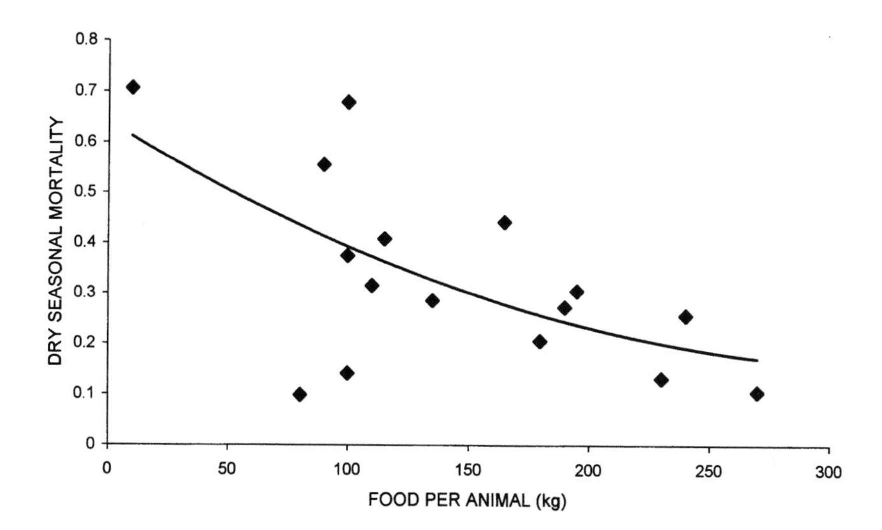

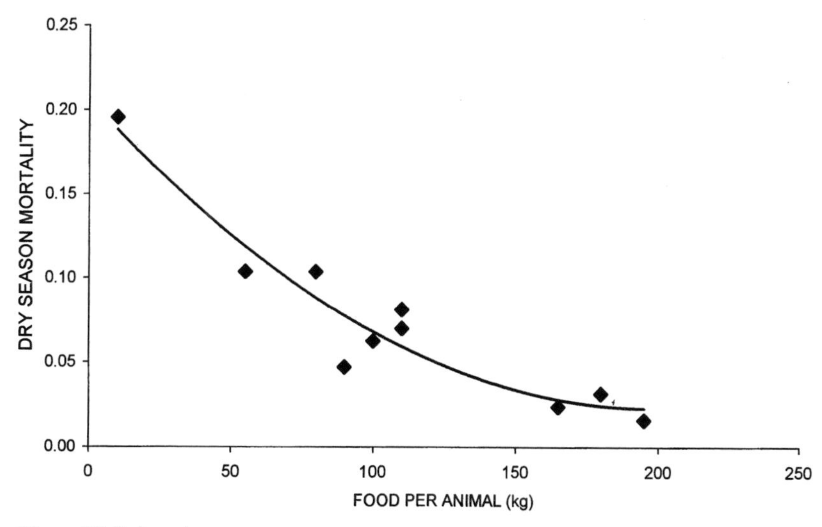

Figure 10.5. (cont.)

For both red deer inhabiting the North Block of the Isle of Rum, and Soay sheep occupying the Village Bay area of Hirta Island in the St Kilda archipelago, juvenile mortality rose linearly with increasing population density, but only after a threshold density had been exceeded

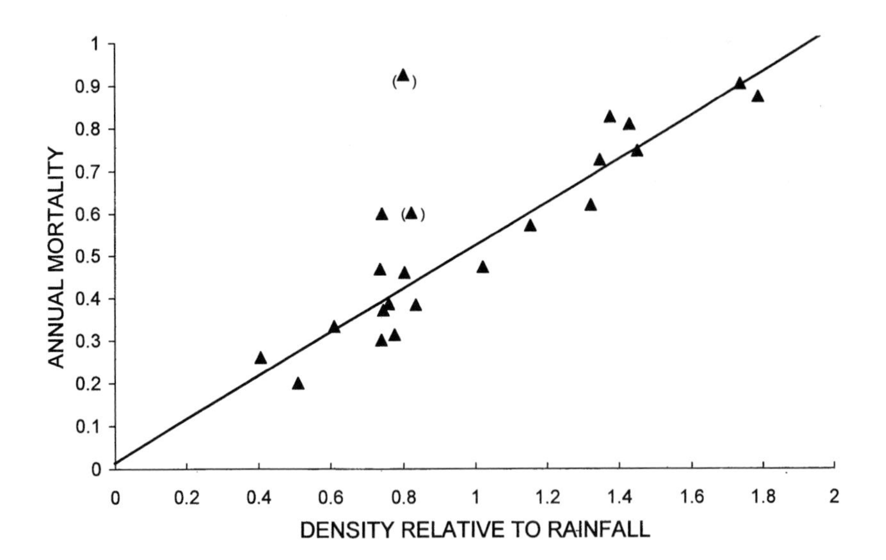

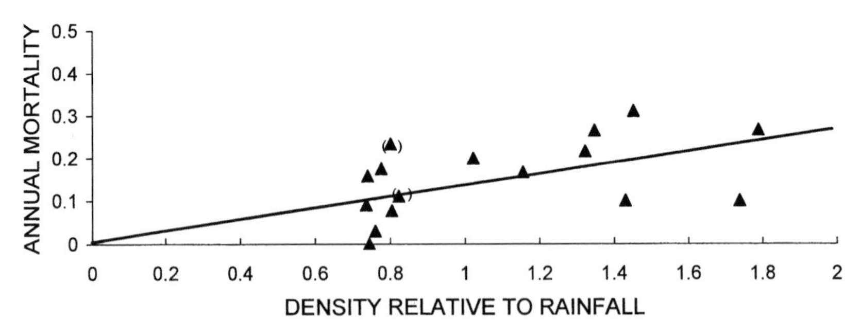

Figure 10.6. Dependence of annual mortality of kudu age segments on population density relative to rainfall. Both annual rainfall and kudu biomass density are normalized relative to their means. Data for the year of the cold spell are bracketed and treated as outliers. (a) Juveniles ($r^2 = 0.882$, $a = 0.014$, $b = 0.760$, $n = 20$, $p < 0.001$). (b) Yearlings ($r^2 = 0.301$, $a = 0.005$, $b = 0.197$, $n = 14$, $p < 0.01$). (c) Prime adult females ($r^2 = 0.168$, $a = 0.027$, $b = 0.082$, $n = 18$, $p < 0.1$). (d) Old adult females ($r^2 = 0.531$, $a = -0.069$, $b = 0.357$, $n = 16$, $p < 0.01$).

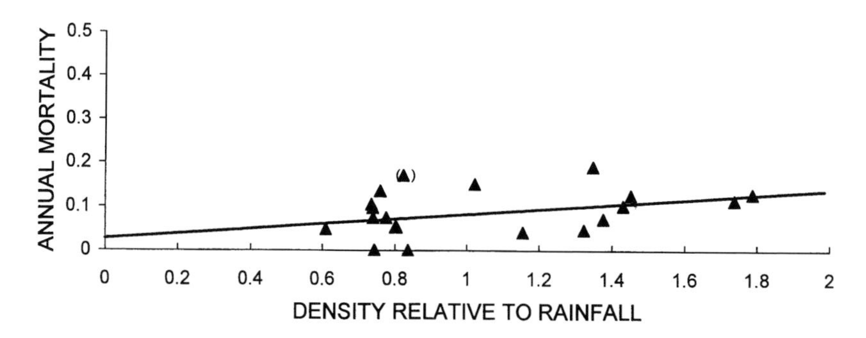

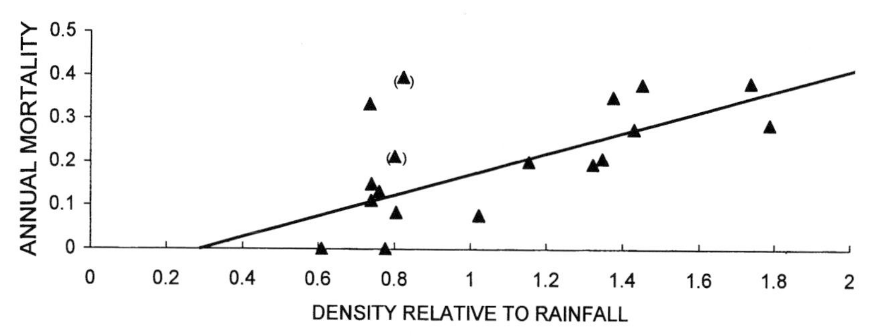

Figure 10.6. (cont.)

(Clutton-Brock *et al.* 1985; Grenfell *et al.* 1992; Milner *et al.* 1999a) (Fig. 10.7a, b). Mortality among adult male sheep was much more sensitive to increasing population density than that among adult females (Fig. 10.7c–d). For both sexes of Soay sheep, susceptibility to mortality rose sharply towards high density levels, precipitating periodic dieoffs. Notably, on these predator-free islands all mortality occurred directly through starvation. Accordingly, mortality losses remained almost zero through the summer period, and under conditions of low density, apart from deaths due to old age.

For African buffalo, the annual mortality of adults increased significantly with rising population density, but not that of calves (Sinclair 1974). However, the wide annual variability in calf survival probably obscured the density-dependent signal. Moreover, the rise in adult mortality could have been influenced by an increasing proportion of old animals with higher specific mortality rates (Loison *et al.* 1999). Studies on reindeer (Skogland 1985, 1990) and mule deer (Bartmann *et al.* 1992) found

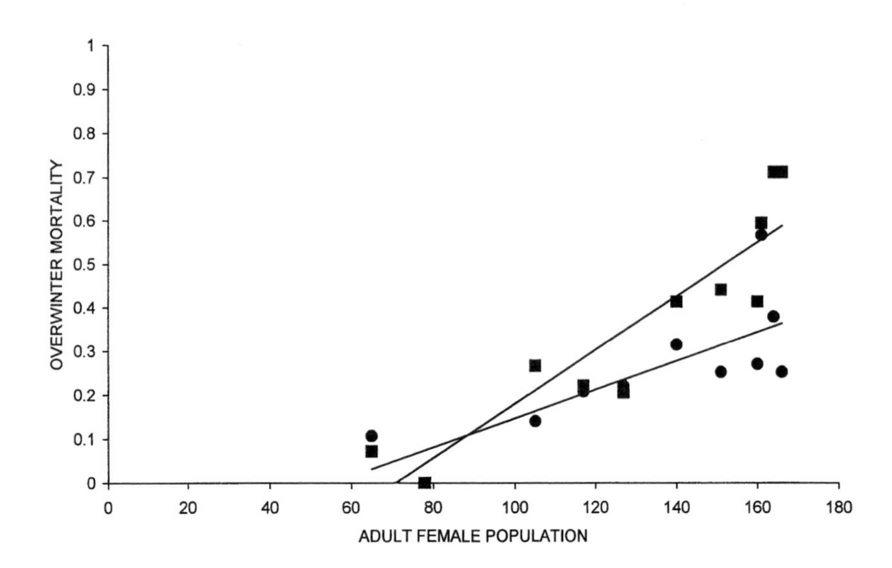

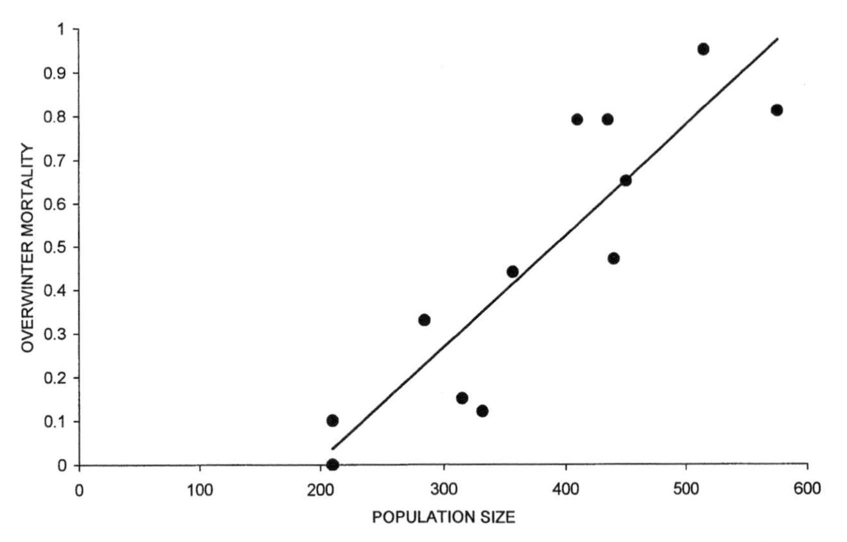

Figure 10.7. Density dependence in the overwinter mortality. (a) Red deer calves in the North Block of the Isle of Rum, Scotland, over 1971–82; squares, males; circles, females. (From Clutton-Brock *et al.* 1985.) (b) Juvenile Soay sheep inhabiting the Village Bay area of the Island of Hirta, controlled for weather influences on survival. (c) Adult male Soay sheep. (d) Adult female Soay sheep. (From Grenfell *et al.* 1992; Milner *et al.* 1999b).

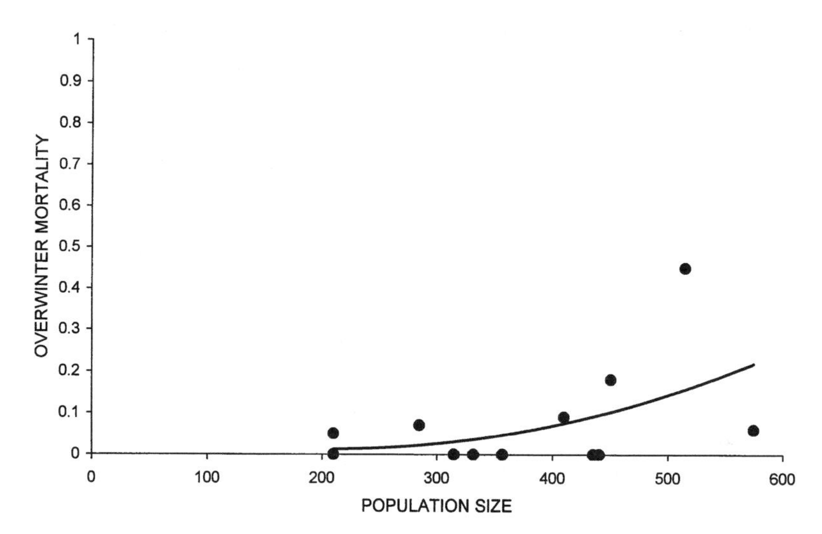

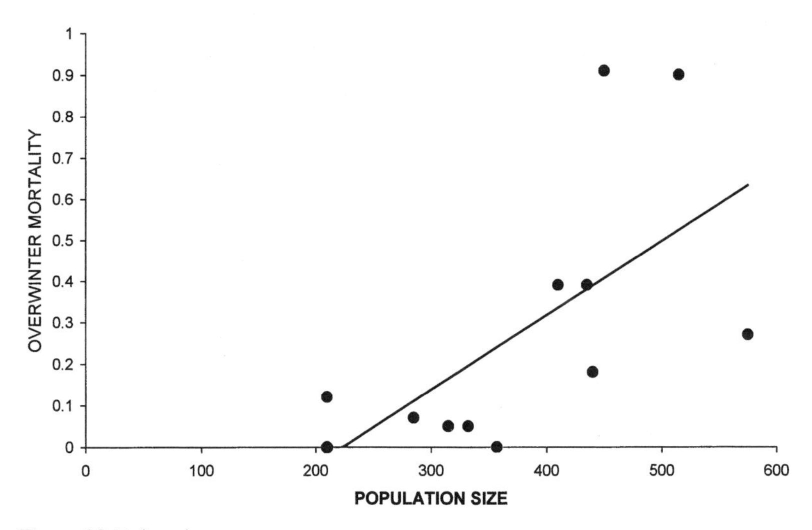

Figure 10.7. (cont.)

juvenile mortality to be dependent on the effects of population density on food availability during winter.

At the population level, losses in biomass through mortality are the aggregated outcome of the mortality rates shown by different population segments. For kudus, the overall pattern of density dependence in mortality may be assessed by building the age-specific mortality functions into an age-class-structured population model (cf. Owen-Smith 2000). The effects of density on net population growth can be revealed by allowing the population to grow from a low initial density towards the level where zero net growth pertains, while holding rainfall constant. The effective loss of population biomass through mortality is the difference between the prevailing population growth rate and the maximum rate exhibited at very low density, after stabilization of the initial age structure. For the female segment, the maximum population growth rate was close to 25% per year, if no density-dependent increase in the fecundity of yearling females was allowed. Assumptions need to be made about how the age-class-specific trends in mortality should be projected below the minimum rainfall:density ratio observed during the study.

If it is assumed that specific mortality rates tend towards zero at very low density, as indicated in Fig. 10.6, the density dependence in overall mortality losses deviates only slightly from linearity (Fig. 10.8a). Alternatively, if it is assumed that the mortality rates for each age class remain at the minimum levels observed during the study under conditions of high rainfall relative to density, the population mortality loss takes on a more strongly curvilinear form relative to density (Fig. 10.8b). This is because the different age classes constituting the population achieve their minimum mortality rates at different population densities. The background mortality rates shown by kudus under low-density conditions of resource superabundance may be presumed to represent 'accidental' predation. Hence, assumptions about how predation losses change with density affect the pattern of density dependence in mortality.

Ratio dependence

In Chapters 2 and 9, the theoretical issue whether the food intake and biomass gain responses should be related directly to food abundance, or to the food supply relative to population density, was raised. The problem is essentially how to link across time frames. In immediate terms, food intake is dependent on current food availability; but the latter is influenced by past consumer density. Moreover, the extent of food depletion varies seasonally, and the gain response is generally non-linear.

The above analysis showed that, for kudus at least, mortality losses are effectively dependent on the resource production:population density

a

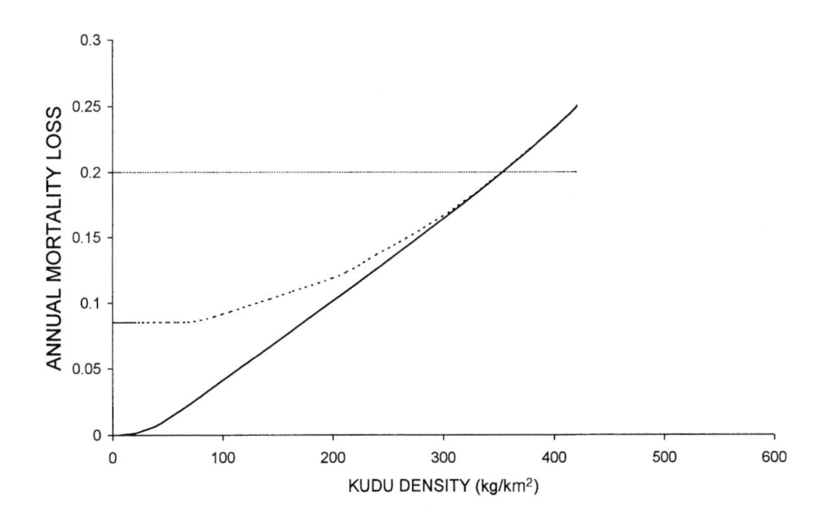

b

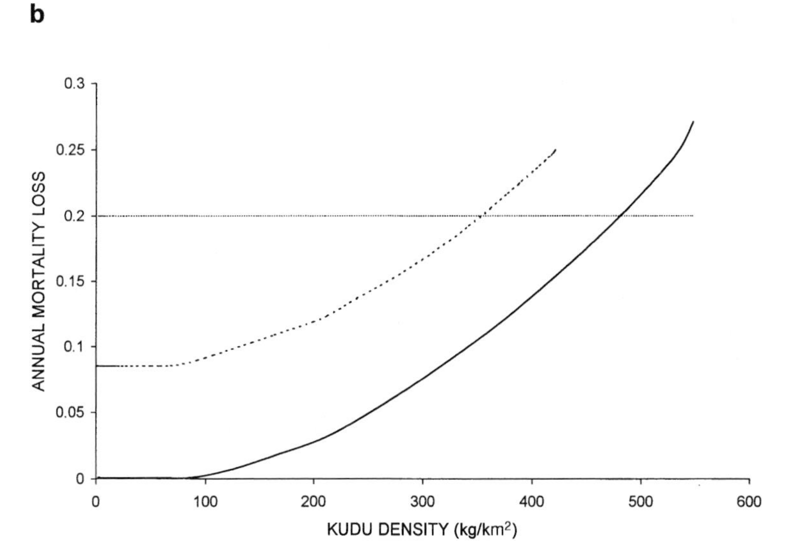

Figure 10.8. Loss in integrated population biomass through mortality relative to increasing population density, estimated from a simulation model incorporating the age-class-specific mortality functions indicated in Fig. 10.6 with different assumptions about minimal mortality levels. (a) Comparison between output assuming (i) minimal mortality rates to be close to zero (prime females 0.01, old females 0.02, yearlings 0.02 and juveniles 0.05) (solid line), and (ii) minimal mortality rates corresponding with those observed for high rainfall relative to population density (prime females 0.07, old females 0.10, yearlings 0.09 and juveniles 0.20, including infertility losses) (dashed line). (b) Comparison between output with predation eliminated by dividing through by maximum survival rates (solid line), compared with pattern from (ii) above.

ratio. Thus ratio dependence is manifested in the mortality ('numerical') response rather than in the intake ('functional') response. This is not surprising, since mortality losses reflect resource gains over some extended period.

Predation

Additive vs. interactive mortality

Bartmann *et al.* (1992) highlighted the distinction between *compensatory* and *additive* mortality, with respect to predation. Predator-induced mortality is additive if healthy animals are killed that would otherwise have survived. However, adaptively hunting predators should preferentially seek prey that are weakened nutritionally, and thus more easily captured and killed. In the latter circumstances the impact of the mortality on the prey population is compensatory, in the sense that animals that would eventually have died from starvation merely die sooner. The mortality loss results effectively from an interaction between the pre-disposing effects of malnutrition, and the risk of predation. Of course, some of the weakened animals may otherwise have survived, had they not had the misfortune to encounter a predator while in a vulnerable state. Predation acts to amplify their susceptibility to mortality.

Pursuit predators, such as African wild dogs and spotted hyenas, generally seek prey that are weakened (Kruuk 1972; FitzGibbon and Fanshawe 1988; Cooper 1990). Bears and coyotes in Yellowstone National Park kill proportionately more underweight elk calves than those with higher birth mass (Singer *et al.* 1997). Predators hunting primarily by ambush, like most felids, may attack healthy animals, although capture success probably depends on the condition of the prey. Most of the wildebeest killed by lions in the Serengeti region were in poor condition, as revealed by low bone marrow fat (Sinclair and Arcese 1995). In the Kruger Park, records of African buffalos killed by lions are greater in years of low rainfall, when animals are weakened, than when rainfall is above average (Mills *et al.* 1995).

For kudus, almost 40% of lion kill records in the Kruger Park fell within the three dry season months (Owen-Smith 1993a). Kudu carcasses uneaten by predators were recorded only under extreme drought conditions. Nevertheless, variability between years in mortality rates was dependent on resource availability as influenced by rainfall (Owen-Smith 1990). Besides lions, kudus are susceptible to mortality from other predators, including leopards, cheetahs, wild dogs and hyenas. An additive component to mortality is indicated by the less than 100% survival shown by kudus during years of high rainfall, even among prime-aged females. Its magnitude appeared to be between 5 and 10% per year for yearlings and adult females, and 10–15% per year for juveniles if allowance is made for a loss of at least 5% prenatally (see Figs. 10.4 and 10.6).

Modelling predation

Predation losses depend basically on the intrinsic risk of predation per encounter with a predator, multiplied by the predator density. If predators hunt unselectively with respect to prey condition, the additive mortality due to predation will diminish as the proportion of animals in the population expected to die anyway from other causes increases. Specifically, the additive mortality loss, Q_A, is given by

$$Q_A = \{q_a - q'(Q_0 + Q_G)\}C, \tag{10.3}$$

where Q_0 and Q_G are as defined in Equation 10.1, C represents the predator (carnivore) density, q_a is the risk of death per predator encounter for healthy animals, and q' is the corresponding risk of death for animals that are nutritionally weakened or senescent.

For simplification, assume that $q' = q_a$, i.e. there is no difference in intrinsic predation risk between healthy and weakened animals. This assumption yields the upper limit to the additive mortality due to predation. The loss in population biomass through additive predation diminishes from its maximum towards zero as the mortality rate pre-disposed by resource deficiencies rises (Fig. 10.9). For illustration, if the encounter-dependent risk of predation is 0.1 over some period, and the expected mortality rate through malnutrition or senescence is 0.2, we have $Q_A = 0.1 - (0.1 \times 0.2) = 0.08$ per encounter. If nutritionally compromised animals are more likely to be killed than healthy animals (i.e. $q' > q_a$), the magnitude of the additive mortality will be reduced even further.

Alternatively, assume that predators kill some healthy animals, but selectively hunt vulnerable prey when the latter are available. Under such conditions, once the proportion of vulnerable prey satisfies predator demand, the additive mortality loss falls to zero (Fig. 10.9). Although all mortality occurs through the agency of predation, the animals killed were doomed to die anyway. In reality, some weakened animals might otherwise have survived, had they not failed to evade the predator pursuit. Hence, the additive mortality component will remain somewhat greater than zero.

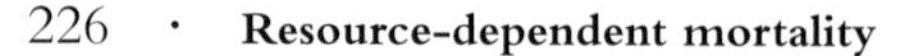

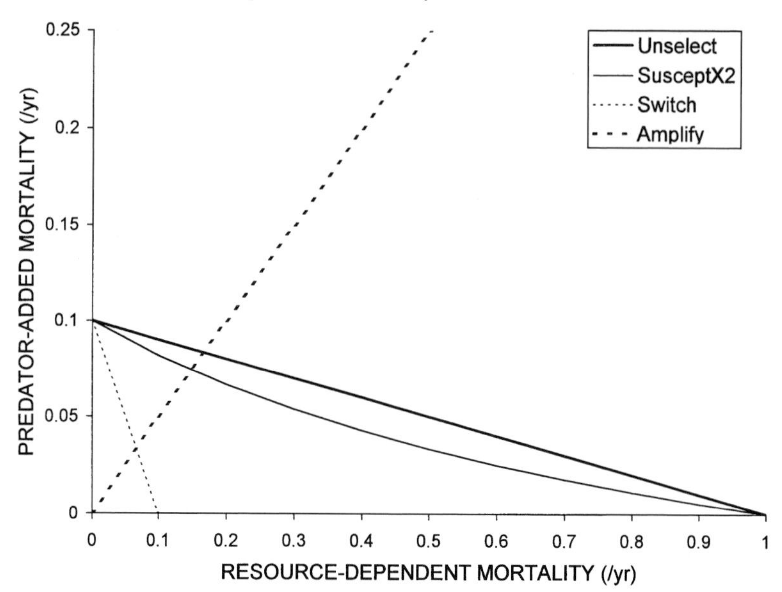

Figure 10.9. Dependence of the additive mortality due to predation on the magnitude of resource-dependent mortality, as influenced by different assumptions about how predators select for prey: (i) predation is unselective for prey condition (bold solid line); (ii) nutritionally weakened animals are twice as vulnerable to being predated as healthy animals (solid line); (iii) predators switch to nutritionally vulnerable prey when these are available (dotted line); (iv) predation amplifies the nutritional pre-disposition to mortality (bold dotted line).

More generally, predation is likely to act interactively with nutritional status, i.e. predators hasten the demise of animals that would have died eventually of malnutrition. The effect is to amplify how steeply mortality rises with diminishing resource gains (Fig. 10.9). However, by removing weakened prey from the population, the resources that these animals would otherwise have consumed become available to others. This could prevent levels of resource depletion leading to deaths directly through starvation from being reached. Thus some compensatory alleviation of mortality is likely to occur through such density-dependent release of resources.

Projected patterns

For the Kruger Park kudus, observed patterns of mortality appeared to be a composite of additive and interactive components. Additive mortality is indicated by the losses that occurred even among prime-aged females in years of high rainfall. The pattern of mortality that might have occurred in the absence of predators can be estimated by dividing the survival rates projected by the regression relations depicted in Figs. 10.4 and 10.6 by the maximum survival rates observed for each age class under conditions of high rainfall relative to density. At the population level, this adjustment leads to zero mortality at low density, while elevating the density at which the population mortality rate just balances recruitment (Fig. 10.8b). Furthermore, the overall form of density dependence in mortality rate becomes somewhat curvilinear, because zero age-specific mortality is now reached at a density level somewhat above zero. This pattern may be compared with the abruptly accelerating density dependence in mortality shown by the Soay sheep population on Hirta, where predators are lacking so that mortality occurs directly through starvation (Grenfell *et al.* 1992) (Fig. 10.7).

Habitat security

Susceptibility to predation can also vary among habitat types. Ungulates that rely upon running speed to escape predators are disadvantaged in thick bush, whereas species dependent upon jumping agility for evasion are more at risk in open vegetation. Moose are handicapped during winter in bottomlands where snow accumulates. Differences in habitat security affect the value of the additive predation risk indexed by q_a.

When food is abundant, animals can restrict their activities to habitats where their vulnerability to predation is low. With rising population density, an increasing fraction of the population may be forced into habitats where food is available, but risk of predation is high. Predation risk has to be balanced against the likelihood of starvation. By restricting habitat use, predation can accentuate resource limitations (Hik 1995; Sinclair and Arcese 1995). Migration allows the Serengeti wildebeest population to occupy habitats with reduced predation risk for part of the year, a crucial factor in their high abundance (Fryxell *et al.* 1988). A similar mechanism may operate among caribou populations (Messier 1995). Hence, differences among habitats in security against predation contribute further to the interaction between resource availability and predation.

State-dependent mortality

Animals die directly from malnutrition after their body reserves have dropped to a critical level. For modelling the dynamics of Soay sheep,

Illius and Gordon (1998) estimated mortality from the extent of body fat reserves. Specifically, mortality losses due to exhaustion of fat reserves were calculated from the mean body fat for each age and sex class, and the distribution of body fat around the mean observed in a population sample obtained in one year. The annual dynamics of body fat was estimated from the energy gain from food consumed, relative to energy expenditures for maintenance, activity and thermoregulation. Allowance was made for the observed seasonal cycle of growth and fattening, and for an upper limit to the extent of body fat reserves. This procedure generated a pattern of density dependence in mortality that took effect above some critical density threshold, somewhat more abruptly than observed for the Soay sheep population on Hirta.

As a metaphysiological counterpart to this physiological model, surplus resources could be subdivided between consequent growth in lean body, and augmentation of stored body reserves (Getz and Owen-Smith 1999). The nutritional deficits occurring at certain stages in the seasonal cycle can then be ameliorated by flows from stored reserves, with consequent deferment of mortality. Mortality risk could then be based directly on body state, as indexed by the magnitude of stored reserves or deficits in lean body mass. However, both of these modelling approaches ignore the spreading of risk that occurs when animals are subject to predation.

Over the seasonal cycle, animals must trade the benefits of storing fat to alleviate starvation-induced mortality against the costs, due to heightened risk of predation, of building and carrying such reserves (Houston *et al.* 1997). In a predictable environment, consumers should store sufficient fat to avoid starvation during the late winter or dry season period, as outlined in Chapter 7. All mortality is then mediated through predation interactively with body condition (McNamara and Houston 1990). However, in stochastic environments some starvation-related mortality will arise when fluctuations in resources or in the thermal regime exceed certain limits.

Although the metaphysiological approach could be extended to accommodate state-dependent mortality through predation, such models become somewhat more complicated. Accordingly, I will resist the challenge of this further elaboration, and retain the flux-dependent formulation of the mortality function defined by Equation 10.2.

Demographic structure

Strictly, the metaphysiological modelling approach rests on a currency of integrated population biomass. However, in this chapter we have recognized the dependence of mortality on individual age, size and sex. Accordingly, we now consider how to accommodate demographic structure as a secondary refinement.

From a metaphysiological perspective, at least four population segments need to be differentiated with regard to their growth potential, relative metabolic requirements and susceptibility to mortality: (1) adult females, with negligible growth potential except via resources diverted to nurture offspring; (2) adult males, with growth potential restricted to the storage component and only genetic contributions to offspring; (3) immature animals of both sexes, which allocate resource surpluses primarily to body growth; (4) juveniles having their growth potential subsidized by maternal investments. The transition from juvenile to immature stage occurs around the time of weaning. Animals become functionally adult around the age of first conception for females, and at the age of sociosexual maturity for males, although growth in body size may continue for a period. Further sub-divisions could be made, e.g. between fetuses supported within the mother's body and infants nourished by maternal milk; between different stages of growth towards maturity; between barren and fecund females; between prime and old adults; and among younger stages with respect to sex. Whether these are necessary depends on how substantially mortality rates and growth potential differ between such segments, and on whether sufficient information is available to specify the differences.

In a demographically structured metaphysiological model, biomass must be transferred between the population segments at appropriate stages. Births are represented by transferring a proportion of adult female biomass to constitute the nucleus of the juvenile segment. Newborn ungulates typically weigh 7–10% of maternal mass, but the mass loss by the mother during parturition is somewhat greater than this because of associated losses in tissues such as the placenta and body fluids. Following parturition, the mother subsidizes offspring growth by providing milk generated from her surplus resource gains, or sometimes even from her lean body mass. Such biomass transfers continue up until the time of weaning. Thereafter the body growth of immatures depends only on resources acquired independently. Fully grown males have no growth potential, but may fluctuate seasonally in body mass through storage and subsequent mobilization of fat reserves, and possibly also through losses in lean body mass during critical periods. Adult females can increase in mass through the growth of fetuses as well as through changes in body reserves. Maternal supplementation of juvenile growth needs to be adjusted for juvenile mortality, so that only the mothers of surviving calves contribute.

Notably, the numerical birth rate is of relevance only so far as it influences the growth potential of the neonates, and their susceptibility to mortality. Many small offspring can in aggregate generate more biomass during their growth to maturity than fewer, larger offspring, but only if the resource regime allows such growth. Restrictions on female fertility (the proportion of females contributing offspring) as a result of inadequate resources may be represented as a loss in the reproductive contribution to the juvenile segment. Offspring mortality can furthermore be elevated if neonates are born underweight (Singer *et al.* 1997; Gustafson *et al.* 1998).

When the age-class-structured model developed for the Kruger Park kudu population (Owen-Smith 2000) was modified to allow for variable growth rates by the juvenile and immature segments, dependent on annual resource availability, the projected rate of increase in population biomass was unaffected. The increased growth in biomass by one segment was offset by a loss in growth capacity in the subsequent demographic stage, if the adult body mass was held constant. If an increase in adult body mass was allowed, the potential biomass gain was counteracted by correspondingly greater biomass losses via mortality of the larger adults, unless the bigger adults survived better. The maximum relative growth rate of the modelled kudu population was identical in both biomass and numerical terms, for a stabilized age distribution. However, increases in the growth rate of immatures could potentially influence the rate of population increase through the consequences for age at first reproduction, i.e. females attaining mature size sooner.

These findings suggest that the resource gain function influences population biomass dynamics largely through its consequences for mortality. Hence, past emphasis in Lotka–Volterra type models on the form of the intake ('functional') response appears misdirected. Instead, more attention needs to be given to the resource dependence of mortality, in interaction with predation and other influences.

Overview

This chapter emphasizes the fundamental dependence of mortality on resource availability, with immatures and ageing adults most susceptible and prime-aged adults resistant. Even if most deaths occur through the agency of predators, variability in mortality may still be largely resource-related. Hence, the additive component of mortality through predation needs to be distinguished from the component that is interactive with nutritional condition. Risk of predation may vary among habitats, influencing the population impact of predation. Interactions between predation, resources and habitat use may arise, with resource deficiencies causing animals to spill into more risky habitats, and predation risk causing them to crowd into depleted habitats. Predators and parasites amplify nutritional susceptibility to mortality.

Although changes in individual growth rate strongly influence the biomass dynamics of immature segments, they have little effect on population biomass dynamics, except by affecting age at first reproduction. Density dependence in structured populations emerges from the mortality patterns of population segments, including minimal mortality experienced at low density. Density dependence is effectively the inverse of resource dependence. For some ungulate species, differences in resource supply may be related to annual or seasonal precipitation, but this does not apply to all species, especially when food quality depends inversely on rainfall.

This chapter completes the analysis of the functional components of the **GMM** model, recognizing, however, that factors influencing physiological costs have been covered only superficially.

Further reading

The book by Burgman *et al.* (1992) presents a thorough outline of population models, particularly with respect to consequences of stochastic environmental variation. Gaillard *et al.* (2000) review patterns of variability in survival rates exhibited by different age classes of ungulates.

11 · *Habitat suitability: resource components and stocking densities*

This chapter initiates the third section of this book, illustrating applications of the **GMM** model to particular issues in herbivore ecology and management. We address first the assessment of habitat suitability, as manifested by the abundance or performance of a herbivore population in a particular region. Wildlife managers usually think of the habitat 'carrying capacity', or maximum population that can be sustained. Livestock ranchers seek the optimal stocking density that would yield the highest production of meat, wool or other products. Caughley (1976b) termed the former 'ecological carrying capacity', the latter 'economic carrying capacity'. Theoretical ecologists symbolize the zero growth density by the constant K in the logistic equation (May 1981), while acknowledging its vague reality. Ecological analysis has focussed largely on the feedbacks regulating populations around some density (Sinclair 1989). Less attention has been paid to the environmental determinants of the density attained.

The basic utility of the 'carrying capacity' concept was questioned by McLeod (1997) for real-world environments where population abundance fluctuates widely over time. Is it represented by the mean density? The peak density attained between disrupting events? Or by some remote upper level, rarely reached (cf. Ellis and Swift 1988)? Densities also differ widely regionally, and change numerically with enlargements in the scale of the area encompassed (Pastor *et al.* 1997).

Many factors contribute to habitat suitability, including not only the availability of suitable food and other *resources*, but also shelter from extreme *conditions*, and security against predators and other *hazards*. However, our focus in this chapter will be on the resource components of habitat suitability. For large herbivores, vegetation structure and composition defines not only the habitat, but also the food resources presented. Vegetation production determines the potential biomass density of herbivores that could be supported, but the actual population attained by specific herbivore species depends on which vegetation components are consumed and how they contribute to supporting metabolism, growth and reproduction at different stages in the seasonal cycle.

For grazing ungulates, recommended stocking densities are assessed on the basis of effective precipitation, range condition and metabolic units of exchange between herbivore species differing in body mass (Grossman *et al.* 1999; Peel *et al.* 1999). For browsing ungulates, a comparable procedure is unreliable, because woody species differ widely in nutritional value as well as in the extent to which they provide forage through the adverse season. Even for grazing livestock, the assumption that range receiving the highest condition score, as conventionally assessed from grassland composition, supports the greatest herbivore production has been questioned (see Tainton 1988; O'Reagain and Mentis 1990). In this chapter, we address first how vegetation components contribute to determining habitat suitability for browsers, then consider the relation between range or 'veld' condition and the productive potential of cattle. This leads us to a generic assessment of the resource components contributing to habitat suitability.

Determinants of browser abundance

Observations

During my kudu study in the Kruger Park, I was impressed by the absence of any resident kudu herds from the region that I traversed between my base and the main study site in the Tshokwane area. During the ten years spanned by the study, I saw kudus only twice in this region, once a group of bulls, and once some wandering sub-adults. A short distance beyond, kudu densities averaged 2–3/km^2. Other browsers, including impala, giraffe and steenbok, were commonly encountered in the region from which kudus were absent. The vegetation there was predominantly umbrella thorn (*Acacia tortilis*) savanna, only subtly different from the mixed knobthorn (*A. nigrescens*) savanna that prevailed over the core study area. In Tanzania's Serengeti Park, where umbrella thorn savanna is widely prevalent, browsing giraffe, impala, eland and gazelles are common, but kudus absent. The feeding studies that we conducted at Nylsvley showed that kudus commonly utilized the vegetation patches where *A. tortilis* and similar species predominated, except during the dry season (Owen-Smith 1993a, 1994). Kudu densities of around 2 animals/km^2 are widely typical of savanna bushveld, but in parts of the Eastern Cape, where succulent semi-evergreen thicket predominates, kudu densities can exceed 10 animals/km^2 (Allen-Rowlandson 1980).

Thus it is apparent that vegetation differences that may seem quite minor can underlie a difference in the population of kudus supported from effectively zero to five times the prevailing average. Features of the browse resource that need to be considered include (1) the rate at which food can be harvested, as affected by structural deterrents like thorns (see Chapter 3), (2) the effective nutritional value of different vegetation components, as influenced by leaf chemistry (see Chapter 6), and (3) the period of retention of foliage through the dry season. Notably, evergreen leaves tend to be more strongly defended by tannins or other allelochemicals than the foliage of deciduous tree species (Coley *et al.* 1985; Cooper *et al.* 1988).

The model

To represent the seasonal dynamics of the browse resource, the 'extended production pulse' model (Chapter 9) is adequate. In this model edible forage, in the form of leaves and supporting shoots, is generated in a brief pulse at the start of the growing season (GS), and persists undepleted through the GS spanning six months. During the dormant season (DS), foliage is subject to attrition through consumption, and perhaps additionally via leaf fall.

From Chapters 2–6, the rate at which herbivores gain biomass from forage consumed while foraging, G_F, depends on (1) the food availability, F, in mass per unit area; (2) the nutritional value of the forage consumed, c, expressed as a conversion ratio from food to consumer biomass; (3) the eating rate obtained while feeding in food patches, I_P, as mass per unit time; (4) the fraction, a, of available food accepted (i.e. consumed), and (5) the search rate s in units of area traversed per unit time. For standardization, rates of intake and gain should be expressed proportionally to consumer biomass. The daily rate of biomass gain, G_D, depends additionally on the fraction of the daily cycle spent foraging, t_F, which may be constrained by digestive capacity. Adapting Equation 3.6 for a single homogeneous food resource, we obtain

$$G_D = \{csaF/(1 + saF/I_P)\}t_F. \tag{11.1}$$

Extending this equation to accommodate multiple food types, we get

$$G_D = \left\{s \sum_{i=1..r} c_i a_i F_i/(1 + s \sum_{i=1..r} a_i F_i/I_{Pi})\right\}t_F, \tag{11.2}$$

where i indexes the food types, and r represents the range of food types included in the diet (cf. Equation 6.4).

The biomass growth potential of herbivores is governed by the net difference between the rate of biomass gain, G, expressed over an appropriate time frame, and the rate of biomass loss through physiological attrition, M_P. Following Equation 7.1, we can represent the *relative growth potential* (RGP) as a dimensionless proportion of the resting (or basal) metabolic requirement, M_{P0}, i.e.

$$\mathrm{RGP} = (G - M_P)/M_{P0}, \tag{11.3}$$

where G and M_P can be expressed either in units of metabolizable energy or its biomass equivalent (thereby avoiding the issue of whether energy or material nutrients such as protein are more limiting).

Allowance must also be made for mortality losses, M_Q. From Equation 10.2, nutritionally dependent mortality is inversely dependent on resource gains relative to metabolic requirements, so that

$$M_Q = z_0 + zM_P/G \tag{11.4}$$

with z_0 and z being curve fitting constants.

In the model, the mortality risk is evaluated from the current resource gain at each time step, and averaged over the year to determine the annual mortality. The annual increase in population biomass is then calculated by multiplying the annual mean RGP by the annual survival rate, i.e. one minus the annual mortality. Because values for the nutritional conversion coefficients c are somewhat arbitrary, for some simulations the maximum potential rate of population growth was set arbitrarily, and the realized population growth rate obtained by subtracting the mortality loss from it. Further details about the model formulation and parameter values used are provided in Appendix 11A.

General output

The first step was to establish the general consequences of heterogeneity in food resources for herbivore performance in a seasonal environment. For this basic model, attrition in food resources during the DS was solely through consumption. The simplest 'habitat' consisted of one medium-quality food, intermingled with a low-quality buffer forage (Table 11.1, Med+Buf). Next, the biomass of the medium-quality food was partitioned evenly among three food types differing in quality, keeping total food production and mean food quality unchanged (Table 11.1, Even4). Conditions with these food types intermingled (variable diet breadth) were contrasted with a situation in which the food types were distributed in discrete patches (alternate patch choices). Lastly, a habitat was constructed with five intermingled food types graded in abundance as well as in quality, such that poor-quality food types were more abundant than high-quality food types (Table 11.1, Graded5).

Table 11.1. *Parameters for food type combinations representing different habitat conditions used in the generalized model*

	Habitat label		
	Med+Buf	Even4	Graded5
Number of food types	2	4	5
Nutritional conversion c	0.6, 0.4	0.7, 0.6, 0.5, 0.4	0.8, 0.7, 0.6, 0.5, 0.4
Starting biomass F (g/m^2)	15, 15	5, 5, 5, 15	2, 3, 5, 8, 12

With just a medium-quality resource type plus a buffer, and herbivore density set fairly high, the relative nutritional gain obtained by the herbivores declined progressively through the DS as the preferred resource become depleted (Fig. 11.1a). Widening the diet to include the buffer resource stabilized the daily food intake somewhat, but the nutritional quality of the buffer was too low to alleviate the widening metabolic deficit. Correspondingly, susceptibility to mortality rose steeply during the late DS (Fig. 11.1b). The overall pattern of density dependence in net population growth rate was quite sharply convex (Fig. 11.1c).

With a wider quality range among food types, herbivores obtained higher nutritional gains during the early part of the DS, by selectively choosing the best food types, but at the cost of later gains (habitat Even4, Fig. 11.1a). However, owing to the selective foraging more forage biomass persisted into the late DS, so that the rise in mortality rate then was not quite as steep as with just two food types (Fig. 11.1b). Both the maximum population growth rate, and the density level for zero population growth, were elevated, relative to the situation with two food types, although the mean food production and quality were unaltered (Fig. 11.1c). Further improvement in these measures resulted when resources were graded in quality and abundance over a wider range (habitat Graded5, Fig. 11.1a–c). This occurred because selective consumption retarded the stage at which animals became forced to depend on the sub-maintenance buffer resource, with consequent rise in mortality. For the situation with patchily distributed resources (i.e. Even4-Patch, Fig. 11.1), nutritional gains and consequent mortality changed abruptly at the stages when consumers switched between resource patches. While the population growth rate at low density was elevated, the sustainable population density was reduced compared with conditions where the same resource types were intermingled. The effect of patchy resources was to allow herbivores to forage selectively in the places where the best quality resources were

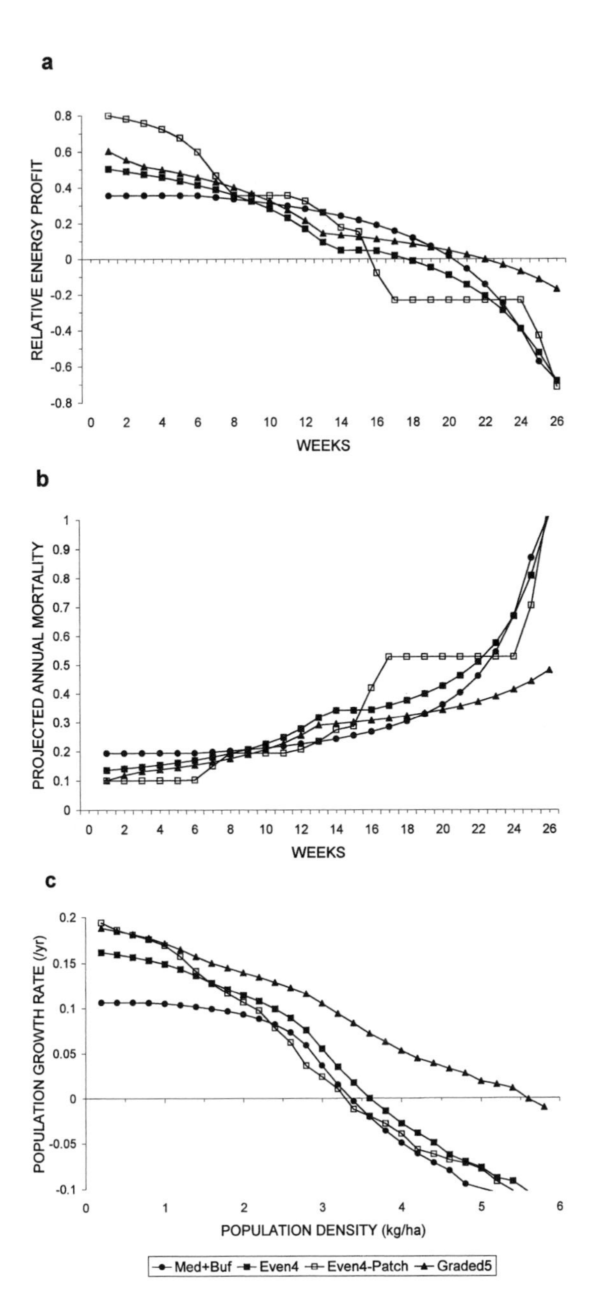

Figure 11.1. Output of general model assessing consequences of resource heterogeneity. Habitat composition as specified in Table 11.1, with Even4-Patch distinguished by having a patchy rather than intermingled distribution of food types. (a) Relative energy balance over the dry season, for high herbivore density of 3 kg/ha. (b) Corresponding weekly mortality expressed as annual rates over the dry season period. (c) Resultant changes in projected population growth rate relative to population density.

concentrated early in the DS, but later animals incurred the penalty for having depleted these patches.

Overall, the availability of multiple resource types tended to make the form of density dependence gently rather than sharply convex. The population level for zero growth was determined largely by the amount of food remaining towards the end of the DS, through its effects on mortality. Accordingly, the zero growth density was rather sensitive to the form of the mortality function.

Application to *Burkea* savanna vegetation

To test the model with real data, it was applied to assess the density of kudus that could be supported by the *Burkea* savanna prevalent at Nylsvley, where our detailed studies on browser feeding ecology were conducted. Following Owen-Smith and Cooper (1987a), the 60 woody species growing within the 210 ha study enclosure were grouped into five distinct food types, differing in effective nutritional value and other features influencing their consumption by kudus and other browsers (Cooper and Owen-Smith 1986; Cooper *et al.* 1988; see also Chapter 6): (1) palatable deciduous unarmed species, favoured as a food source year-round; (2) palatable deciduous spinescent species, varying in acceptability because of the effects of thorns or spines on intake rate; (3) relatively palatable evergreen species, neglected during the GS but favoured throughout the DS; (4) unpalatable deciduous species, neglected except when new leaves were present, and (5) unpalatable evergreen species, accepted only towards the end of the DS. Although individual species of forb (i.e. herbaceous plants excluding grasses, plus shrublets) varied quite widely in their nutritional properties, as a category forbs were generally high in nutrients, and hence were amalgamated into a single high-quality food category for the model.

Unpalatable woody species made up most of the 60 g/m^2 of edible foliage available within the height reach of kudus during the GS. Evergreen species constituted only a small fraction of the woody vegetation, but contributed much of the 2 g/m^2 of woody browse remaining by the end of the DS in September (Fig. 11.2). Forbs started declining in availability earliest during the DS, partly through withering and partly owing to consumption. Unpalatable deciduous woody species tended to retain foliage longer than palatable deciduous species. The DS decline in foliage on evergreen species was due solely to consumption. Kudus did not consume shoots lacking leaves, with rare exceptions.

The simplified vegetation parameters used to represent the *Burkea* savanna are given in Table 11.2. The biomass conversion coefficient, c,

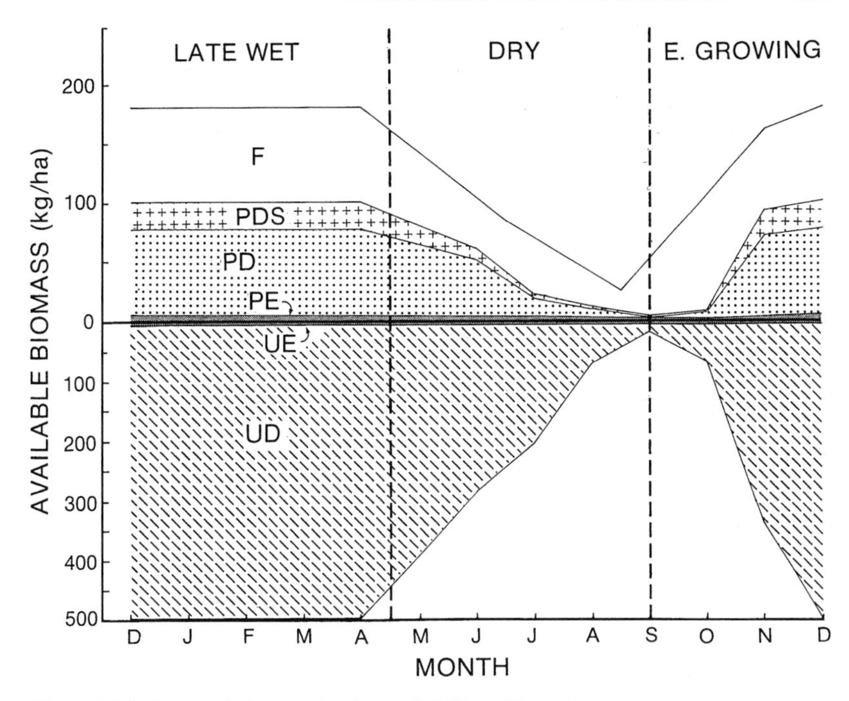

Figure 11.2. Seasonal changes in the availability of broad resource types documented in the Nylsvley study area for *Burkea* and *Acacia* savanna types combined: F, forbs; PD, palatable deciduous unarmed woody species; PDS, palatable deciduous spinescent woody species; UD, unpalatable deciduous woody species; PE, relatively palatable evergreen woody species; UE, unpalatable evergreen species. (From Owen-Smith and Cooper 1987a; Owen-Smith 1994.)

for each food type was derived from estimates of dry matter digestibility. Because the unpalatable deciduous category was heterogeneous, it was sub-divided into two components. The component labelled UnpDec2, together with unpalatable evergreens, constituted the low-quality buffer. Food types were ranked according to the observed frequencies with which they were consumed.

Despite the vegetation simplification, the model output replicated the estimated energy balance of the kudus through the DS (Owen-Smith 1994), provided allowance was made for the observed monthly differences in daily foraging time (Fig. 11.3a). The discrepancies apparent during the early and late dry season periods can be explained by model omissions. The model did not allow for the reduced foraging effort shown by the kudus in the early dry season, involving reductions in the fraction of available food accepted as well as in daily foraging time. Also not taken into account was the small flush of forbs that occurred in August. Finally,

Table 11.2. *Food type parameters used in the model to represent the Nylsvley* Burkea *savanna*

	Food type					
	Forbs	PalDec	PalSpi	PalEvg	UnpDec1	UnpDec2
Nutritional conversion c	0.72	0.65	0.70	0.58	0.50	0.35
Initial biomass F (g/m^2)	3	8	1.5	1.5	1.5	19
Eating rate I_P (g/min)	5	5	2.5	5	5	5
Leaf attrition rate (/week)	0.15	0.15	0.15	0	0.15	0.15
Onset of leaf fall (week)	4	10	8	0	14	14

the beginning of the new leaf flush on certain unpalatable deciduous species during September elevated the actual nutritional gain of the kudus above that projected by the model.

For the *Burkea* savanna vegetation, as approximated in the model, the model output projected a zero growth level in herbivore biomass density of about 3 kg/ha (Fig. 11.3b). This closely matched the actual kudu density in the 210 ha study enclosure (five 100–150 kg kudus), as well as local kudu densities typical of savanna bushveld (2–3/km^2, population mean body mass *ca.*140 kg).

Having thus established the basic capability of the model to project realistic population levels, we may ask further how different vegetation components contribute towards supporting this population density. Of how much value is the vast amount of unpalatable foliage that is mostly little utilized? A preliminary attempt to answer these questions, by calculating changes in short-term rates of nutritional gain, was made by Owen-Smith and Cooper (1989). The **GMM** model enables the overall annual gain to be estimated not only in nutritional terms, but also allowing for mortality.

The model output suggested that eliminating half of the palatable deciduous woody component (the dietary staple) made a greater difference to the zero growth density than reducing the abundance of the forbs (the high-quality component) by a similar proportion (Fig. 11.3b). This is because 50% of palatable deciduous foliage constituted a larger absolute amount than 50% of forb biomass, and the amount of food produced largely determined the density sustained. Eliminating all evergreen

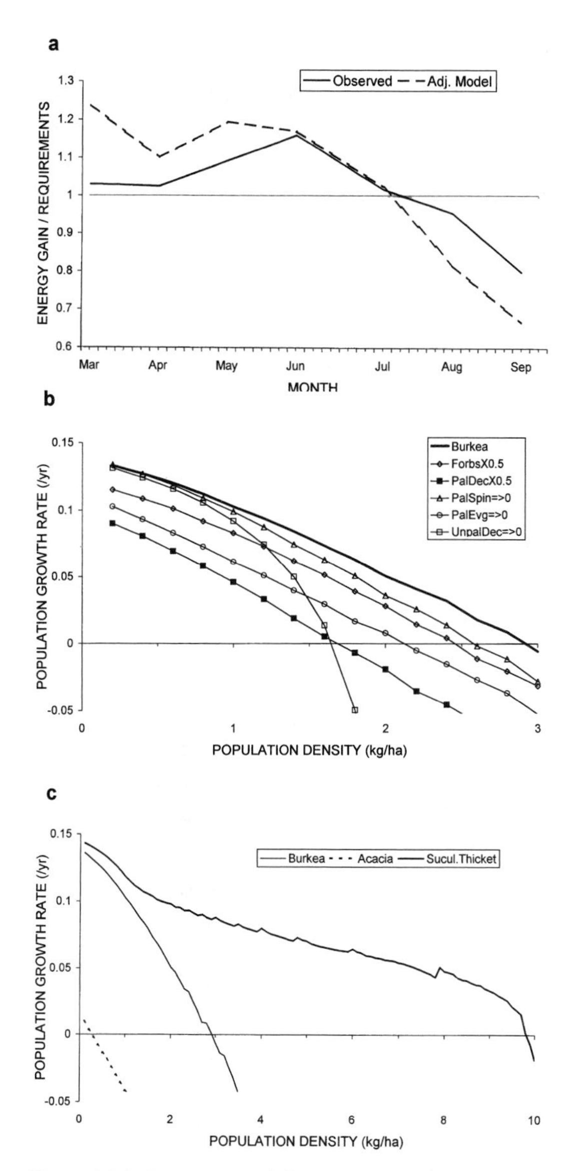

Figure 11.3. Browser model output. (a) Observed energy balance of kudus over the dry season for the Nylsvley *Burkea* savanna (Owen-Smith 1994), compared with model output adjusted for varying daily foraging time. (b) Projected population growth rate relative to kudu biomass density for the *Burkea* savanna habitat, and adjustments comprising (i) reducing abundance of forbs by half, (ii) reducing the abundance of palatable deciduous (PalDec) woody plants by half, (ii) eliminating all palatable spinescent (PalSpi) species, (iv) eliminating all palatable evergreen (PalEvg) species, and (v) eliminating all unpalatable deciduous (UnpalDec) species. (c) Density dependence in population growth rate for *Burkea* savanna (Table 11.2) compared with that for *Acacia* savanna (Table 11.3a) and for succulent thicket vegetation (Table 11.3b).

foliage had the same effect on total food biomass as a 50% reduction in forb biomass, but exerted a somewhat stronger influence on the zero growth density. This emphasizes that resources persisting through the dry season largely control the herbivore population levels that can be supported, through the consequences for resource-dependent mortality at this time of the year. Similarly, although the unpalatable browse component (both deciduous and evergreen) contributed only a small amount to herbivore diet during the late DS, eliminating the buffering effect of this component made as great a difference to the potential population level as removing half of the palatable deciduous staple (but, notably, with five times as much change in overall food biomass). The slow eating rates obtained from spinescent woody species yielded a rate of nutrient gain well below kudu maintenance requirements, but removing this minor component of *Burkea* savanna reduced the zero growth density by a not insignificant amount.

Application to other vegetation

Whether by lucky circumstance or parameter accuracy, the model successfully predicted the observed kudu densities supported by *Burkea* savanna vegetation. A more demanding test is to assess how closely it projects the widely discrepant densities observed in other vegetation types, with appropriate parameter adjustments. Vegetation measures available for the *Acacia* patches present in the Nylsvley study area were used to represent umbrella thorn savanna (Table 11.3a). Palatable spinescent trees were sub-divided between two components differing in the eating rates they yielded. Forbs were somewhat more abundant in the *Acacia* vegetation, and the overall food quality higher, than that offered by *Burkea* savanna. The total browse biomass presented was a little less than that offered by *Burkea* savanna, and the evergreen component was lacking. For succulent thicket vegetation, empirical data were lacking. Arbitrarily it was assumed that overall food abundance and quality were similar to that in *Burkea* savanna, but with forbs absent and all woody components evergreen (Table 11.3b).

Model output for the *Acacia* savanna indicated close to zero capacity to support a kudu population (Fig. 11.3c). This was due largely to the absence of an evergreen component to carry kudus through the dry season, coupled with low eating rates obtained from the predominantly spinescent trees. Findings thus confirmed that a kudu population could not persist year-round solely in *Acacia* habitat as defined in Table 11.3a. However, a kudu population could seasonally make use of this habitat provided animals had access to other vegetation during the dry season period.

Table 11.3. *Food type parameters used in the model to represent the Nylsvley* Acacia *savanna and Eastern Cape succulent thicket*

(a) Acacia *savanna*

	Food type				
	Forbs	PalDec	PSpiA	PSpiB	UnpDec
Nutritional conversion *c*	0.75	0.65	0.70	0.72	0.50
Initial biomass F (g/m^2)	6	0.5	4	12	3
Eating rate I_P (g/min)	5	5	2.5	1.5	5
Leaf attrition rate (/week)	0.15	0.15	0.15	0.15	0.15
Onset of leaf fall (week)	4	10	8	8	14

(b) *Succulent thicket*

	Food type				
	EvgA	SpiEvg	EvgB	EvgC	EvgD
Nutritional conversion *c*	0.65	0.7	0.60	0.5	0.40
Initial biomass F (g/m^2)	8	8	8	8	8
Eating rate I_P (g/min)	5	2.5	5	5	5
Leaf attrition rate (/week)	0	0	0	0	0
Onset of leaf fall (week)	0	0	0	0	0

For the evergreen thicket approximation, the model output indicated that a herbivore density of around 10 kg/km^2 could be supported, closely matching the observed kudu density in this vegetation type. This high population density was due largely to the year-round persistence of the predominantly evergreen foliage (Fig. 11.3c).

Overall assessment

The model output emphasized how minor resource components bridging critical periods could be crucial in sustaining browser populations. Regions with similar resource production could support very different population levels, depending on the seasonal persistence of food resources. Regional heterogeneity in habitats could enable populations to persist, and seasonally utilize vegetation types within which they could not persist year-round. In the Kruger Park, some kudu groups foraged in open acacia savanna during the wet season, but contracted their range either to adjoining rocky hills or to riparian thickets during the dry season (Owen-Smith 1979; du Toit 1988).

A tall browser like a giraffe can survive where kudus cannot by accessing higher levels of forage on trees. Moreover, the feeding action of giraffe allows animals to strip leaves from the branch tips of acacia trees, so that their food intake rate is less influenced by small leaf size (Pellew 1984). Additionally, being larger than kudus they range over a wider area, encompassing a greater variety of habitat types (du Toit and Owen-Smith 1989). Impalas, which are somewhat smaller than kudus, obtain an adequate rate of food intake relative to their body size from umbrella thorn trees (Cooper and Owen-Smith 1986). Furthermore, being adapted morphologically as mixed feeders, their diet encompasses a wide range of plant types, including grass as well as browse.

In north temperate regions, herbivores cope with winter food deficits by storing substantial fat reserves. However, fat stores can be costly (Chapter 7). The associated costs may be higher in the tropics than in temperate regions through (1) negative consequences for thermoregulation, (2) heightened predation risk, or (3) lesser opportunities to acquire energy surpluses during the growing season (Geist 1974).

Range condition and grazer stocking densities

Observations

Range condition assessment is commonly based on grass species composition in terms of species that either increase or decrease in abundance in response to heavy grazing pressure (Tainton 1988; Hardy *et al.* 1999). 'Decreaser' species are tufted perennials that are favoured by cattle, and thus decline in representation in swards subject to heavy stocking levels. The 'Increaser' category encompasses two types. 'Increaser I' species are represented by fairly tall tufted grasses of low nutritional value that become prevalent in certain regions when the grassland is too lightly grazed. 'Increaser II' species, which become predominant under conditions of heavy grazing, are fairly disparate in their attributes. Included are fibrous species neglected by grazers, short or decumbent grasses offering inadequate rates of intake, and annuals.

Because short grasses and annual species can be highly nutritious, 'veld' judged to be in intermediate or poor condition may sometimes yield a better-quality diet, and hence higher potential animal production, than grassland apparently in better condition (O'Reagain and Mentis 1990). Grasslands in northern Australia that were dominated by palatable perennial grasses gave lower live mass gains when cattle were stocked at low density, but gave better gains at higher stocking densities, than apparently poorer range where annual grasses and forbs were prominent and annual herbage production less (Ash *et al.* 1996) (Fig. 11.4). Their cultivation of

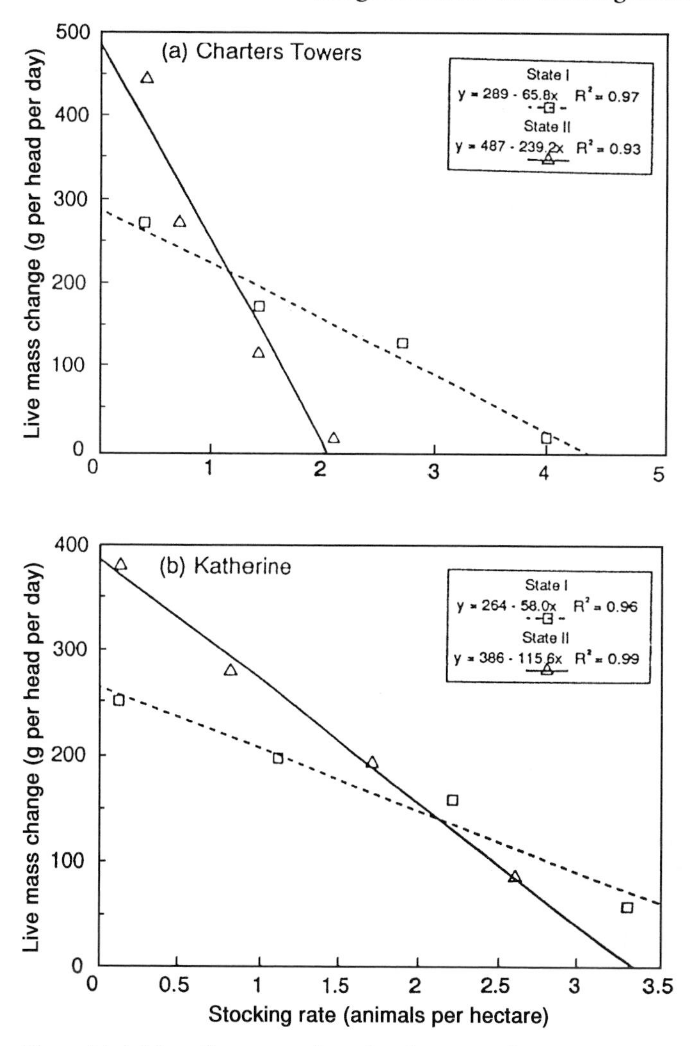

Figure 11.4. Mean live mass gains of cattle averaged over the year in relation to stocking density, comparing effects of grassland condition in two study areas in northern Australia. Good-condition grassland (dashed lines) has predominantly perennial 'decreaser' species. Moderately degraded grassland (solid lines) has a high proportion of 'increaser' perennial grasses, annual grasses, forbs and legumes. (a) Sites near Charters Towers, northern Queensland. (b) Sites near Katherine, Northern Territory. (From Ash *et al.* 1996, Reprinted with permission from Elsevier Science).

'grazing lawns' has been recognized as contributing to the high abundance of wildebeest and other grazing ungulates in the Serengeti ecosystem in Tanzania (McNaughton 1979a). The Umfolozi Game Reserve in KwaZulu–Natal, where I conducted my white rhino study, has continued to support a high abundance and diversity of grazing ungulates, despite the conversion of much of its area to a short grassland mosaic through the grazing pressure of white rhinos (Owen-Smith 1988).

The improved animal nutrition that may be obtained from 'Increaser II' grasses at certain times of the year must be counterbalanced against the nutritional deficits that may arise from inadequate forage reserves during the dormant season for plant growth. Previously I modelled the effects of veld condition on the performance of adaptively foraging herbivores confined within grazing camps for short periods of a few weeks (Owen-Smith 1991). This temporal restriction allowed grass re-growth to be ignored. The **GMM** framework enables us to evaluate the nutritional balance of animals over the complete annual cycle, allowing for compensatory vegetation growth.

The model

For the purposes of this assessment, a fairly detailed representation of the seasonal growth dynamics of grass resources was required. Changes in food quality as well as in standing biomass, as a consequence of consumption as well as growth phenology, needed to be assessed through both the growing and dormant seasons. However, for this agricultural context the herbivore population could be represented simply as an even-aged cohort of growing animals.

The starting foundation was provided by the seasonal grass production model developed in Chapter 8 and used in Chapter 9 to assess the resource-related density dependence of a herbivore population. In previous applications of this model, herbivores restricted their foraging to the resource patch type that yielded the highest nutritional gain at each time step. The formulation now needs to be modified to accommodate adaptive adjustments in herbivore diet selection among food types intermingled within a grassland habitat, defined by its composition in terms of these grass types. Hence, the diet breadth model applies (Chapter 6), with food types ranked according to their effective value. The herbivores are assumed to be young growing steers, with performance expressed via their growth in body mass. At each weekly iteration of the model, the herbivores adjusted their diet selection so as to maximize either their intake rate, or their rate of digestion, of digestible dry matter, whichever was more limiting. Nutritional quality differed among grass types as determined by concentrations of digestible dry matter, and varied seasonally for each species owing to changes in the ratio of green leaf (biomass) to standing dead material (necromass) (O'Reagain *et al.* 1995; O'Reagain & Owen-Smith 1996). Diet quality was influenced also by the stem proportion in the ingested material, which was low early in the GS, but increased as grass matured towards the mid GS. When green leaf biomass was high, herbivores were strongly selective for green biomass over necromass. This selectivity diminished until, when little biomass remained, animals consumed both green and dead leaves somewhat indiscriminately.

The food intake rate obtained while foraging decreased with lowered forage biomass, and hence grass height, from some maximum rate attained beyond a particular biomass level (Chapter 3; O'Reagain *et al.* 1996). However, animals compensated for diminished food biomass by feeding for longer within patches, thereby accepting a higher fraction of the forage resources contained therein (Chapter 4). The model calculated the daily food intake and digestible dry matter gain each week, and converted the latter into a relative growth potential (RGP) based on digestible dry matter, i.e. the effective energy yield. The RGP was transformed into a daily mass gain, and corresponding weekly mortality rate. The annual food and digestible DM intake, mass gain and mortality rate were calculated by summing, or averaging (as appropriate), the weekly values. The specific parameter values and functional relationships used in the model are given in Appendix 11B and its accompanying table.

Grassland condition was defined by the relative abundance of four grass types. These types, exemplified by typical African species, are: (1) staple good-quality, high-forage-bulk species, e.g. *Themeda triandra*; (2) medium-quality, fairly high-bulk species, e.g. tall *Eragrostis* or *Sporobolus* spp.; (3) poor-quality, high-bulk species, e.g. *Aristida* or *Hyparrhenia* spp., and (4) high-quality, low-growing species, e.g. *Urochloa, Cynodon* or lawn-forming *Sporobolus* spp. The parameter values representing these grass types are given in Table 11.4a. Following convention, veld in good condition has a high proportion of 'Decreaser' grasses corresponding with type (1) above, whereas veld in poor condition contains predominantly 'Increaser II' grasses of either type (3) or type (4). Veld in medium condition contains an intermediate mix of these grass types (Table 11.4b).

Table 11.4. *Parameters for the four grass types, and corresponding grassland compositions, used in the veld condition assessment model*

	Representative grass type, and grazing response category			
	Themeda Decreaser	*Eragrostis* Increaser IIb	*Aristida* Increaser IIc	*Urochloa* Increaser IIc
(a) *Grass type parameters*				
Growth rate g_V (per week)	0.5	0.5	0.4	0.75
Saturation biomass v_{max} (g/m^2)	400	300	500	200
Nutritional value c (digestible DM)	0.7	0.65	0.6	0.75
Stem proportion	0.1	0.15	0.2	0.05
(b) *Grassland composition representing each veld condition, by relative basal area of the above grass types, and approximate veld condition score*[a]				
Good	0.80	0.16	0.02	0.02
Medium	0.30	0.60	0.05	0.05
Poor-A	0.05	0.30	0.60	0.05
Poor-B	0.05	0.30	0.05	0.60
Veld condition score	100	50	25	25

[a] *Source*: After Tainton (1981).

Model output

Initially a simplified seasonal cycle was used, with 26 weeks of constant rain followed by 26 weeks with no rain. With a moderately high stocking density of one 300 kg steer per hectare, the digestible dry matter intake, and hence RGP, increased to a peak during the early GS, then dropped to lower levels later in the GS as indigestible stem as well as standing necromass accumulated in the sward (Fig. 11.5a). The RGP declined further during the DS as green leaf became depleted, with herbivore diet being expanded to encompass lower-quality grass types. Veld in good condition yielded lower gains during most of the GS than either veld in medium condition, or veld in poor condition with nutritious but low-bulk grass types predominating. This was because even small amounts of high-quality food ameliorated the digestive capacity constraint. However, veld in good condition supplied adequate energy for growth throughout the DS, whereas veld in poorer condition yielded sub-maintenance intake towards the end of the DS after staple grass types had been depleted.

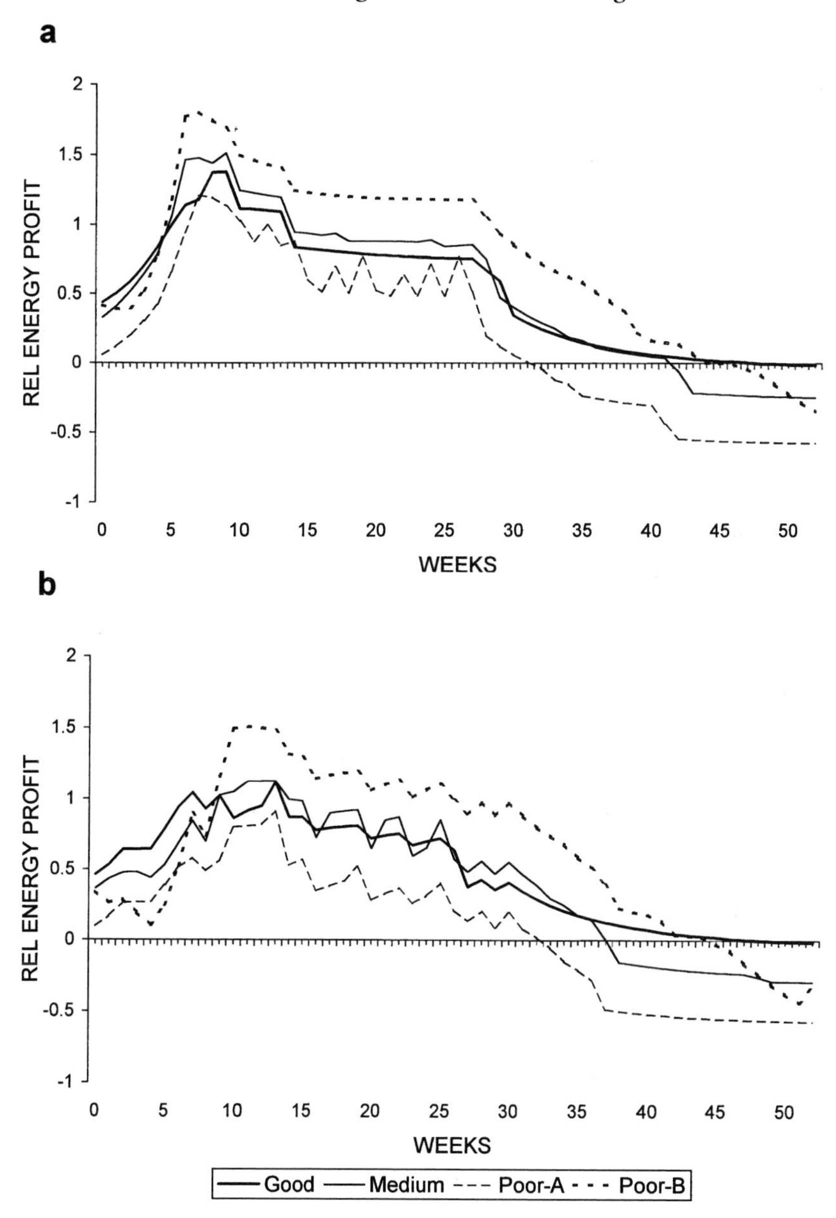

Figure 11.5. Weekly nutritional gains over the annual cycle for grassland with different veld condition scores, from grazer model output. (a) Conditions of constant weekly rainfall through 26 weeks of the growing season followed by no rain for 26 weeks through the dry season. (b) Conditions with rainfall varying erratically among weeks, following pattern depicted in Fig. 8.4c.

The model was then modified to allow for seasonally variable rainfall, following a set pattern with rainfall occurring in certain weeks but not in others. Grass growth was suppressed in weeks without rain. Under these conditions, the intake rate obtained from veld dominated by lawn-forming grasses was only marginally adequate early in the GS, during which period rainfall was erratic, regrowth limited and carryover of necromass from the previous season insufficient (Fig. 11.5b). Conditions of weekly varying rainfall also produced an earlier onset of submaintenance levels of intake during the DS for grassland in less-than-good condition.

In terms of potential growth rate per animal, relative to stocking density, the alternative versions of poor veld diverged in the pattern of density dependence generated: convex where nutritious but low-growing grasses predominated, and somewhat concave where high-bulk but low-value species prevailed (Fig. 11.6a). For good veld, gain per animal increased initially with increasing stocking density, owing to a suppression of necromass buildup, but declined with further increases towards very high stocking densities. Veld in medium condition yielded better animal gains than good-condition veld at low stocking densities, but lower gains at high stocking densities. The best animal growth potential overall was obtained from poor veld dominated by nutritious short grasses. However, under such conditions consumers were susceptible to high mortality when forage ran out towards the end of the DS. Hence, net gains allowing for mortality losses were no better than those from good-condition veld if the stocking density was high (Fig. 11.6b).

The maximum animal production achieved per hectare was similar for veld in good condition and for veld in poor condition dominated by nutritious short grasses, but occurred at a lower stocking density for the latter (Fig. 11.6c). The onset of a decline in production per hectare was somewhat abrupt for veld in good condition, because of its fairly uniform grass composition. This could create practical problems in setting the optimal stocking density where rainfall, and hence grass production, varies substantially between years. However, the problem could be alleviated by setting aside some areas as fodder banks for the bad years, albeit at the cost of productive capacity during the good years.

The daily food intake and biomass gain responses to changing food abundance inherent in this model, compiled from various grassland compositions, are illustrated in Fig. 11.7. The effective half-saturation level for food intake is about 50 g/m^2, and that for biomass gain about 70 g/m^2, depending on the assumed maximum values for these rate processes.

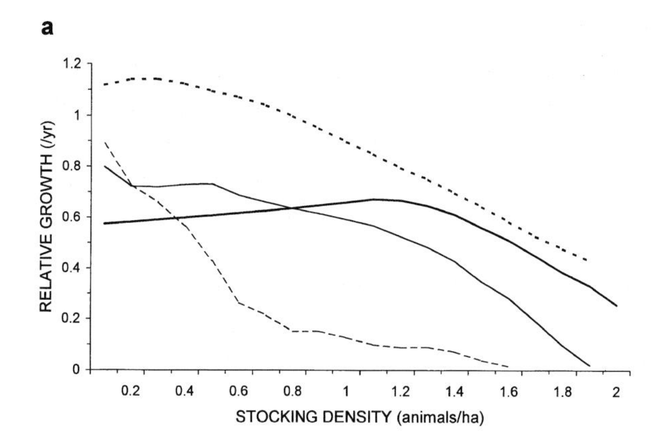

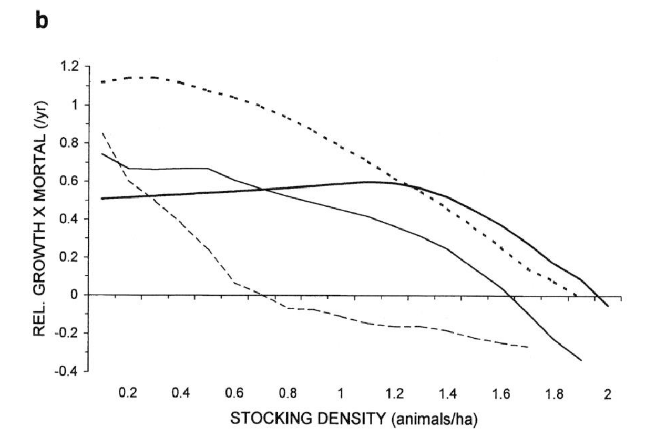

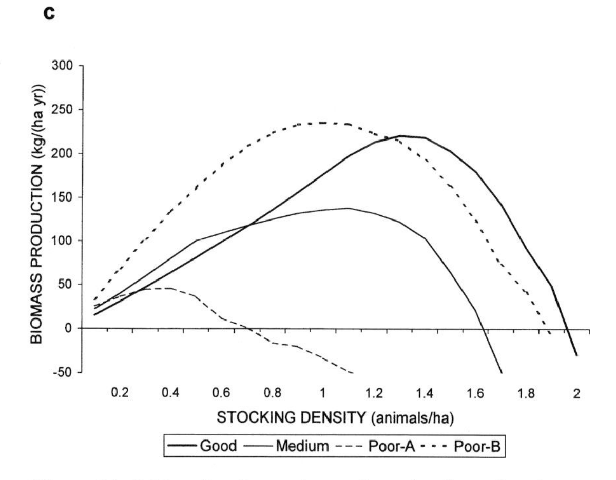

Figure 11.6. Density dependence in animal productive potential for grassland representing veld in different condition, from grazer model output for conditions of variable weekly rainfall. (a) Density dependence in relative growth per animal. (b) Density dependence in the product of relative growth and mortality losses. (c) Density dependence in animal biomass production per unit area.

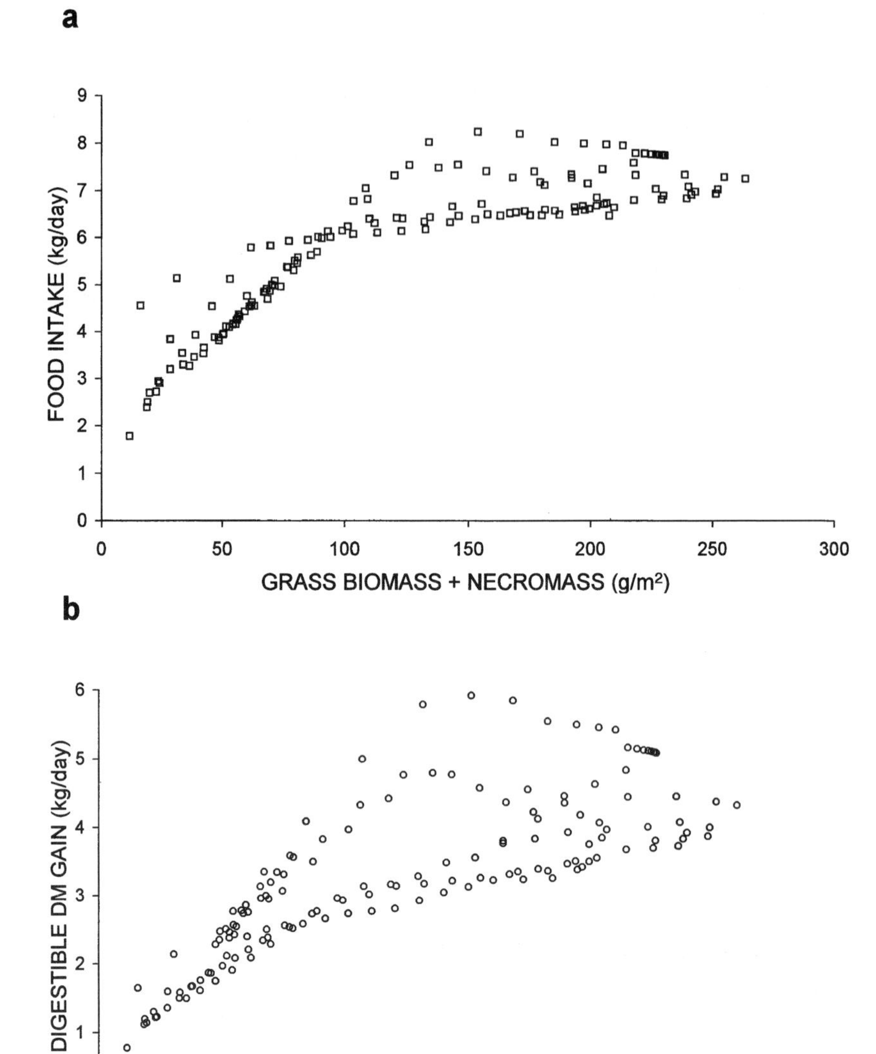

Figure 11.7. Form of the food intake and biomass gain responses inherent in the grazer model, compiled from various grassland compositions relative to total grass biomass plus necromass. (a) Food intake response. (b) Nutrient gain response in terms of digestible dry matter.

Overall assessment

The modelling exercise demonstrated the tradeoff that arises between high forage quality during the GS, affecting productive potential per animal, and high forage bulk persisting into the DS, determining stocking potential. What range condition is best depends on the objectives of the manager, the stocking level maintained, and the type of herbivore stocked. Large grazers such as cattle are rather more dependent upon the high bulk and moderate nutritional value of grasses like *Themeda triandra* than smaller grazers like sheep and certain antelope species. The latter are favoured somewhat more by high-quality grass types than are cattle, even if these grasses offer only moderate bulk.

The model projects an optimal stocking density of somewhat over one steer per hectare for grassland with an annual forage production potential of around 600 g/m^2. This happens to be the stocking rate at which annual consumption amounts to about 50% of annual grass production, a common rule of thumb for setting the upper limit to permissible stocking densities. The model does not take into account consumption by insect herbivores, variability between years in grass growth and hence stocking potential, or landscape variability in the occurrence of grassland types. All of these factors would reduce the actual stocking potential somewhat below that for the idealized environment used in the model.

The model also did not consider the long-term effects of grazing on grassland composition, which is also an important consideration in setting stocking levels relative to veld condition (Tainton 1988). A shift from a good veld condition, with abundant grasses like *Themeda triandra*, to medium condition with less nutritious tall grasses assuming greater prominence, brings a loss in productive capacity. However, if high-quality lawn-forming grasses later prevail, the productive potential could improve. This depends on the forage height and DS carryover such grasses offer for the specific type of herbivore stocked. Lack of adequate food during the late DS could be alleviated by promoting a patch mosaic of grassland types, the benefits of which were modelled in Chapter 9.

Generic resource types

The consequences of specific resource contexts for herbivore population performance have been illustrated above. Now we need to consider more generally how particular kinds of resources contribute to supporting herbivore populations. Generic resource types may be distinguished

a

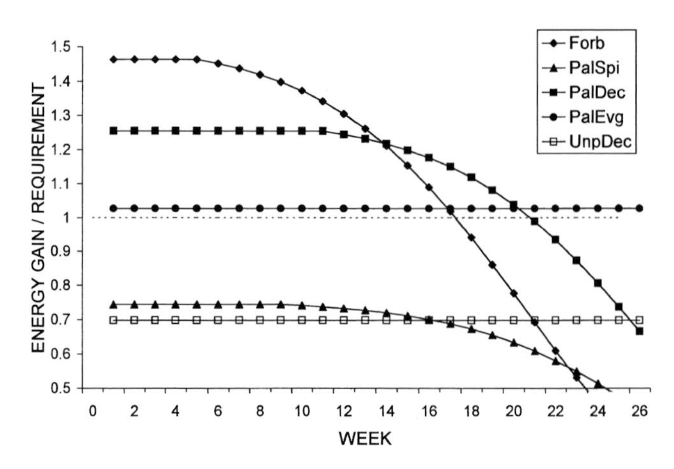

ENERGY GAIN / REQUIREMENT
WEEK
Forb
PalSpi
PalDec
PalEvg
UnpDec

b

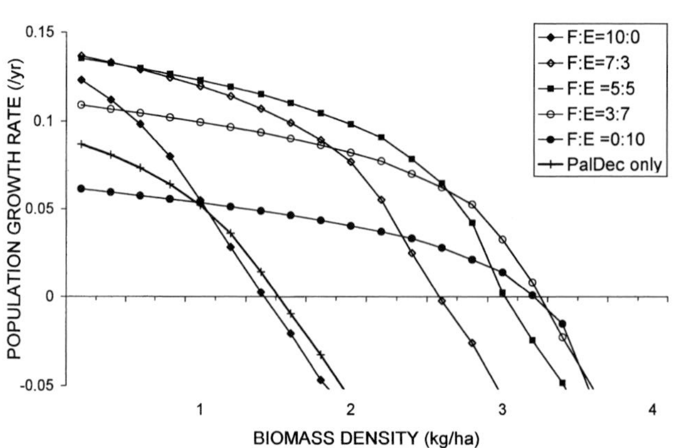

POPULATION GROWTH RATE (/yr)
BIOMASS DENSITY (kg/ha)
F:E=10:0
F:E=7:3
F:E =5:5
F:E=3:7
F:E =0:10
PalDec only

c

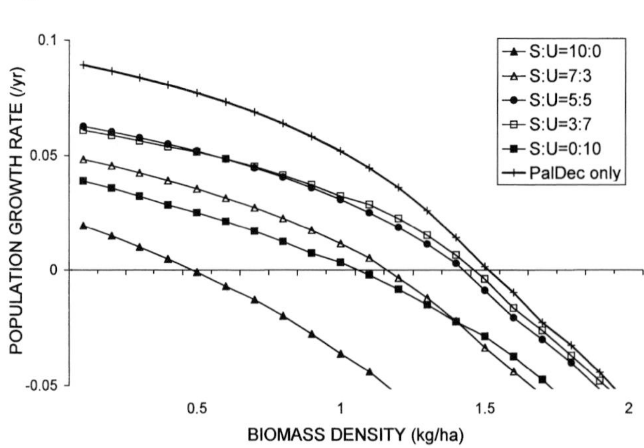

POPULATION GROWTH RATE (/yr)
BIOMASS DENSITY (kg/ha)
S:U=10:0
S:U=7:3
S:U=5:5
S:U=3:7
S:U=0:10
PalDec only

functionally in terms of differences in nutritional value, growth potential, potential intake rate offered, and capacity to retain forage through the DS. I propose the following functional classification of vegetation resources into five categories:

1. *Quality* resources elevating productive potential, e.g. forbs for browsers and grasses offering high-quality, low-fibre but adequate bulk for grazers (e.g. buffalo grass, *Panicum* spp., for cattle, lawn-forming *Sporobolus* spp. for wildebeest);

2. *Staple* resources providing an adequate-quality diet for most of the year, e.g. palatable deciduous woody plants for browsers, *Themeda* and similar medium–tall grasses for cattle, more nutritious *Panicum*, *Cynodon* or *Urochloa* species for smaller grazing ungulates;

3. *Restricted-intake* resources that are commonly high in quality, but yield inadequate intake rates, which can improve diet quality if complemented by more rapidly ingested resources, e.g. spinescent woody plants for large browsers, lawn-forming grasses for large grazers;

4. *Reserve* resources sustaining herbivores through much of the DS, e.g. the more palatable evergreen woody plants for browsers, *Themeda* for many grazers;

5. *Buffer* resources that may help alleviate starvation during critical periods, e.g. unpalatable deciduous and evergreen woody plants for browsers, unpalatable grasses or even browse components for cattle.

Figure 11.8a illustrates the seasonal resource gains provided by each of these generic resource types in isolation, for a browsing ungulate. High-quality resources such as forbs yield greatest energy gains during the GS, but generally do not maintain such gains through the DS. Accordingly,

Figure 11.8. Browser model output for generic food types, alone and in combination. (a) Relative energy balance through the dry season for zero stocking density, i.e. decline in food abundance due only to inherent attrition. Each uniform food type has a standard abundance of 20 g/m^2, backed by a low-quality food buffer ($c = 0.25$, $F = 5$ g/m^2). Parameter values for each food type are given in Table 11.5. (b) Density dependence in population growth rate for different combinations of forbs relative to evergreens, retaining a standard abundance of palatable deciduous and buffer components. Overall food abundance (excluding the buffer) = 20 g/m^2, half made up by the staple palatable deciduous category. (c) Density dependence in population growth rate for different combinations of low intake (spinescent) food types relative to unpalatable deciduous food types offering elevated intake rates, retaining the standard abundance of palatable deciduous and buffer components.

Table 11.5. *Parameters for generic food types used in the model, based on browse components*

	Food type					
	HQual	RestrInt	Staple	Reserve	BufferD	BufferE
Nutritional value (digestible DM)	0.75	0.70	0.60	0.50	0.40	0.35
Basic biomass F (g/m^2)	25	25	25	25	25	25
Eating rate I_P (g/min)	5	5	2.5	5	7.5	7.5
Leaf attrition rate (/week)	0.15	0.15	0.15	0	0.15	0
Onset of leaf fall (week)	4	8	10	0	14	0

because dry season deficits exert the greatest leverage on mortality, staple food types sustain higher stocking densities despite being less nutritious. Reserve resources yield a marginally positive energy balance, but slightly negative population growth rate taking mortality into account. Both restricted-intake and buffer resource types give sub-maintenance nutritional gains, and so are inadequate for supporting a population in isolation. Nevertheless, they can make an important contribution to stocking densities in combination with other resource types.

Next we consider the best combination of these resource types for promoting herbivore performance. Specifically, how important are high-quality resources, relative to dry season reserves? What mix of high-quality, staple and reserve food types yields the maximum productive potential? What is the value of restricted-intake resources, relative to low-quality buffers? Answers to these questions were again sought by using the browser model.

Augmenting the staple food type with both high-quality forbs plus reserve evergreens brought a substantial improvement in resource gains year-round, and hence in population performance, compared with a situation with palatable deciduous species alone (Fig. 11.8b). Overall the contribution of reserve resources to stocking density appeared to be more substantial than that of high-quality resources that became depleted during the DS. However, this analysis did not allow for the important contribution that high-quality resources could make towards offspring production during the GS.

Replacing some of the staple resource with a combination of restricted-intake and buffer resources lowered permissible stocking densities (Fig. 11.8c). However, the consequences could be minor where the low intake rate of a high quality resource is balanced by a higher than average intake rate from a lower-quality resource. Nevertheless, the buffer resource made a greater contribution to stocking density than the restricted-intake resource, by maintaining a food supply through the critical DS period.

For browsers, the period over which leaves are retained into the dry season can vary substantially between years, depending on prior rainfall conditions. The effect of such variability in leaf retention on herbivore biomass dynamics depends on the vegetation composition. In forb-rich habitats with a limited availability of evergreens, a shift by *ca.* 3 weeks in the commencement of leaf fall could cause the herbivore density supported to vary by about 40% (Fig. 11.9a). Where evergreens predominate and forbs are less prevalent, the zero growth density appears somewhat more resistant to variability in foliage retention by the deciduous woody and forb components (Fig. 11.9b).

Overview

This chapter assessed how not only the annual vegetation production, but also seasonal fluxes in resource availability, contribute to habitat suitability in terms of herbivore population performance. For browsing ungulates in savanna regions, the population supported is strongly dependent on the presence of bridging resources to provide forage during bottleneck periods in the late dry season, and relatively less influenced by the productive capacity of the vegetation in the wet season. This is because savanna trees are predominantly deciduous, and hence lose their foliage during the course of the dry season. Even chemically defended and thus unpalatable evergreens may make an important contribution, by serving as a buffer against starvation during critical periods. The contribution of tree species with highly nutritious foliage may be restricted by the low rates of intake they yield, as a result of physical defences such as thorns. However, the influence of these factors depends on the body size and feeding adaptations of the particular species of browser concerned. The exercise also illustrated how vegetation heterogeneity can contribute towards maintaining herbivore populations.

For grazing ungulates, the stocking densities that can be supported depend largely on the presence of productive perennial grasses that retain

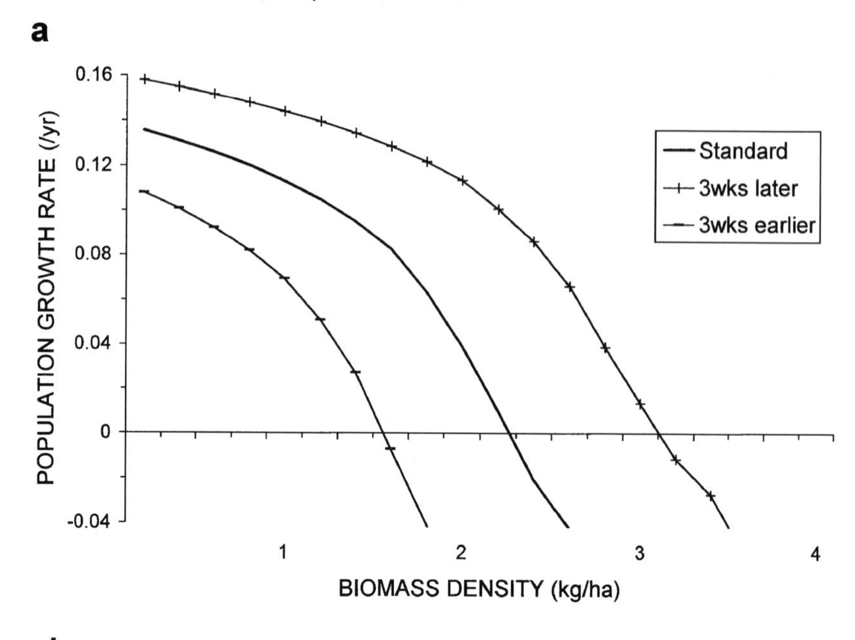

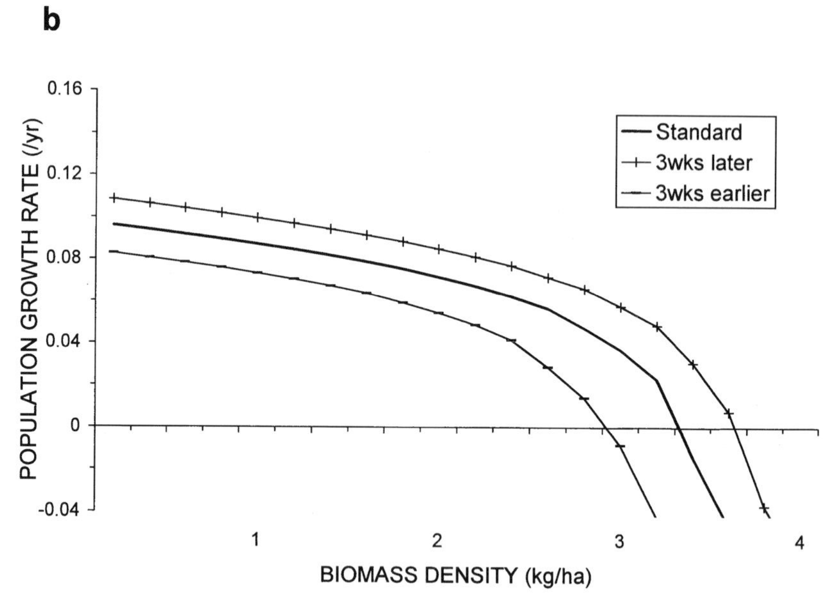

Figure 11.9. Effects of changes in duration of leaf retention on herbivore population performance, from browser model output. (a) Forbs abundant relative to evergreens, and standard palatable deciduous and buffer components as in Fig. 11.8, adjusting the onset of leaf loss three weeks earlier or later relative to standard conditions. (b) Habitat with evergreen foliage much more abundant than forbs, otherwise as in (a).

adequate nutritional value to support animals through the dry season. These 'Decreaser' grasses are the primary component of grassland judged to be in good condition. Nevertheless, at low stocking densities higher productive outputs may be obtained from cattle grazing grassland with a predominance of less bulky but highly nutritious 'Increaser II' grass species. Grassland with a mixed composition may generally yield the best all-round performance. Smaller grazers are dependent more on the high-quality grass types, and less on the high-bulk species, than cattle.

A generic suite of resource types was proposed, taking into account not just production, quality and intake rates yielded, but also the use of these resources at different stages in the seasonal cycle. A favourable habitat requires an appropriate combination of these resource types.

Other components of habitat suitability besides resources have not been considered. Predation risk can greatly influence habitat suitability, as recognized in Chapter 10. The presence of other species that compete for similar resources can also modify habitat suitability, as will be assessed in the next chapter.

Further reading

McLeod's (1997) critical assessment of the carrying capacity concept provides a useful overview. Techniques of veld condition assessment as applied in South Africa are summarized by Tainton (1999). More general reviews of principles of grazing livestock management are provided by Heitschmidt and Stuth (1991) and Heady and Child (1994).

Appendix 11A. Structure of the browser habitat suitability model

The model was written as a TrueBASIC program in two versions. The simple version merely calculated nutritional gains and associated measures on a weekly iteration. Mortality losses were estimated after transferring this output to a spreadsheet. The elaborated version transformed weekly gains into estimates of annual mortality for different densities. The acceptance coefficient for available food biomass was set at 0.25, the maximum observed DS value for kudus (Owen-Smith 1997). The daily foraging time t_F was fixed at 0.45 of the 24 h day (the observed year-round mean

Table 11A. *Parameter values used in the browser and grazer habitat suitability models*

Parameter	Function	Value	
		Browser	Grazer
Body mass M (kg)		100	300
Rumen capacity (kg)	(i) $M \times 0.022$	2.2	—
	(ii) $M \times 0.027$	—	8.1
Turnover rate of rumen digestible matter (/day)		2.2	1.0
Passage rate from rumen (/day)		0.35	0.3
Resting metabolic requirement (digestible DM, kg/day)	(i) $M^{0.75} \times 0.38/(18 \times 0.83)$	0.8	—
	(ii) $M^{0.75} \times 0.34/(18 \times 0.83)$	—	1.65
Active metabolic requirement (digestible DM, kg/day)	$1.8 \times$ resting	1.5	3.0
Search rate (m^2/min)	(i) $M^{0.67} \times 0.9$	20	—
	(ii) $M^{0.67} \times 0.67$	—	30
Mortality intercept z_0		-0.9	-1.0
Mortality coefficient z		1.2	1.15
Maximum eating rate (g/min)	$M \times 0.09$	—	27
Food biomass for maximum eating rate (g/m^2)	$I_P \times 11$	—	300
Ungrazable biomass (g/m^2)		—	10

for kudus). The mortality function was adjusted so that nutritionally related mortality was zero when resource gains exceeded maintenance requirements by one third, and 30% per year when resource gains just met maintenance requirements. Additive mortality due to predation plus animals dying from senescence amounted to 10% per year, diminishing in its overall magnitude as nutritionally dependent mortality increased. To calculate the annual population growth rate, annual mortality was subtracted from a maximum potential recruitment rate of 30% per year. No age structure was considered.

In the general model food types were assumed to be intermingled, so that the diet encompassed the combination of these food types maximizing the rate of nutrient gain. Food types were arbitrarily ranked in order of their inclusion in or exclusion from the diet. An alternative formulation considered food types dispersed in discrete habitat patches, with herbivores selecting the best patch type each week. The available food biomass and effective density of herbivores within the chosen patch was adjusted to allow for this localized concentration of consumption.

For each weekly iteration, the biomass F of each food type was reduced first through consumption, and then through attrition due to leaf fall or withering. The standard weekly rate of biomass attrition was set to 0.15, equivalent to a halving of the foliage retained every four weeks. The week in which leaf attrition commenced varied among food types.

Mortality susceptibility depended on weekly resource gains. To calculate annual mortality, an average weekly mortality rate was first calculated for the DS. The mortality in the first week of the DS, before resources had become depleted, was assumed to represent the GS mortality rate. The annual mortality rate was the average of the GS and DS mortality. The population growth rate was determined by subtracting this mortality loss from the potential population recruitment rate. Parameter values are listed in Table 11A.

Appendix 11B. Structure of the grassland–grazer model for veld condition

The model, written in TrueBASIC, was parameterized for a steer weighing 300 kg. The rate of food intake obtained while foraging followed Equation 3.6, i.e. $I_F = saF/(1 + saF/I_P)$, where F = available food biomass, weighted for selection against necromass, a = acceptance coefficient for this biomass, s = searching rate as area per unit time, and I_P = within-patch eating rate in mass per unit time. To convert this to a daily food intake, I_F was multiplied by a fixed daily foraging time proportion, set at to 0.45 of the 24 h cycle, close to the upper limit observed for cattle (O'Reagain *et al.* 1996). The relative growth potential of the animals was calculated in energetic units, converting digestible dry matter into its energetic equivalent relative to basal requirements.

The eating rate I_P reached a maximum above a certain forage biomass, and declined linearly with decreasing biomass (and hence grass height) below this level. Parameter values were based on the maximum intake rates in relation to grass height recorded by O'Reagain *et al.* (1996). The acceptance fraction a diminished linearly with increasing grass biomass, from a value of 0.9 when grass biomass was one fifteenth of the biomass yielding maximum eating rate, to 0.25 when grass biomass was at the value yielding the maximum eating rate. The selection coefficient against necromass was $1/3$ when green biomass exceeded the level where maximum eating rate was attained, $1/2$ when green biomass was less than this but more than half the maximum eating rate level, and 1.0 below the latter biomass. Effectively, the standing necromass was reduced by this factor in its relative contribution to available forage biomass and intake. The indigestible stem fraction in the food ingested was zero before week 10 in the growing season, half of the maximum species-specific value up to 14 weeks, and at the maximum species value (as indicated in Table 11.4) for the remainder of the year. Food types were ranked in value based on either the intake rate while foraging, or the digestion rate, of digestible dry matter, whichever was more limiting in the current iteration (from Owen-Smith 1994).

The daily digestive capacity was determined for a fixed rumen contents mass, multiplied by the turnover rate of digestible dry matter plus the passage rate of material from the rumen. Mortality parameters were set so that starvation mortality amounted to 15% per year if resource gains equalled the active requirement, and zero when gains were 15% above this requirement. The realized growth potential overall was the product of the rate of mass gain per unit of herbivore biomass and the annual survival rate (one minus mortality). The productive potential per unit area was the product of this growth potential and the stocking density.

Grass growth followed the model specified by Equation 8.2, for each grass type independently. Contributions to available food biomass were weighted by the proportional basal area of each grass type. The digestibility of necromass was 0.75 times that of green biomass.

Additional parameter values are given in Table 11A. With this mechanistic model, the half-saturation level for food intake rate was not preset, but emerged from how eating rate, forage acceptance and digestive capacity changed with grass standing crop (i.e. biomass + necromass). For the specified parameter values, consumption rate declined linearly below a food biomass of 300 g/m^2, projecting a half-saturation level of around 150 g/m^2. However, digestive capacity truncated the daily food intake towards higher food biomass, and acceptance increased with diminishing food biomass. Plotting the model output relative to changing grass standing crop over the seasonal cycle, for different grassland compositions, shows that the maximum daily food intake was attained for a grass standing crop of around 150 g/m^2 (depending on conditions), with the effective half-saturation value being around 45 g/m^2 (Fig. 11.7a). However, under certain circumstances the half-saturation value could be as low as 20 g/m^2, specifically when animals fed on remnants of tall grass remaining in a mostly depleted sward. The half-saturation value for daily nutrient gain (as digestible DM) was somewhat higher, around 70 g/m^2 (Fig. 11.7b).

12 · *Resource partitioning: competition and coexistence*

The importance of competition in structuring herbivore species assemblages is widely surmised, but rarely demonstrated. One view is that past competition over evolutionary time has resulted in niche divergence, such that competition has only a minor influence on extant populations (Owen-Smith 1985), or at least is manifested only intermittently (Owen-Smith 1989). Sinclair (1985) suggested that risk of predation has an overriding effect on species associations. However, Prins and Olff (1998) considered competition to be pervasive within grazing ungulate assemblages, such that species of closely similar size rarely coexist.

Part of the problem is that competition among large herbivores arises largely indirectly via vegetation modification or 'sward capture' (Murray and Illius 1996, 2000), rather than through overt interference. Smaller ungulates have the potential to out-compete larger species by depressing vegetation biomass below that needed to meet the greater absolute food requirements of the latter (Illius and Gordon 1987). On the other hand, the vegetation impacts of the bigger species can alter vegetation structure such that habitat conditions are changed for other species. However, the habitat modification need not be detrimental. A reduction in grass height could improve food access and dietary quality for smaller species better adapted to exploit short grasslands, leading to interspecific facilitation rather than competition (Vesey-Fitzgerald 1960; McNaughton 1976; Prins and Olff 1998). Nevertheless, despite the short-term gains in nutritional intake that may result for these small species, consequent increases in population abundance have not been observed (Sinclair and Norton-Griffiths 1983).

Competition may be obscured when the presence of one species inhibits another from occupying a habitat it would otherwise have exploited (Rosenzweig 1981). Herbivore species alter their habitat use and diet selection seasonally, so that competition may be effective only during particular times of the year (Owen-Smith 1989). Competition could be accentuated during the late dry season, when resources are most sparsely available, or diminished during this period if species then rely on distinct food refuges (Jarman 1971).

The potential for interspecific competition has important implications for managers of multi-species ranches and wildlife reserves. If competition is ignored and each herbivore species is stocked at its independent 'carrying capacity', resource overuse and diminished animal performance could result. If stocking is restricted to species with distinct food and habitat preferences, the benefits of facilitation and fine-scale resource partitioning are foregone.

To assess the deferred effects of resource exploitation on population abundance, the consequent nutritional gains, and hence population growth potential, must be evaluated through the complete seasonal cycle, taking into account adaptive changes in habitat use. In this chapter I will consider firstly how potential resource competition among herbivore species differing in body size affects the respective abundances that they might attain, in different environments. Next, I will evaluate how ecomorphological distinctions affecting resource use might ameliorate competition. Finally, I will examine the tradeoffs that can exist between facilitation and competition over the seasonal cycle.

Lotka–Volterra model

The conceptual foundation for assessing competition is provided by the Lotka–Volterra equations, found in every ecology textbook. These represent the competitive impact of each species on the population of another, by incorporating a competition coefficient into the respective logistic growth equations:

$$\begin{aligned} \mathrm{d}N_1/\mathrm{d}T &= r_1 N_1(1 - (N_1 + \alpha_2 N_2)/K_1); \\ \mathrm{d}N_2/\mathrm{d}T &= r_2 N_2(1 - (N_2 + \alpha_1 N_1)/K_2), \end{aligned} \qquad (12.1)$$

where N_1, N_2 = abundances of species 1 and 2 respectively, r_1, r_2 = respective intrinsic growth rates, K_1, K_2 = equilibrium densities, for each species in isolation, and α_1, α_2 = competition coefficients representing the unit effect of each species on the population of the other.

If $\alpha = 1$ in both cases, individuals of each species are identical in their competitive impacts on each population, so that chance decides which species prevails. For $\alpha_1 > 1$, species 1 is the superior competitor and is expected to displace species 2, unless predation or other influences restrict the abundance of species 1. If the α coefficients are both fractional, the two species can coexist at reduced abundance compared with what they would have reached in isolation. A negative α coefficient would imply facilitation, i.e. the presence of one species elevates the abundance of the other. Because the interaction is defined phenomenologically, the population consequences could arise via sharing a mutual predator (so-called 'apparent competition'), rather than through resource overlap.

In seasonally variable environments, the degree of resource overlap changes through the seasonal cycle. To assess the consequences, we need to turn to an appropriate version of the **GMM** model. This will reveal that the meaning of the α coefficients is not so clear as is implied by the Equations 12.1 above.

Competitive habitat partitioning

To accommodate adaptive responses by individuals of each species to changes in habitat suitability brought about by the presence of the other, we draw upon the analysis of competitive habitat selection developed by Rosenzweig (1981, 1991). This rests on the concept of an 'ideal free' distribution, attained where each individual within a population acts to enhance its own benefits (Fretwell 1972). At low density, individuals should occupy the best habitat, i.e. that offering highest survival and reproductive success. However, as population density rises, the fitness benefits of this habitat become diluted, e.g. through interference with food access. At some density, it becomes advantageous for some individuals to settle in an intrinsically inferior, but less crowded, habitat.

Rosenzweig extended this concept to interspecific competition. He defined the 'isoleg' (point of equal choice) as the population level at which two habitats become equally favourable. With two potentially competing species present, effective habitat suitability depends on their joint abundance. Two alternative contexts may exist: (1) the species prefer different habitats, but are inhibited from spilling over into the alternative habitat by the presence there of the competing species; (2) both species prefer the same habitat, with the inferior competitor becoming relegated to a second-rate habitat.

These two situations can be depicted graphically in the phase space formed by the relative density of each species (Fig. 12.1). The competitive interaction causes the isolegs, which would otherwise be perpendicular to each axis, to become inclined (Fig. 12.1a). This occurs because each species must reach a higher density before spillover into the habitat

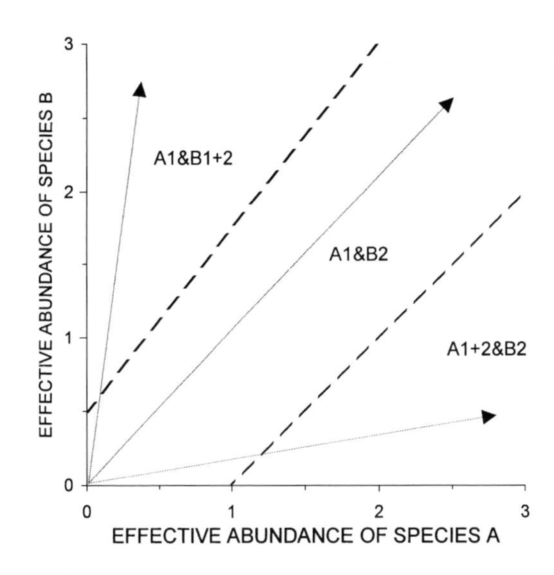

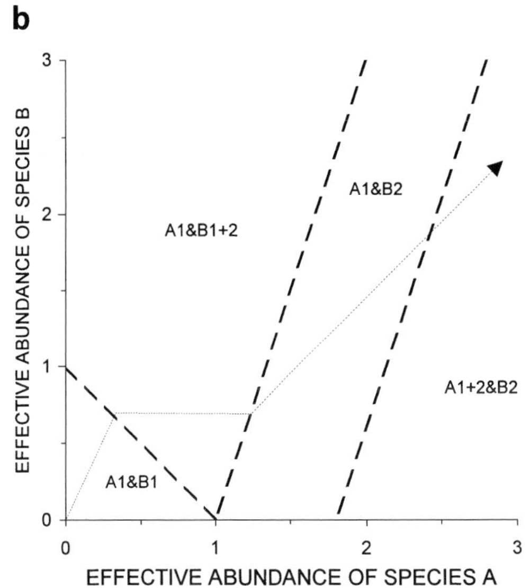

Figure 12.1. Graphical representation of competitive habitat occupation by two consumer species in relation to the population abundance of each. Dashed lines indicate isolegs defining regions of differing habitat occupation (e.g. A1 signifies that species A occupies habitat type 1 only, B1+2 signifies species B occupies habitats 1 and 2). Vectors show how habitat occupation would change as effective abundance is depressed, owing to declining resource availability. (a) Distinct habitat preferences. (b) Shared habitat preferences. (Adapted from Rosenzweig 1981, 1991.)

becomes advantageous, when the competitor is present. With shared rather than distinct habitat preferences, the location of the isolegs is displaced, and there is a small region at low density where both species coexist in the mutually preferred habitat (Fig. 12.1b). The isolegs partition the phase space into relative density regions where the two species either overlap in one or other habitat, or segregate into different habitats.

To anticipate the consequences of seasonally changing resource availability, we embellish the diagram with one additional feature. The effect of seasonally decreasing resources is the same as that of increasing density relative to constant resources. This is indicated by the vectors, which suggest how species that overlap in habitat use at times of the year when resource abundance is high may segregate into distinct habitats when resource availability becomes diminished.

The second step introduced by Rosenzweig was to superimpose zero growth isoclines (ZGIs) on the phase diagram. As noted above, mutual competition reduces the density at which the population growth rate of each species becomes zero, as represented by the α coefficients in Equations 12.1. Habitat segregation causes the ZGIs to become warped because, in the region where both species occupy different habitats, their densities do not respond to changes in the abundance of the other (Fig. 12.2).

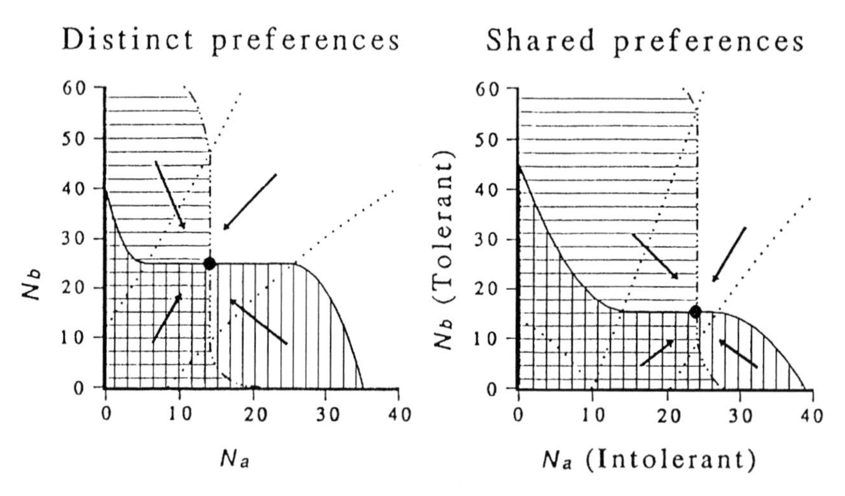

Figure 12.2. Zero growth isoclines (solid lines) superimposed over isolegs for habitat choice (dotted lines) in the phase space defined by the relative abundance of two competing species. In the region where the species are segregated into distinct habitats, the zero growth isoclines are perpendicular to each axis. Stable coexistence occurs where the isoclines intersect. Arrows indicate trajectories of population change towards this joint equilibrium. (From Rosenzweig and Abramsky 1997, with kind permission of Kluwer Academic Publishers).

The ideal free distribution concept applies most simply to situations where competition arises immediately through interference with foraging. Where competition occurs primarily through resource depression, as is usually the case for large herbivores, the consequences are deferred. Species may aggregate in the short term while resources are plentiful, and separate into alternative habitats only after resources have been depleted below some threshold availability (Chapter 9). Moreover, some species may be more effective at exploiting sparse resources, and others better able to use abundant resources, so that they select the same habitat at different stages of resource depletion (Tilman 1986).

Exploitation competition model

For this model, basic habitat suitability will be indexed by the *relative growth potential* (RGP) offered by the resources contained in the habitat, so that the performance of each herbivore species can be evaluated in comparable terms. The RGP, defined by Equation 7.1, is the net difference between the rate of biomass or energy gain obtained while foraging, G, and physiological costs, M_P, normalized relative to the basal metabolic requirement, M_{P0}:

$$\mathrm{RGP} = (G - M_P)/M_{P0}. \tag{12.2}$$

For simplicity, M_P will be assumed constant, i.e. physiological expenditures do not differ between habitats. Differences in mortality risk between habitats will likewise be ignored (but could easily be incorporated). Accordingly, habitat suitability is governed solely by the rate of resource gain, G. Modifying Equation 3.6, the daily resource gain, G_D, obtained from a particular habitat is asymptotically dependent on the food availability, F, expressed as a mass per unit area, that this habitat offers, i.e.

$$G_D = \{csaF/(1 + saF/I_P)\}\, t_F, \tag{12.3}$$

where s = area searched per unit time, a = acceptance coefficient, c = food conversion coefficient into herbivore biomass, I_P = rate of food intake obtained while feeding within food patches, and t_F = daily foraging time as a proportion. Furthermore, digestive or metabolic processing constraints may restrict the daily foraging time, and hence the maximum daily gain (Chapter 5).

Hypothetical trends in RGP with changing food availability are depicted for four habitat types differing only in the food quality they present in Fig. 12.3a. The habitats can be characterized as preferred, alternative,

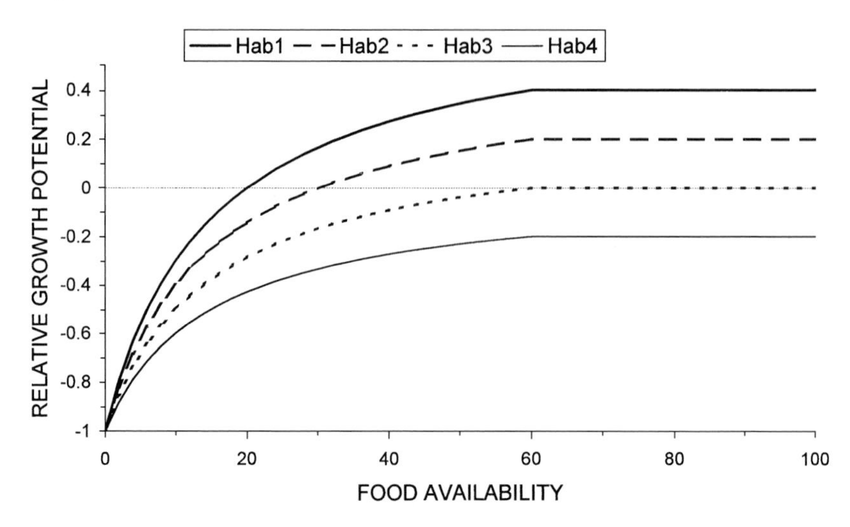

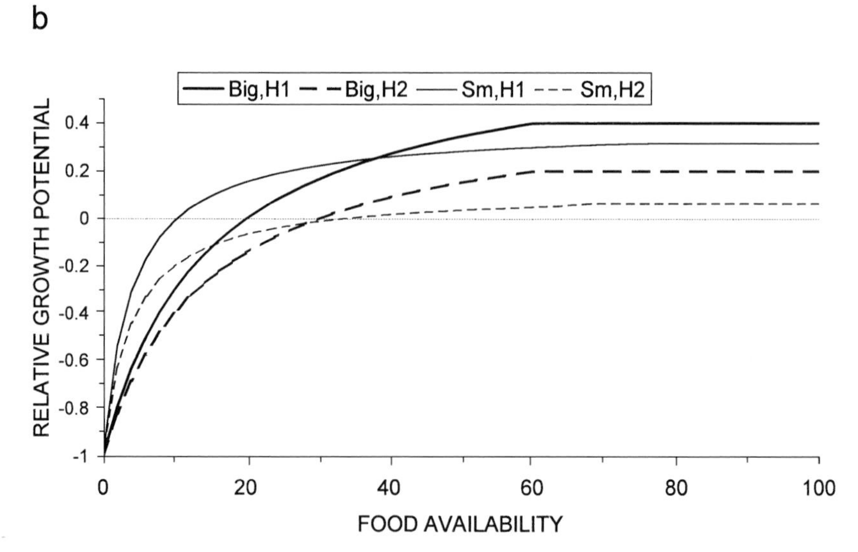

Figure 12.3. Comparative resource gain functions for the alternative habitats used in the model, expressed as dimensionless relative growth potential. (a) Basic gain functions for four habitats differing only in food quality, labelled 1–4 from best to worst. (b) Comparative gain functions for the two best habitats for two herbivore species differing in size and hence in the initial slope and ultimate elevation of their respective gain functions. (c) Comparison among habitats for the big herbivore, following a reduction in effective eating rate in the best-quality habitat. (d) Comparison between the best two habitats for two herbivores differing in size, after adjustment (c).

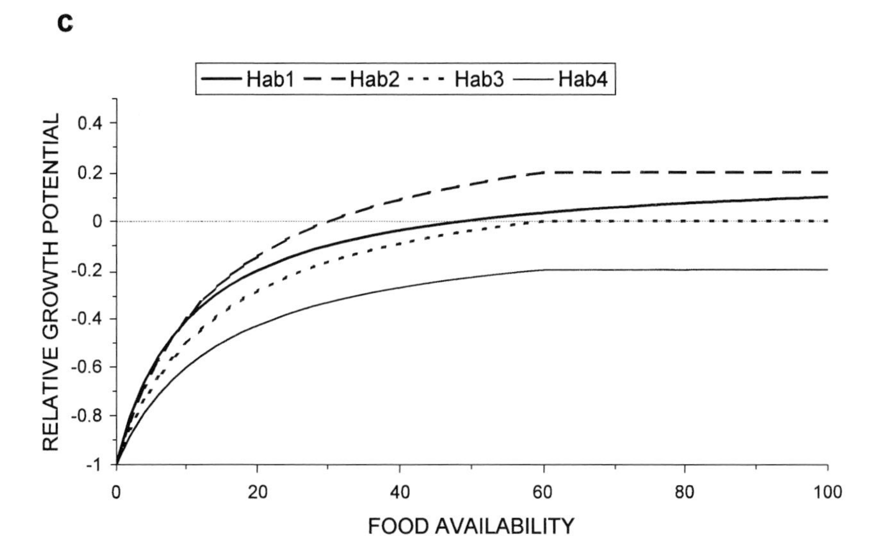

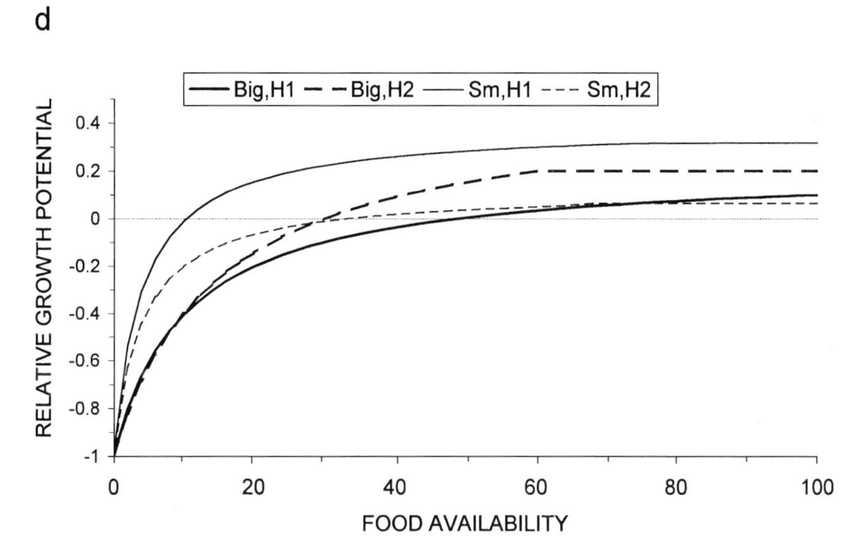

Figure 12.3. (cont.)

reserve, and buffer, with the buffer yielding sub-maintenance gains under all conditions. The habitat type offering the highest-quality food is likely to be generally preferred, while this food remains sufficiently abundant. However, small herbivores may be relatively more effective at exploiting sparse food, but less effective when food biomass is high, than big herbivores (Illius and Gordon 1987). To represent this, the respective resource gain functions, expressed in dimensionless RGP units, can be distinguished by (a) the asymptotic level, set by the food intake rate I_P relative to basal metabolic requirements M_{P0}, and (b) by the abruptness of the initial rise at low food availability, set by search rate s relative to the eating rate I_P (Fig. 12.3b). Hence, relative competitive superiority shifts with changing resource abundance.

Alternatively, we may recognize that food of the highest quality commonly yields a restricted rate of intake. For grazing ungulates, this might occur where short grasses are the most nutritious. For browsers, it can arise where tree species associated with fertile soils have thorns limiting bite size and hence eating rate. As a result, the habitat offering the best food quality could be less favourable in the growth potential that it offers than another habitat yielding a higher rate of food intake, particularly for larger species of herbivore (Fig. 12.3c). The outcome would be distinct rather than shared habitat preferences for herbivores differing in body size (Fig. 12.3d).

To model alternative scenarios, four habitats will be distinguished by their effective values for c, I_P and maximum F, as defined in Equation 12.3, as well as by their relative extent in the environment. Two competing herbivore species will be labelled suggestively 'big' and 'small', based on distinctions in their resource gain functions. Both are assumed to be grazers, so that vertical distinctions in resource access can be ignored. Only dormant season (DS) conditions when resources are non-renewing need be considered. Food availability is progressively depleted through consumption only, i.e. inherent attrition is ignored. At each time step, the model calculates the daily rate of resource gain, G_D, and its RGP equivalent, and identifies the best habitat, for each of the competing species in turn. Herbivores concentrate their feeding exclusively in the habitat yielding the highest resource gain, within each time step. Lastly, the mean RGP is calculated over the entire DS. A RGP averaging zero is assumed to give zero net population growth, obviating the need to estimate mortality. Further details of the model and parameter values used are given in Appendix 12A.

Shared preferences

With habitat types distinguished only by resource quality, as in Fig. 12.3a, the two herbivore species share the same preference ranking of habitats, despite the difference in their respective resource gain functions, shown in

Fig. 12.3b. Initially, habitats yielding lower resource gains were assumed to be more extensive than richer habitats. The model was run first for each species in isolation. The output showed how herbivores progressively shifted their habitat occupation from the best habitat towards the poorest habitat over the course of the DS (Fig. 12.4a, b). Because the RGP is truncated by digestive satiation at high resource abundance, habitat switching rather than broadened habitat use occurred when resources in the most favorable habitat became reduced below the critical level. With the second-best habitat being more extensive, it took longer to reduce its resources to the critical level for habitat transfer than it did for the best habitat. At a later stage, consumers periodically returned to the initially better habitats to depress further the sparse resources remaining. Eventually the herbivores turned to the buffer habitat, despite its poor food quality. As a result of the changing habitat occupation, the animals experienced diminishing nutritional gains over the course of the DS, with abrupt transitions associated with habitat shifts. For the specific parameter values used in the model, the small species persisted in the best habitat for longer than the big species, and did not exploit the buffer habitat at any stage.

For the competitive situation with both herbivore species present, the resource gain from the mutually preferred habitat remained favourable for the small species at the resource level at which it became advantageous for the big species to switch to the alternative habitat. By continuing to depress the availability of resources in the best-quality habitat, the small species made it unprofitable for the big species to return to this habitat (Fig. 12.4c). In this way, the small species was the superior competitor for high-quality habitat. On the other hand, by reducing food availability in other habitats, the big species reduced the resource gain that the small species obtained when it eventually transferred to alternative habitats (Fig. 12.4d).

Through competition, both species incurred a marked reduction in growth potential through the DS compared with what they obtained when present alone (Fig. 12.4c, d). The small species suffered most. The initial feeding period by the big species in the best habitat caused the small species to abandon this habitat sooner than it would otherwise have done. The competitive impact of the big species was exacerbated because it attained a higher biomass density than the small species, as a result of its ability to use a wider range of habitat types effectively.

Zero growth isoclines (ZGIs) were obtained by fixing the population density of one species, then adjusting the density of the other species until the mean RGP of the latter over the DS became zero. The intersection of the two ZGIs indicates the relative densities of the two species for coexistence. For the specific parameter values used in the model, the abundance of the big species was reduced by less than 10% when the small

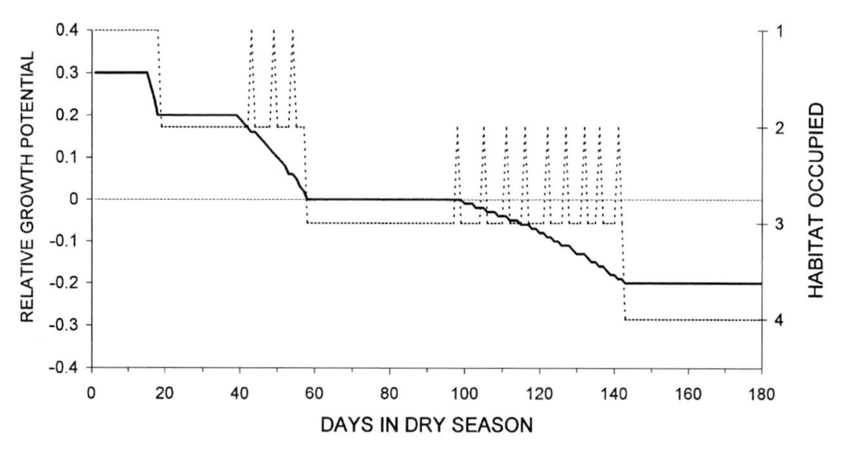

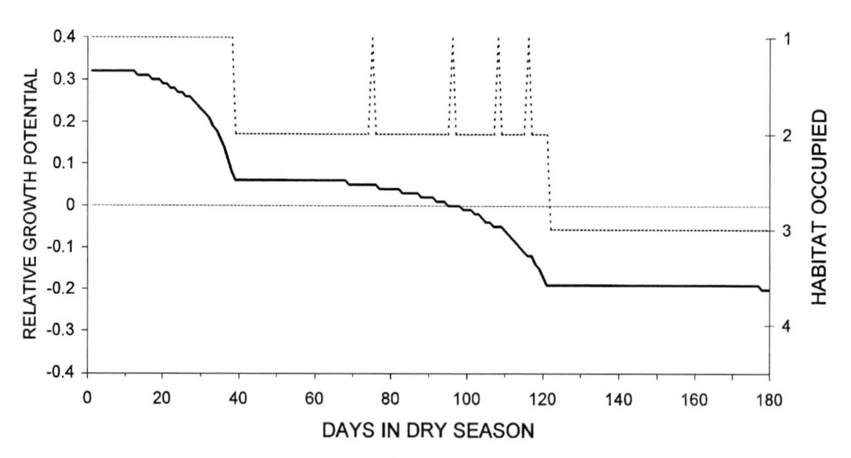

Figure 12.4. Comparative resource gains (bold line) and habitat occupation (dotted line) for modelled big and small herbivore species through the dormant season. Each species is stocked at its respective zero growth density. (a) Big species in isolation. (b) Small species in isolation. (c) Big species in competition with the small species, compared with its gains when alone. (d) Small species in competition with the big species, compared with its gains when alone.

c. BIG, COMPETING

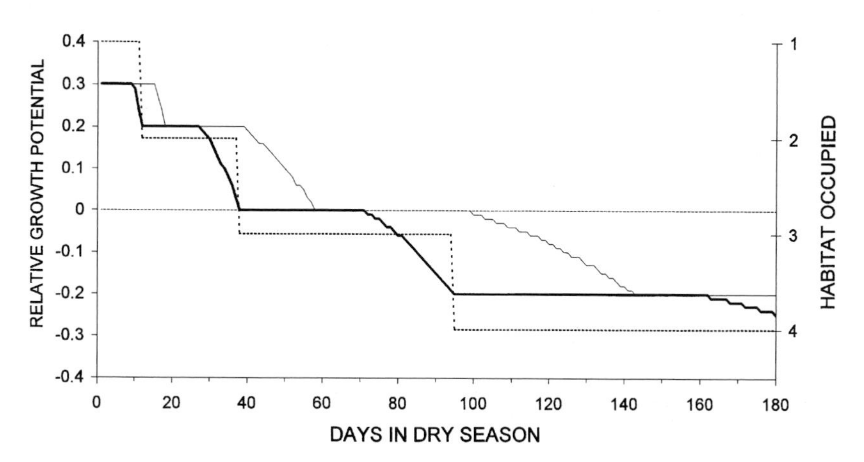

d. SMALL, COMPETING

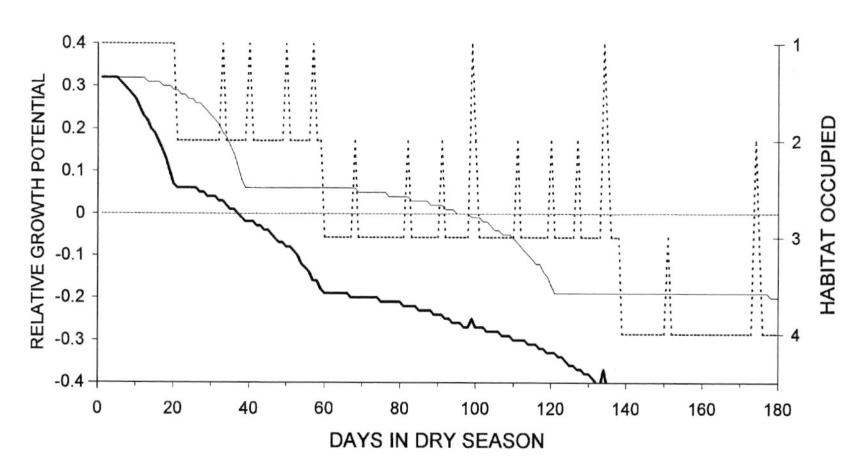

Figure 12.4. (cont.)

species was present (Fig. 12.5a). However, the small species reached only about one quarter of the density that it would have reached in isolation. This implies that the big species was the superior competitor. But α coefficients determined from the reduction in biomass density below the population attained by each species in isolation, per unit biomass of the competitor, suggest the opposite (see Fig. 12.5a). In other words, a unit mass of the small species had more effect on the abundance of the big species than vice versa. The big species attained a higher overall biomass

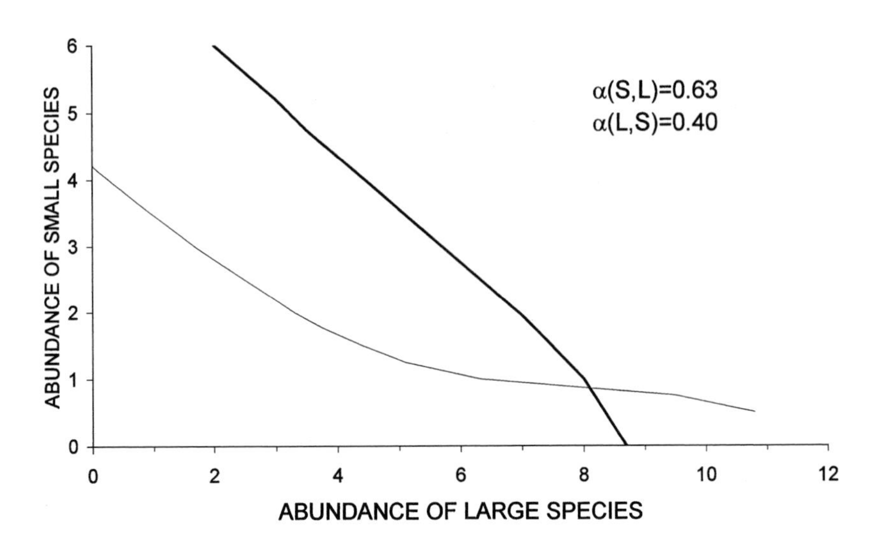

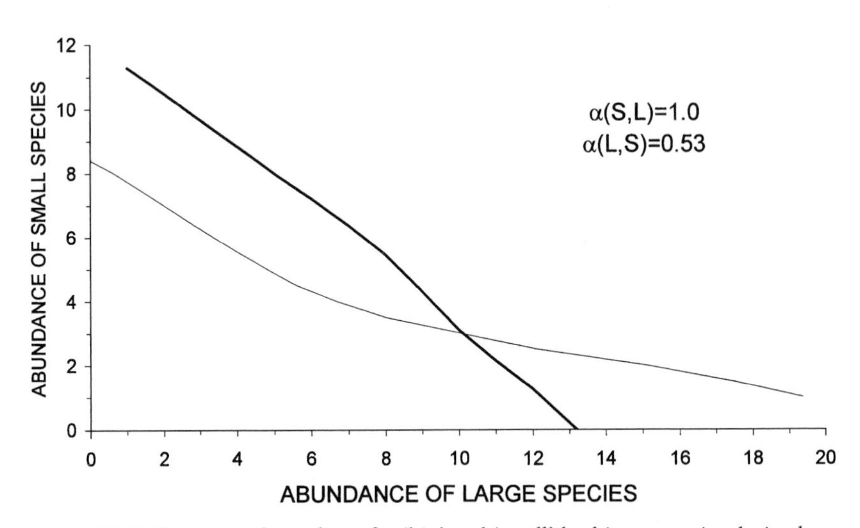

Figure 12.5. Zero growth isoclines for 'big' and 'small' herbivore species derived from the model. Alpha coefficients are indicated for the effects per unit biomass of each species on the abundance of the other. (a) Shared habitat preference, poorer-quality habitats more extensive (1:2:4 area ratio). (b) Shared habitat preference, better-quality habitats more extensive (4:2:1 area ratio). (c) Distinct habitat preference, poorer-quality habitats more extensive (1:2:4 area ratio). (d) Distinct habitat preference, better-quality habitats most extensive (4:2:1 area ratio).

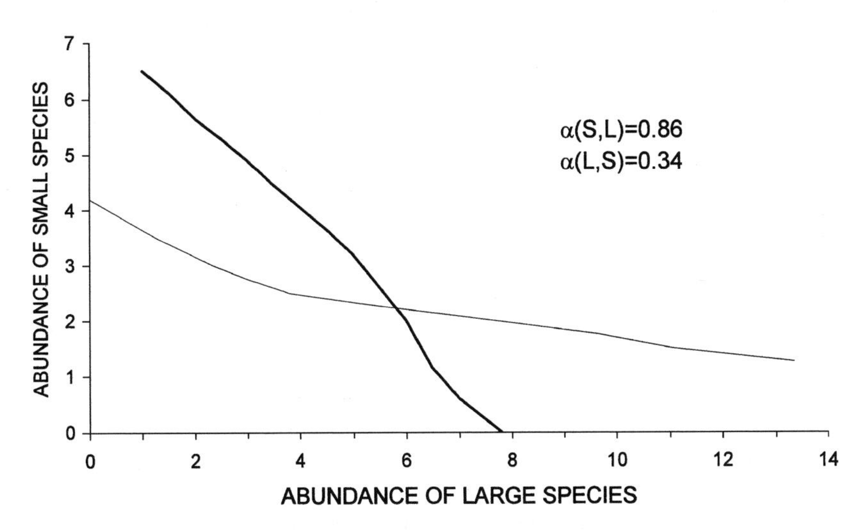

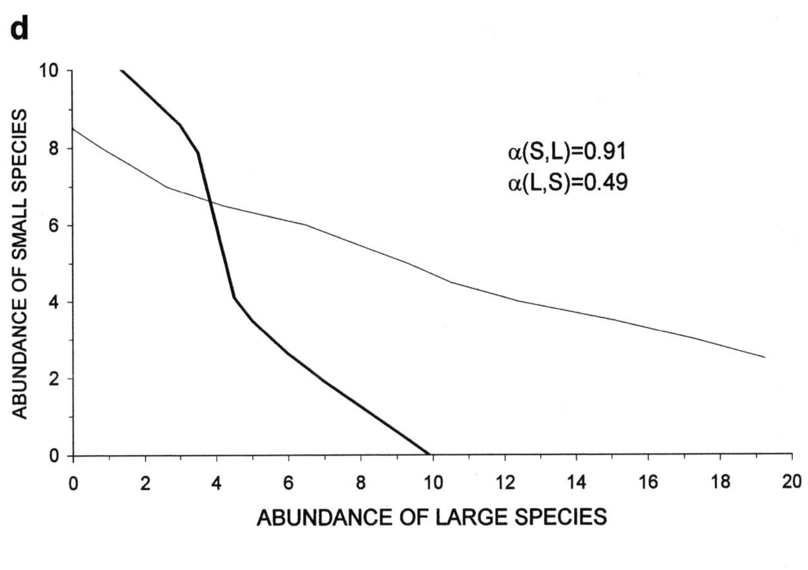

BIG SMALL

Figure 12.5. (*cont.*)

in the environment, because it was better able to exploit poorer-quality habitats. Hence it ended up having a greater impact on the population level of the small species, through a swamping effect.

Interestingly, in the region of joint population stability, the small species would be little affected by variations in the abundance of the big species, because its ZGI is almost parallel to the *x* axis. In contrast, changes in the abundance of the small species have a marked influence on the abundance of the big species, as reflected by the α coefficient.

When the relative habitat extent was changed so that better quality habitats were more extensive, both species attained higher densities, but the small species benefited most (Fig. 12.5b). The α coefficient for the biomass-specific effect of the small species on the abundance of the big species increased to 1.0. Nevertheless, the density of the small species was reduced more by the presence of the big species than vice versa.

In the extreme situation, where the best-quality habitat occupies all of the landscape, the small species, which is better able to survive on sparse resources, would displace the big species from the region. At the other extreme, if only poor-quality habitat was available, the small species could not persist. Environmental heterogeneity facilitates coexistence, even when species have common habitat preferences. The competitive balance among herbivore types depends on the habitat distribution in the environment.

Distinct preferences

To model a situation where two species have distinct habitat preferences, the resource gain functions depicted in Fig. 12.3c were used for the big species. Thus, whereas the small species attained highest gains in the best quality habitat, the big species achieved its best performance in the habitat type ranked second in quality. The relative suitability of the other two habitat types remained as in Fig. 12.3a.

The form of the resultant ZGIs did not deviate greatly from those for shared preferences with the same habitat area distribution (Fig. 12.5c, d). However, at joint population stability the density of the small species was reduced somewhat less, and that of the big species somewhat more, than for the corresponding shared preference situation. If the habitat offering the highest quality was made the most extensive, the small species became clearly the superior competitor and attained a higher biomass density at zero growth than the big species (Fig. 12.5d). The big species was effectively excluded from using the best-quality habitat, unless it could secure food unavailable to the smaller species (e.g. foliage at higher levels on trees for browsers). This occurred even though the α coefficient for the effect of the small species on the big species differed little from that for shared habitat preferences.

Overall assessment

The outcomes described above are dependent to some degree on the specific parameter values chosen. Only exploitation competition under DS conditions was considered. Effects on relative habitat suitability that may arise from the depletion of specific food types within each habitat patch were not considered. Nevertheless, the findings are suggestive of some of the confounding influences on competitive outcomes: (1) the impact of competition on the population levels of each species may not be revealed by the biomass-specific α coefficients; (2) the outcome can depend on the habitat type distribution, for both shared-preference and distinct-preference situations; (3) the regional abundance of a species can influence its competitive impact.

Current theory assumes that smaller species of herbivore are competitively superior through their ability to subsist on sparser resources (see, for example, Prins and Olff 1998). This capacity is manifested only in the higher-quality habitat patches that the small species can effectively exploit. Larger species can competitively outweigh smaller species through their ability to use a wider range of resources, including some of the food present in the patches favoured by smaller species. Moreover, the supposed competitive superiority of small herbivores is not supported by observations. For example, Fritz *et al.* (1996) reported from Zimbabwe that impala were displaced from their preferred habitat types by high cattle numbers, although in this case other causes of the habitat displacement were not excluded (e.g. disturbance by herdsmen).

Size difference for coexistence

Body size affects all parameters in the resource gain function, including most notably metabolic costs. As noted by Bell (1971) and Jarman (1974), small herbivores tend to be selective feeders seeking high-quality plant types and parts, whereas large species can effectively utilize lower-quality but more abundant resources. In consequence, larger ungulates may utilize a wider range of habitat types (du Toit and Owen-Smith 1989). However, if species are too similar in body size, the distinction in habitat use may be insufficient to allow coexistence (Prins and Olff 1998). How different must competing herbivore species be in body mass in order to coexist?

We need thus to specify how relevant parameters change allometrically with body mass. Across a wide range of species, metabolic requirements are proportional to body mass raised to the power 0.75 (Peters 1983). Maximum eating rate was found to scale with $M^{0.71}$ (where M = body mass), i.e. closely in line with metabolic needs (Shipley *et al.* 1994). The area searched while foraging depends on step length multiplied by path width scanned, and so should theoretically scale with $M^{0.67}$. Step rate, like other basic rates, should decrease with $M^{-0.25}$ (Calder 1984). Hence, the area scanned per unit time should increase according to body mass raised to the power 0.42 (i.e. 0.67 − 0.25). Notably, the half-saturation biomass determining how steeply intake rate rises at low food availability is determined by the fraction I_P/sa (cf. Equation 3.7). If a is a constant unchanged by body size (which is questionable), the half-saturation level should be proportional to $M^{0.29}$ (i.e. 0.71 − 0.42). Accordingly, an increase in body mass by a factor of ten should almost double the half-saturation level for intake rate.

Lastly, we must establish how digestive capacity varies with body mass. Theoretically, as a volume rate it should be related to $M^{0.75}$. This relation was supported by Illius and Gordon (1992). This means that a decrease by a factor of ten in body mass should almost double daily food intake relative to body mass. However, empirical observations indicate that the relative food intake per day rises somewhat less steeply with diminishing body size, according to $M^{0.90}$ (Owen-Smith 1988; Meissner and Paulsmeier 1995). The latter relation was used in the model. Specific details of the model formulation are given in Appendix 12B.

The model enables ZGIs to be calculated for pairs of species differing by varying amounts in body mass. The body mass of the larger species was fixed at 250 kg. The body mass of the smaller species was increased from 25 kg (i.e. a mass ratio of 10) to 200 kg (mass ratio 1.25). It was assumed that the two species exhibit shared habitat preferences, and that poorer-quality habitats are both more extensive and more productive of vegetation biomass per unit area than better-quality habitats.

With a wide difference in body mass, the smaller species was restricted to about one tenth of the population density that it could have attained in isolation, but at this density it was resistant to changes in the abundance of the larger species (Fig. 12.6a). However, when the body mass ratio was diminished to about 2:1, the ZGIs closed up, and the slope of the ZGI for the smaller species became more inclined in the region of joint coexistence (Fig. 12.6b–c). This outcome implies a rather delicate equilibrium, with a small increase in the abundance of the larger species resulting in extirpation of the smaller species. The biomass that the smaller species attained in the presence of the larger competitor, relative to its biomass density in isolation, decreased sharply when the body mass ratio became less than 2:1 (Fig. 12.6d).

Although merely the outcome of the specific parameters used, the body mass ratio of greater than two required for stable coexistence is intriguingly similar to that suggested by Hutchinson (1959) for limiting morphological similarity among competitors (see also May 1973). Whether Hutchinsonian ratios have any real meaning has been contentious. Prins and Olff (1998) identified a regularity in body mass ratios among co-occurring grazers in African ungulate assemblages. I found that browsers exhibited a body mass ratio of about 2–3 among syntopic species, but could not discern a similar pattern among grazers (Owen-Smith 1989).

Ecomorphological distinctions

Other morphological distinctions besides body size differences can influence resource use patterns by mammalian herbivores. Grazers and browsers differ in dentition, digestive anatomy and physiology (Hofmann 1973, 1989), and in relative liver and salivary gland sizes (Hofmann 1989; Robbins *et al.* 1995). Among grazers, those species that prefer short grass, such as wildebeest, have relatively wide muzzles and incisor breadths compared with species of similar body mass that feed usually on somewhat taller grass (Bell 1971; Owen-Smith 1982; Gordon and Illius 1988).

In a previous analysis, I used a modified diet breadth model (Chapter 6) to evaluate the consequences of anatomical distinctions for resource gains, and hence habitat preferences, among large herbivore species (Owen-Smith 1985). Because this article appeared in an obscure publication, I will summarize some of its findings. The analysis was based on the resource gain function for multiple resource types, adapted from Owen-Smith and Novellie (1982) (cf. Equation 6.4):

$$G_F = s \sum_{i=1..r} c_i a_i F_i / (1 + s \sum_{i=1..r} a_i F_i / I_{Pi}), \qquad (12.4)$$

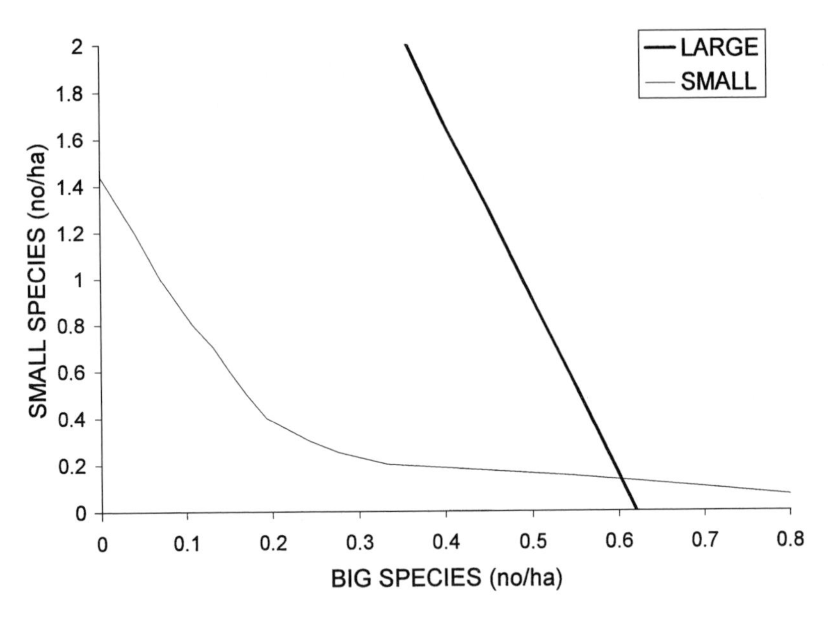

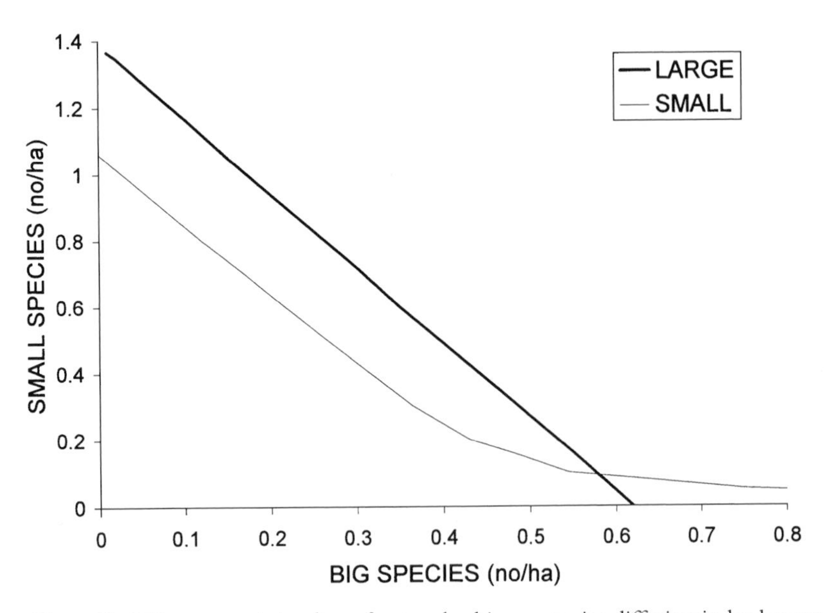

Figure 12.6. Zero growth isoclines for two herbivore species differing in body mass by different ratios, indicating the competitive impact of the bigger species on the abundance of the smaller species at equilibrium and vice versa. Body mass of the

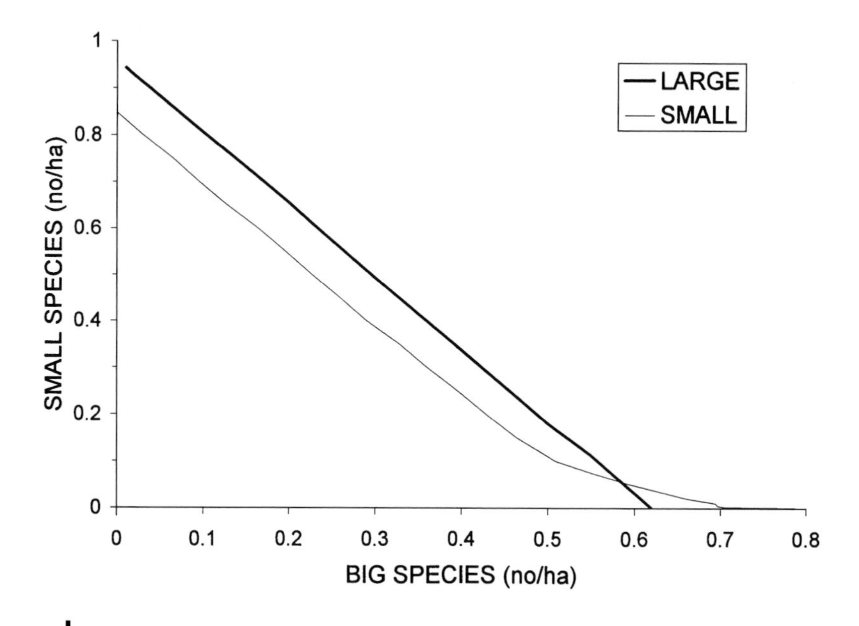

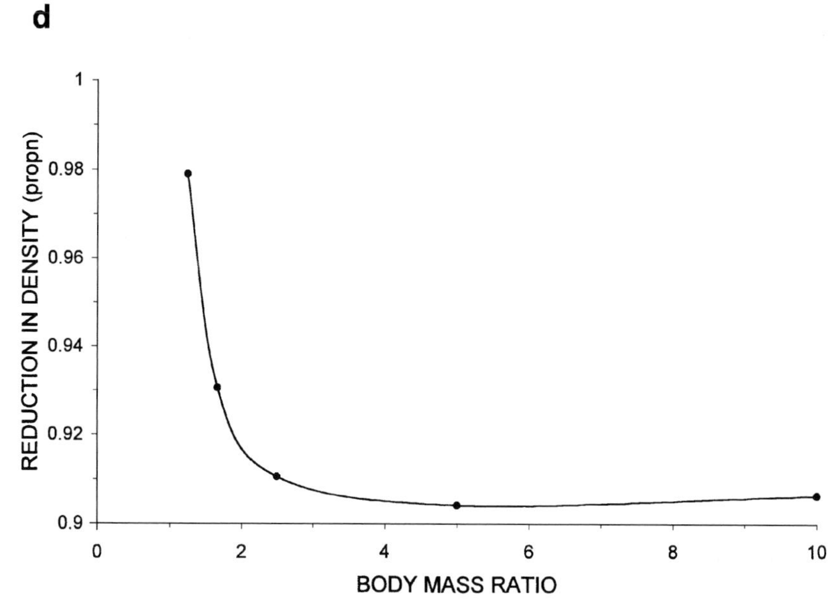

Figure 12.6. (*cont.*) big species was set at 250 kg. (a) Body mass of smaller species = 25 kg. (b) Body mass of smaller species = 100 kg. (c) Body mass of smaller species = 150 kg. (d) Corresponding reduction in equilibrium density of the smaller species as the body mass ratio is changed.

where symbols have the same meaning as in Equation 12.3, for food types labelled $i = 1..r$. To transform the rate of resource gain while foraging, G_F , into the daily gain, G_D, it must be multiplied by the proportion of the day available for foraging. In applying the model, protein and energy were considered separately as potentially limiting nutrients, relative to metabolic requirements.

The digestive capacity constraint was calculated, relative to diet quality, by using a simplified version of Equation 5.1:

$$D = d_F(k_s C_s + k_p), \qquad (12.5)$$

where $D =$ digestive processing capacity (mass per unit time), $d_F =$ mass of digesta in the main fermentation chamber (the rumen for ruminants), $C_s =$ cell solubles content of the diet, as a proportion, $k_s =$ fractional digestion rate for the cell solubles component, and $k_p =$ fractional passage rate of material from the fermentation chamber.

Food gains were converted into digestible energy yield, assuming the cell solubles fraction to be completely digestible and the cell wall component to be partly digestible. The extent of digestion was reduced according to the passage rate, following Mertens (1977). Digestible protein was calculated by subtracting metabolic faecal and urinary protein losses. The 'profit factor', for either energy or protein, was calculated as the ratio between nutrient gains and metabolic requirements, with the latter including energetic expenditures for foraging activity. Daily foraging time and other parameters were held constant. The most profitable diet was determined from the intersection between the daily intake and digestive capacity constraints. For low-quality diets, protein rather than energy sometimes became the effectively limiting nutrient.

Influence of gut anatomy

Hofmann's (1973) detailed anatomic measurements suggested that grazing ruminants have a somewhat greater rumen capacity, but slower passage rate of indigestible material from the rumen, than browsers. Through more prolonged digestion, grazers should be better able than browsers to degrade the cellulose component of the cell wall, which constitutes a higher proportion of grasses than tree leaves. Furthermore, hindgut fermenters, such as equids, have a relatively fast passage rate, reduced somewhat less by high fibre content, compared with ruminants of similar size (Bell 1971; Janis 1976).

To represent these distinctions in the model, the rumen contents mass of grazers (d_F in Equation 12.5) was assumed to be 1.5 times that of browsers. This estimate was based on measurements of rumen contents mass for wildebeest (a grazer) and kudu (a browser) reported by Giesecke and van Gylswyk (1975) under dry season conditions. Since no information on digestive passage was available, the passage rate (k_p) of a browser was estimated arbitrarily to be 1.5 times that of a grazer. A hindgut fermenter was assumed to have half the effective gut capacity, but five times the passage rate, compared with a grazing ruminant of similar size. This is obviously a vast exaggeration (Illius and Gordon 1992), but was necessary to reduce digestive efficiency sufficiently in the model. In reality, the greater digestive efficiency of ruminants depends mainly on their ability to re-masticate the digesta (chew the cud). The highest-quality food types, supposedly representing forbs and woody plant foliage, were arbitrarily excluded as potential food for grazers.

With these assumptions, the model output indicated that a medium-sized grazing ruminant such as a wildebeest should be substantially more effective at extracting energy from a grass diet than a similar-sized browser such as a kudu (Fig. 12.7a). An equivalent-sized non-ruminant, suggestively labelled 'zebra' in Fig. 12.7, should gain less energy from better-quality food types than a 'wildebeest', but be somewhat more tolerant of poor-quality forage. Accordingly, 'zebra' should be habitat generalists, accepting a wider range in grassland quality than 'wildebeest'. Nevertheless, non-ruminants must obtain a relatively high daily food intake to meet their metabolic requirements, because of their poor digestive efficiency. This need is supported by observations (Duncan *et al.* 1990).

Hofmann's functional interpretation of his anatomical observations has been contested (Gordon and Illius 1994). Additional factors might influence the dietary distinction between grazers and browsers. Notably, grazing ruminants have relatively smaller livers and salivary glands, and hence a lesser capacity for handling toxins and tannins, than browsers (Hofmann 1989; Robbins *et al.* 1995).

Influence of oral morphology

Certain grazing ruminants have wide mouths relative to their body size, whereas other grazers have mouths of intermediate width, and browsers generally have narrow muzzles (Bell 1971; Owen-Smith 1982, 1989; Gordon and Illius 1988; Perez-Barberia and Gordon 2001). Mouth width can be indexed by either muzzle width, or the breadth of the incisor

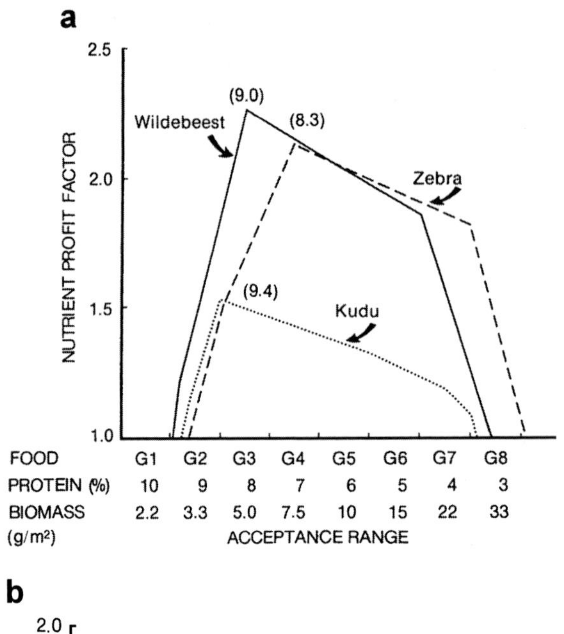

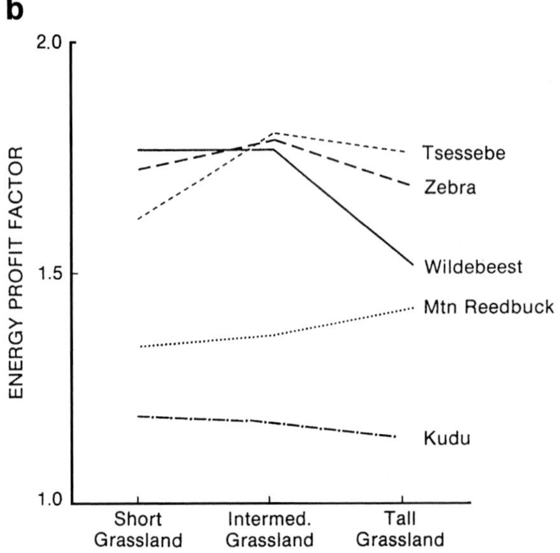

Figure 12.7. Modelled effects of morphological differences among herbivore species on nutritional gains (expressed as either energy or nutrient profit, whichever is more limiting). (a) Effects of digestive anatomy distinctions on nutritional gain relative to diet breadth for a grazing ruminant ('wildebeest'), grazing non-ruminant ('zebra') and browsing ruminant ('kudu') of similar body size. (b) Effects of differences in mouth width relative to body size on the effective energy gains obtained by modelled grazing ruminants in different grassland structures. 'Wildebeest' and 'zebra' have a relatively wide mouth, 'tsessebe' and 'kudu' a relatively narrow mouth, and 'mountain reedbuck' a small mouth due to small body size. (From Owen-Smith 1985.)

tooth row. Functionally, a relatively wide mouth should enable animals to enhance the bite mass they can prehend, especially for short grass swards where bite depth is restricted. The downside is that wider mouths may make animals less adept at separating leaf material from surrounding stem when feeding in taller grass swards. The potential for selecting specific plant parts, such as individual leaves, depends also on the absolute bite size, and hence on body size.

To represent these mechanisms in the model, a wide-mouthed 'wildebeest' was allowed a bite mass 1.5 times larger, relative to its body mass, than a narrow-muzzled 'tsessebe' (with hindsight, waterbuck would have been a better example). For each grassland habitat, the bite mass constituted entirely by green leaf was set at a certain value. If the bite mass exceeded this amount, green leaf was diluted increasingly by indigestible stem.

The model output suggested how such effects could influence foraging performance in different grassland types (Fig. 12.7b). As expected, a wide-mouthed 'wildebeest' does best in short grassland, whereas a more narrow-muzzled 'tsessebe' (or waterbuck) is most successful in medium-height grassland. A small-bodied grazer like a mountain reedbuck seems most effective at extracting nutrients from fairly tall grassland, where large leaves allow a large bite free from stem. Despite its wide mouth, a non-ruminant 'zebra' appears less influenced by differences in grassland height than a similar-sized 'wildebeest', owing to its greater digestive tolerance for fibre. A narrow-mouthed browsing kudu performs poorly in all grassland types, mostly on account of its digestive inefficiency on a high-fibre diet. Differences in grazing performance will lead to corresponding grassland habitat preferences.

Observations provide some support for this model, although not in a simple and direct way. Topi (conspecific with tsessebe) did not obtain a smaller bite size than wildebeest while grazing short grass swards, but instead had their intake rate restricted by a slower bite rate (Murray and Brown 1993; Murray and Illius 2000). Apparently topi spent extra time seeking out the bite size they needed. However, most of the trials used tufted grasses so that wildebeest did not benefit from having a wider bite. When instead a mat-forming grass species was presented, wildebeest obtained a bite size 2.6 times that of topi, although the young wildebeest used for the trials were only 25% heavier. Tall swards slowed down the biting rate of wildebeest much more than they did that of topi, possibly because the wildebeest sought green leaves but took somewhat longer to separate these from stems than did topi. Wildebeest were able to maintain a positive energy balance with a sward height of 2 cm, whereas topi required a height of more than 3 cm to satisfy their energy requirements. Hartebeest, with a similar muzzle width to topi, exhibited a smaller bite size and hence lower intake rate than topi, but appeared adept at selecting leaf material from short senescent swards. The reduced metabolic requirement of hartebeest may compensate for their lower intake rate (Murray 1993).

Based on a theoretical analysis, Wilmshurst *et al.* (2000) concluded that the grass biomass selected for feeding should increase allometrically with increasing herbivore size. This expectation was supported by observations on grazing ungulates in the Serengeti ecosystem, where Thomson's gazelle was the smallest species. In contrast, my diet breadth model suggested that 'mountain reedbuck', representing the smallest grazer, should perform best on relatively tall swards. Observations show that mountain reedbuck do indeed favour tall grassland habitats, and select tall rather than short grass species for feeding (Irby 1977; Oliver *et al.* 1978). Again, specific feeding mechanisms override allometry. Mountain reedbuck can obtain a pure leaf, and therefore high-quality, diet by selectively plucking green leaves from tall grass swards. Gazelles rely instead on the leafy structure of the short grasses that predominate on extremely fertile soils in the Serengeti region.

At the other extreme, the exceptionally wide mouths typifying the two largest grazing ungulates, white rhino and hippopotamus, enable them to feed very effectively in short grasslands (Owen-Smith 1988). Thus they can compete successfully with smaller-sized ruminants that are more efficient at digesting grass (Eltringham 1974, 1980).

Tradeoffs between facilitation and competition

Facilitation occurs where consumption by one species enhances resource gains for another species. Originally this phenomenon was highlighted in the context of a grazing succession, with herbivore species replacing one another in sequence following reductions in grass height through grazing and trampling during the early dry season (Vesey-Fitzgerald 1960; Bell 1970). In particular, through exploiting taller grassland larger species promoted shorter grass that was more accessible to smaller species. In addition, the nutritional quality of the grass re-growth following grazing can be substantially higher than that presented by taller, mature swards. Hence, grazing facilitation can also arise during the course of the growing season. For example, a substantial enhancement of energy flow to gazelles feeding in areas where grass had been grazed short by wildebeest concentrations was recorded in the Serengeti region of Tanzania (McNaughton 1976).

Facilitation may also take place between seasons. In Scotland, winter grazing by cattle improved the availability of green grass in spring for red deer, with consequent increases in the calving success of the deer (Gordon 1988). In North America, forage quality for elk during the subsequent winter was improved by cattle grazing during spring, such that the local abundance of elk increased almost four-fold (Anderson and Scherzinger 1975). However, Hobbs *et al.* (1996a,b) demonstrated experimentally that grazing by elk in winter could lead to reduced growth by cattle using these areas during spring and early summer. While there was a weak enhancing effect of elk grazing on the nutritional quality of cattle diet, at high density elk consumed over half of the standing crop of dead perennial grass, thereby reducing the amount of forage subsequently available to the cattle. However, the experimental setup, which entailed confining animals within fenced enclosures, precluded habitat separation.

Long-term facilitation may develop where one species transforms habitat conditions such that they become more broadly favourable for other species. Through opening savanna habitats via their destruction of trees, elephants improve conditions for grazers favouring open conditions, but to the detriment of browsers requiring more densely wooded vegetation (Owen-Smith 1988). By promoting short-grass grasslands at the expense of taller grassland, white rhinos may expand grazing habitats for other species favouring short grass. However, white rhinos also graze effectively on short grass, and thus compete with these species for available forage during the dry season when resources are non-renewing.

Despite widespread evidence for seasonal grazing facilitation, the expected consequences in terms of increased abundance by the species benefiting have not been observed. The rise in wildebeest numbers in the Serengeti region have not brought about an increase in the abundance of gazelles. Instead, the Thomson's gazelle population has either remained static, or declined slightly (Sinclair and Norton-Griffiths 1983; Borner *et al.* 1987). To evaluate the population consequences of seasonal facilitation in resource gains, relative to competitive impacts, the **GMM** model needs some further elaboration.

Facilitation model

To change the competition model into a facilitation model, two basic adjustments are needed. First, nutritional gains must increase as forage biomass is decreased, for the species benefiting from the grass height reduction. Secondly, the temporal frame must be expanded to cover the full annual cycle, so that gains in one season can be balanced against consequences at other times of the year.

Accordingly, the gain response functions for alternative habitats species shown in Fig. 12.3 were changed for the smaller herbivore species into the form depicted in Fig. 12.8a. Parameter values were adjusted such

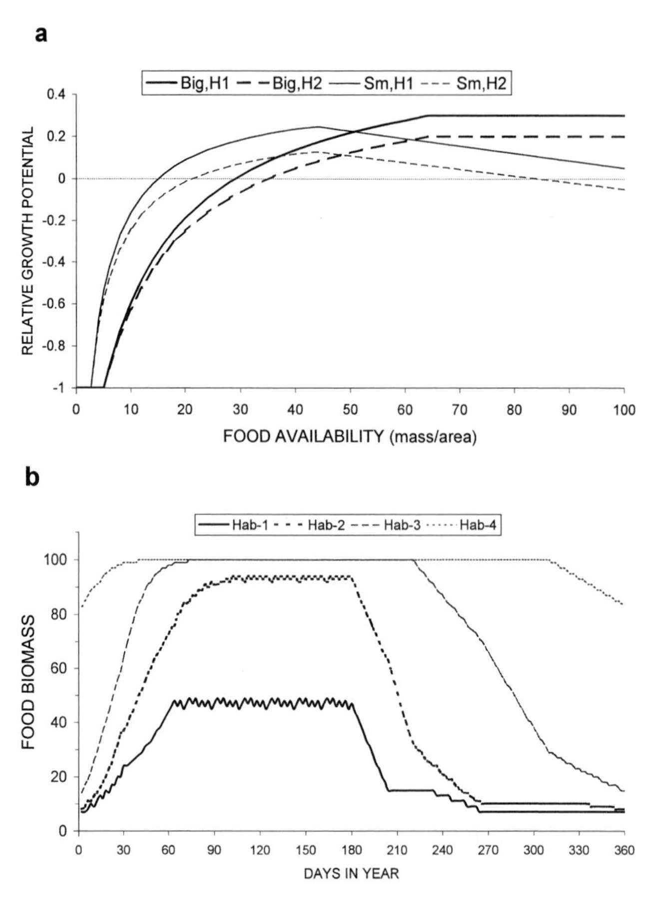

Figure 12.8. Assumed resource gain functions and output from the facilitation model. (a) Comparative gain functions for big and small herbivore species from two habitats differing in resource quality, with gain by the small species declining with increasing forage biomass above a threshold level. (b) Annual regime of resource availability in the four habitats corresponding with situation of joint population stability. (c) Annual variation in relative growth potential of the small herbivore species for situation of joint population stability, compared with that exhibited in isolation at near-zero density. (d) Potential initial rate of population increase by the small herbivore species in relation to increases in the biomass density of the big species.

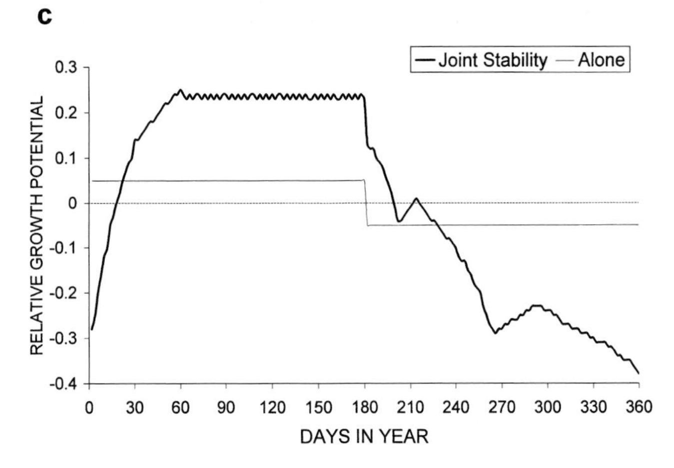

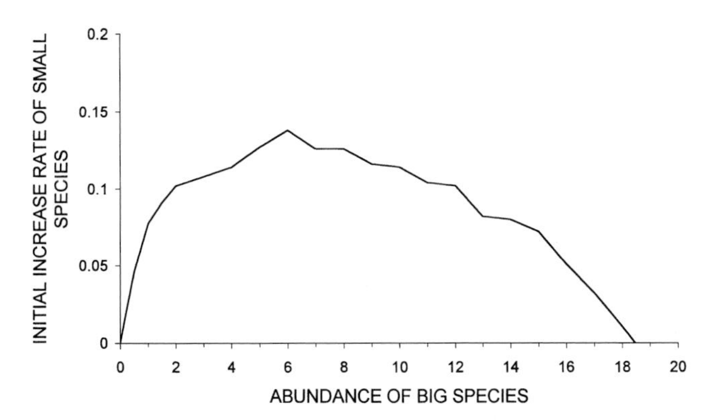

Figure 12.8. (cont.)

that the growth potential of the small species was only marginally positive at maximum forage biomass levels, but increased substantially to peak at around 40% of this maximum. Furthermore, the relative rates of resource gain obtained from alternative habitats were set for the big species such that the latter switched from the best habitat to the next-best at about the resource level where the small species achieved highest gains. Thus these conditions represented strong seasonal facilitation. Notably, in the second-best habitat the small species obtained a negative net resource gain at high forage biomass levels, but could survive and grow if forage biomass was reduced sufficiently. The model also included a seasonal reduction in food quality, to prevent the small species from performing better in the dormant season than during the growing season. As a result, the small species also obtained a negative resource gain from the best habitat during the DS, at high forage biomass levels. The seasonal alternation was represented by a stepwise transition, with forage biomass increasing logistically during the 180 day growing season (GS), and decaying solely through consumption during the 180 day DS. To prevent the grass biomass from being reduced too close to zero during the DS, thereby suppressing re-growth during the following growing season, an ungrazable reserve was specified, as indicated in Fig. 12.8a. Further details of the model formulation are given in Appendix 12C.

Model output

From the model, zero growth isoclines were obtained for both herbivore species, as in the competition model. Firstly, a situation was considered in which habitats offering higher food quality were less extensive than those presenting poorer-quality resources. The model output showed that increasing the abundance of the big species enhanced energy flow to the small species during the GS, by drawing foraging biomass down to the level where the small species achieved maximum gains. Subsequently during the DS, forage biomass was reduced progressively through the combined feeding impacts of both herbivore species. Despite the seasonal facilitation, the abundance of the small species was not elevated above the maximum population that it could have obtained in isolation (Fig. 12.9a). All that happened was that the competitive impact of the big species was ameliorated. At joint stability, where the isoclines crossed, the abundance of the small species was reduced by only about 20% below its potential maximum. The α coefficient for the biomass-specific impact of the big species on the abundance of the small species was only 0.13, compared with 0.40 for the competition-only situation depicted in Fig. 12.5. The impact of the small species on the abundance of the big species remained strong, because it consumed much of the high-quality food that the big species would otherwise have been able to eat in the best habitat. However, despite the α coefficient for this interaction being substantially greater than 1.0, coexistence occurred.

Notably, where facilitation occurs the combined biomass of the two herbivore species at the point of joint stability exceeds the population biomass of the big species supported in isolation (Fig. 12.9a). This was not the case for the purely competitive situation depicted in Fig. 12.5. Of course, other mechanisms could also elevate the productivity of multi-species assemblages, e.g. complementary resource use between grazers and browsers.

a

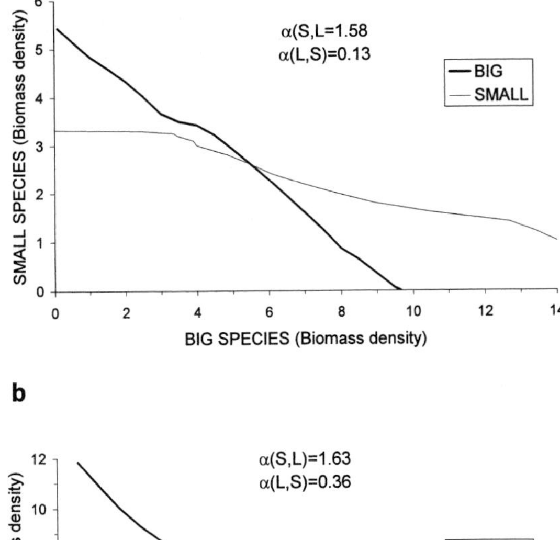

b

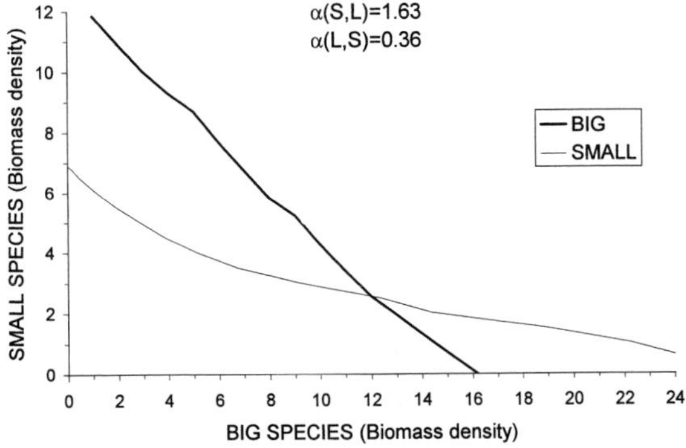

c

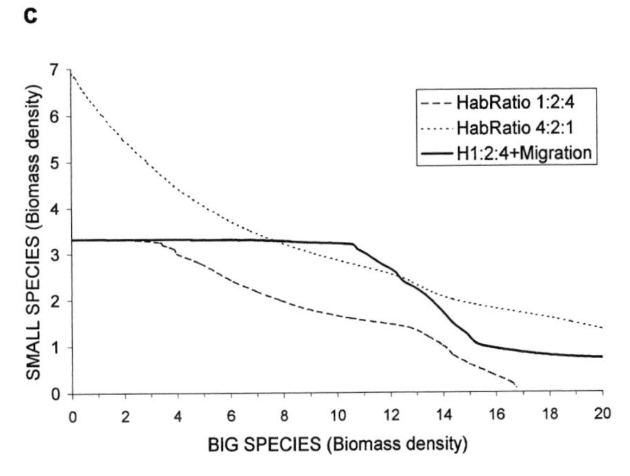

Figure 12.9. Zero growth isoclines derived from output of the facilitation model.(a) Poorer-quality habitats are more extensive (1:2:4 ratio). (b) Better-quality habitats are more extensive (4:2:1 ratio). (c) Isoclines for small species for situation where the big species migrates out of the area during the dormant season, compared with basic non-migratory situations for different habitat ratios.

Figure 12.8b shows the annual forage biomass regimes in the four habitats for the population levels where both herbivore types were jointly stable. This reveals how forage biomass in the best habitat remained at the optimum level for the small species through most of the GS, but became reduced to low levels through the grazing impact of the small species during the DS. Forage availability in the second-best habitat was likewise greatly depressed during the course of the DS, forcing the big species to rely more on the reserve and buffer habitats. Notably, the effect of the reduced forage biomass persisted through the early part of the growing season. At sufficiently high population levels, the small herbivore completely suppressed grass re-growth during the early GS. Such 'sward capture' made the best habitat effectively unavailable to the big herbivore species. Less extreme sward capture occurred even at the population levels conferring joint stability.

The strong enhancement of nutritional gains by the small species during the GS was counterbalanced by the negative deficits that it incurred during the DS, compared with what it might have obtained at low density in isolation (Fig. 12.8c). The higher the density of the big species, the less the forage biomass carried through into the DS. Thereby grazing facilitation during the one season becomes counteracted by competition consequent upon resource depletion during the DS. Moreover, during the DS forage biomass did not remain for long at the level favoured by the small species, because of the progressive depression of food availability.

Altering the environment such that the best-quality habitat was most extensive raised the potential abundance of the small species in isolation, but not at joint stability (Fig. 12.9b). Improving conditions in this way merely intensified the competitive impact of the big species, which also favoured the best habitat. Thus potentially favourable changes in habitat composition, which may develop over longer time periods as a result of the grazing impacts of the big species, may also not lead to population-level facilitation. Raising the quality of the best habitat, perhaps through enhanced nutrient cycling, would simply cause the big herbivore to delay its switch to the second-best habitat until resources in the best habitat had been reduced to a lower level, thereby also intensifying competition.

In the Serengeti region, the competitive impact of the vast wildebeest population is suppressed somewhat because these animals migrate away from the short grass plains before the end of the wet season, leaving remaining grass resources entirely to the gazelles and a few other resident herbivores. However, in the model, even entirely eliminating DS competition from the big species failed to generate a facilitatory increase in the population level attained by the small species. All that happened was that the competitive impact of the big species was alleviated over a wider range in its abundance (Fig. 12.9c).

For the specified parameter values, the population growth potential of the small species at low density was effectively zero in the absence of the big species, because of its inefficiency in exploiting high-forage-biomass conditions. Increases in the abundance of the big species improve forage availability for the small species, enabling it to achieve a positive population growth (Fig. 12.8d). From this perspective, the big species facilitated the small species by enabling it to invade an environment where it could not initially establish a viable population. Following initial establishment, the population of the small species could maintain itself through its own grazing impacts on vegetation biomass.

Overall assessment

The modelling exercise suggests why population-level consequences rarely follow from feeding facilitation, which is invariably restricted seasonally in its occurrence. Competition is pervasive, wherever species overlap in resource use. The greater the share of resources consumed by one species, the less remains to support the population of another species through the critical period when resources are non-renewing. Nevertheless, seasonal facilitation of foraging gains can ameliorate the population-level effects of competition. This could be manifested either through improved reproductive success, or via greater body reserves carried through into the dormant season. Supported by heterogeneity in available resource types, such facilitatory interactions help promote coexistence among species differing in body size and other features, and may enhance overall secondary production. A population-level facilitation may occur transiently, where the grazing impacts of one species enable another to invade a region where it could not establish successfully in isolation.

Overview

Modelling emphasized the overwhelming importance of habitat heterogeneity for species coexistence in seasonally variable environments. This was revealed only because allowance was made for adaptive shifts in the habitats occupied by particular species at different stages of the seasonal cycle. With competition expressed primarily via resource depression, herbivore species may overlap in habitat use during the growing season, but segregate into distinct habitats during the course of the dormant season when resources become reduced in abundance and quality. Consequences of distinctions in body mass and other morphological features for foraging performance contribute towards coexistence, and may cause species to favour different habitats.

Smaller species are potentially superior competitors within high-quality habitats, through their ability to survive on sparse resources. Larger species are better able to satisfy their nutritional needs in poor-quality habitats, provided resources are sufficiently abundant. Through being able to utilize a wider quality range of resources, larger species build up higher abundances than smaller species, and may competitively overwhelm the latter through the weight of their consumption. The relative competitive impact of large and small species may be influenced by the habitat distribution in the environment. Competition coefficients defined in terms of competitive impacts per biomass unit do not represent the density reduction that can result from the competitive interaction.

Grazing facilitation through enhanced resource access by smaller species, consequent upon grass height reduction or expansion in the area covered by high-quality habitats, may alleviate competition. Enhancements of densities above those reached in isolation are unlikely to occur, because gains during one season are counteracted by competitive resource depletion at a later stage in the seasonal progression. Other improvements in habitat conditions that benefit both species merely intensify competition.

Habitat distinctions in security against predation, and in physiological stress, were not considered in this chapter. These could readily be incorporated within the **GMM** model, to explain additional aspects of niche separation.

Further reading

Rosenzweig and Abramsky (1997) review applications of isoleg analysis to competitive habitat occupation by rodents. Broad reviews of factors influencing competition and coexistence in African ungulate assemblages are given by Murray and Illius (1996) and Prins and Olff (1998). Schmidt *et al.* (2000) outline a model of competitive habitat selection under temporal heterogeneity, focussed on broader-scale distributions between source and sink habitats with rodents in mind. Ziv (2000) outlines an alternative model suggesting how habitat specificity scales with body size.

Appendix 12A. Habitat competition model based on resource gain functions

The model is a TrueBASIC program. The herbivore species are defined by the parameter values presented in Table 12A(a), for the assumed habitat parameters given in Table 12A(b). A daily iteration was used. On each day the habitat type currently yielding the highest nutritional gains was identified, for each herbivore species in turn. Resource gains were calculated from either the intake rate function, adjusted for daily

Table 12A. *Herbivore and habitat parameters used for the basic habitat competition model*

(a) *Herbivore parameters (in arbitrary units)*

	Herbivore	
Parameter	big	small
Relative intake rate I (mass/time)	7.5	10
Relative search rate s (area/time)	0.5	2.5
Daily feeding time t_F (proportion)	0.5	0.4
Digestive processing capacity (% of biomass per day)	3.0	3.8
Digestive efficiency (proportion)	1.0	0.82
Relative metabolic requirements M_P (proportion)	1.5	1.5
Maximum daily metabolic gain G (proportion)	1.3	1.4

(b) *Habitat parameters*

	Habitat			
Parameter	(1)	(2)	(3)	(4)
Relative area extent	1	2	4	8
Relative nutritional quality c (proportion)	0.7	0.6	0.5	0.4
Maximum biomass production	100	100	100	100

foraging time, or the digestion capacity constant, whichever was more limiting. An upper limit to the maximum biomass gain per day, relative to metabolic requirements, was also set. The amount consumed was subtracted from the previous food biomass in the selected habitat, adjusted for relative habitat area. No resource renewal was allowed, nor any vegetation attrition besides that due to consumption. The daily nutritional gains relative to metabolic requirements were then averaged over the 180 day dry season. Gains were converted to relative growth potential following Equation 12.2. It was assumed that zero net gain over the dry season represented zero net population growth.

Appendix 12B. Habitat competition model based on allometric relations

This TrueBASIC program is an elaboration of the above model, with parameters represented as functions of body mass rather than as constants. The allometric equations, and representative values for two herbivore species weighing 250 and 25 kg, respectively, are given in Table 12B.

Table 12B. *Allometric equations and representative values for herbivore parameters used in the habitat competition model based on relative body mass*

		Herbivore body mass (kg)	
Parameter	Equation	250	25
Maximum intake rate I (g/min)	$0.35\,W^{0.71}$	17.6	3.4
Search rate × acceptance fraction sa (m^2/min)	$0.08\,W^{0.42}$	0.81	0.31
Daily feeding time t_F (proportion)		0.4	0.4
Digestive processing capacity (kg/day)	$0.05\,W^{0.88}$	6.44	0.85
Basal metabolic requirement M_{P0} (metabol. DM kg/day)	$0.3\,W^{0.75}/15$	1.26	0.22
Active metabolic requirement M_P (metabol. DM kg/day)	1.8 × basal	2.27	0.40
Maximum metabolic processing capacity	multiple of active requirement	1.4 ×	1.5 ×
Mortality function intercept (z_0)		− 1.0	− 1.0
Mortality function slope (z)		1.3	1.3

Allowing an ungrazable food biomass of 5 g/m^2, the half-saturation level was 25 g/m^2 for the large species and 16 g/m^2 for the small species.

Daily resource gains may be truncated by either metabolic processing capacity when food quality is high, or digestive processing capacity when food quality is low. The potentially reduced digestive efficiency of the smaller herbivore is assumed to be counterbalanced by increased dietary selectivity, and so is ignored. The mortality function was set so that zero growth potential over the dry season was associated with a nutritionally dependent mortality rate of about 0.35 per year. Averaged over the whole year, and allowing additive mortality due to predation or senescence of about 0.05 per year, this would just counterbalance a biomass growth potential of 0.225 per year. This level of resource gain was assumed to give zero annual population growth. Habitat parameters were as shown in Table 12A, except that resource production was assumed to vary negatively with resource quality. The standing biomass in each habitat at the start of the dry season was assumed to be 50, 75, 100 and 150 g/m^2, respectively.

Appendix 12C. Facilitation model

The starting basis was the TrueBASIC model outline in Appendix 12A. To generate a humped resource gain function most simply for the small species, the digestive capacity constraint was made a linearly declining function of increasing forage biomass, as depicted in Fig. 12.8a. The time frame was also extended to encompass the full annual cycle, with vegetation biomass accumulating logistically during the 180 day growing season, then declining through consumption alone through the 180-day dormant season. Iteration was at 2 day intervals. To generate the carryover effect of vegetation remaining at the end of the dormant season into the start of the next growing season, the model was run over two annual cycles, with only the output from the second year used. Incorporation of an ungrazable reserve biomass helped suppress the tendency for vegetation to be held at its low starting biomass through the growing season. Small adjustments were made in parameter values for the small herbivore species: (i) search rate s reduced from 2.5 to 2.0, to elevate the half-saturation biomass slightly; (ii) digestive capacity increased from 3.8 to 4.0, but reduced by 0.25 per increment of 100 units in food availability; (iii) digestive efficiency of the small herbivore increased from 0.82 to 0.875. The ungrazable reserve was set at 2.5 units of forage biomass for the small herbivore and at 5 units for the large herbivore. The food quality of the best habitat, in terms of the conversion coefficient c, was reduced from 0.7 to 0.65. Furthermore, food quality c for all habitats was lowered by an amount of 0.05 during the dry season.

13 · *Population dynamics: resource basis for instability*

Populations of large mammalian herbivores are notorious for their propensity to 'irrupt' to high abundance, followed by a crash to lower density in a degraded environment (Caughley 1970, 1976a; McCullough 1997). The archetypal example is provided by mule deer inhabiting the Kaibab region of Arizona (Rasmussen 1941). Better-documented cases exist for reindeer occupying islands off Alaska (Scheffer 1951; Klein 1968), and for white-tailed deer in mainland as well as island situations in North America (McCullough 1997). Persistent oscillations in abundance have been shown by feral Soay sheep inhabiting islands in the Hebrides off Scotland (Clutton-Brock *et al.* 1991, 1997; Grenfell *et al.* 1992, 1998). The moose population occupying Isle Royale in Lake Superior has periodically increased to high abundance, followed by substantial dieoffs, despite the arrival of wolves as predators (Peterson 1999). Several ungulate species have exhibited severe population crashes in a large private nature reserve in South Africa, precipitated by drought conditions in a situation where water supplies had been augmented (Walker *et al.* 1987). Drought-related dieoffs have been widely documented for free-ranging livestock, especially in communal grazing systems (McCabe 1987; Scoones 1993; Hatch and Stafford Smith 1997).

However, most ungulate populations show no more than minor fluctuations in abundance from year to year, although long-term trends or even cycles may be apparent in association with changes in vegetation, climate or other environmental features. The key feature distinguishing unstable dynamics is periodic severe mortality reducing population abundance by half or more within a brief period. For this to occur, mortality must encompass not only the vulnerable young and old, but also prime-aged adults. As a result, several years elapse before the population regains its former abundance.

Elimination of predator regulation is widely invoked as promoting population instability, justifying routine culling of herbivore populations in protected areas where predators are lacking (Brooks and Macdonald

1983; Huff and Varley 1999). This perception has led further to the widespread culling of megaherbivore populations, i.e. species including elephants, hippos and rhinos that are subject to little natural predation, even in vast national parks (Laws and Parker 1968; de Vos *et al.* 1983; Martin *et al.* 1989; Whyte *et al.* 1998). McCullough (1997) identified habitat improvements such as supplementary feeding, and confinement within restricted areas, as additional factors pre-disposing ungulate populations to irruptions and crashes. Nevertheless, some ungulate populations remain relatively stable even when such conditions pertain.

The propensity of large herbivore populations towards unstable dynamics was captured by Caughley (1976a) in his classical model of interactive herbivore–vegetation systems. Rosenzweig (1971) identified further a 'paradox of enrichment' in such models, whereby an increase in the productive potential of the resource destabilizes consumer dynamics (see also Crawley 1983). These concepts provide the starting foundation for this chapter. We explore further how features of the resource base might promote or suppress tendencies towards instability in herbivore populations, in environments that are spatially and temporally variable. These concepts are then applied to a specific example that has been studied in great detail, the dynamics of the Soay sheep population inhabiting Hirta Island in the St Kilda archipelago. Finally, concepts and modelling approaches are generalized to apply to other species in other contexts.

Interactive herbivore–vegetation dynamics

Basic model

Caughley's (1976a) model was deliberately vague about the time frames represented. In particular, he did not distinguish between the annual production of vegetation components feeding the herbivore population, and the longer-term dynamics of plant populations in response to herbivore impacts, although the latter was implied. However, it is the grazing impacts of herbivores over the seasonal cycle that may cause vegetation biomass to expand or contract between years, with consequent effects on herbivore population dynamics. A logistic production function for vegetation, as was used by Caughley, is most appropriate for within-season growth dynamics (see also Roughgarden 1997). How vegetation abundance and composition changes between years may involve somewhat different processes, which we will not model directly. The model will thus address how the annual growth dynamics of vegetation with a particular composition influences the stability of a herbivore population dependent on it and influencing vegetation growth.

Following Equation 8.1, we have

$$dV/dT = g_V V(1 - V/v_{max}), \qquad (13.1)$$

where V = vegetation biomass, g_V = maximum relative growth rate of vegetation, and v_{max} = maximum standing biomass that the vegetation can attain. Strictly, V represents only the above-ground component of vegetation, available for consumption and driving biomass increase through photosynthesis.

The rate of consumption of vegetation biomass by herbivores is assumed to follow an asymptotically saturating Michaelis–Menten function, i.e.

$$I = i_m F/(f_{1/2} + F), \qquad (13.2)$$

where I = relative intake rate, i_m = maximum potential intake rate, F = forage biomass, and $f_{1/2}$ = half-saturation biomass, i.e. the forage biomass at which intake rate drops to half of its maximum. Notably, the forage biomass is somewhat less than the vegetation biomass because not all vegetation is available for consumption. Specifically, $F = V - v_u$, where v_u is the ungrazable vegetation biomass at which I becomes zero (following Noy-Meir 1975).

Following Equation 2.28, changes in herbivore biomass, H, result from the difference between the relative rates of biomass gained from food consumed, G, and biomass lost to metabolic attrition, M_P, and mortality, M_Q :

$$dH/dT = (G - M_P - M_Q)H. \qquad (13.3)$$

Like the intake rate I, the relative rate of resource gain G is assumed to be hyperbolically saturating, modified by the conversion coefficient from vegetation into herbivore biomass c:

$$G = c i_m \{F/(g_{1/2} + F)\}. \qquad (13.4)$$

Notably, the half-saturation level for herbivore biomass gain, $g_{1/2}$, may deviate from that controlling intake rate, $f_{1/2}$, depending on how food quality varies with changing food abundance. Effectively, c then represents the maximum conversion coefficient at high forage biomass, and the relationship between $g_{1/2}$ and $f_{1/2}$ indicates how rapidly forage quality diminishes as food biomass is decreased. This effect could be represented more explicitly by partitioning the forage biomass F into components differing in quality, as will indeed be done as a later elaboration.

For simplicity, the metabolic biomass loss, M_P, will be assumed constant, i.e. $M_P = p$ where p = relative rate of physiological attrition due to basal metabolism, activity costs and other metabolic processes. Mortality losses include deaths through senescence or additive predation (Chapter 10), as well as mortality as a result of nutritional deficiencies (including interactive predation). Following Equation 10.1, the expected loss in biomass through mortality is inversely dependent on resource gains, relative to physiological costs. Hence

$$M_Q = z_0 + zp/G, \tag{13.5}$$

where z_0 and z are curve-fitting parameters determining how sharply mortality rises with diminishing resource gains. However, mortality losses cannot fall below some minimum level q_0 set by inevitable deaths through senescence, additive predation and accidents.

Parameter values were chosen to be roughly representative of a cow-sized herbivore grazing a grassland (Table 13.1).

Non-seasonal conditions

As a control baseline, the model was applied initially to a non-seasonal environment with continual vegetation growth. Iteration was weekly. With other parameters at their standard settings, a value for c of 0.65 was needed to enable the herbivore population to increase. Increasing c slightly to 0.67 gave a 28% annual rate of increase in herbivore biomass; raising c to 0.70 increased the potential herbivore growth rate to 76%. With $c = 0.75$, herbivore biomass could almost triple over one year. The sensitivity of the inherent growth rate of the herbivore population to slight changes in the resource conversion coefficient reflects the compounding effect of small biomass gains iterated weekly.

Because of the steeply saturating form of the intake response, the model generated unstable dynamics even for $c = 0.67$, when $g_{1/2}$ was equated with $f_{1/2}$ and equal to $v_{max}/20$ (Fig. 13.1a). In reality food quality c is likely to decline as vegetation biomass becomes depleted, because consumers have less opportunity to be selective when food is sparse. This was represented by setting $g_{1/2}$ to be $1.5 \times f_{1/2}$. In effect, food quality was lowered by a factor of 0.8 when $F = f_{1/2}$, which seems reasonable. With this adjustment, the model generated a smooth approach towards

Table 13.1. *Parameter values used for the herbivore–vegetation model*

Parameter	Value	Justification
v_{max}	500 g/m^2 (dry mass)	Typical peak standing crop biomass for regions with rainfall 500–1000 mm
g_V	0.2 per week	Range of variation for grasses is 0.1–0.8 per week, but high values probably unsustainable over a season
v_u	10 g/m^2	Small residue of leaves able to generate growth but inaccessible to herbivores
d_g	0.01 per week	Upper value for death rate of leaves of grasses
i_m	2.5% of live mass per day or 17.5% per week	Representative upper limit for cattle with body mass 400–500 kg
$f_{1/2}$	$0.05 \times V_{max}$	Steeply saturating intake response
$g_{1/2}$	$1.5 \times f_{1/2}$	Represents decline in food quality by factor of 0.8 when $V = f_{1/2}$
p	1.42% of live mass per day or 10% per week	Basal metabolism = 30 MJ per day for 450 kg cow, and energy content of mass lost = 25 kJ per g dry mass, or 8 kg per g live mass, giving mass loss 0.8% per day; maintenance metabolism = 1.77 × basal metabolism
c	range 0.5–1.2	Represents outcome of digestion, assimilation and growth efficiency, plus transformation from plant dry mass to animal live mass; dormant season c = 0.9 × growing season c
q_0	0.07 per year	minimum mortality for potential longevity of 15 years
z_0	-1.25	adjusted relative to z so that nutritionally related mortality = 25% per year when G = maintenance requirement, and 0% per year when $G = 1.2 \times$ requirement
z	1.5	—

equilibrium for $c = 0.67$, but not for higher values of c (Fig. 13.1b, c). For $c = 0.75$, a tripling of $g_{1/2}$ relative to $f_{1/2}$ was necessary to achieve an equilibrium herbivore biomass approached with only minor ripples (Fig. 13.1d). This represents a decline in effective food quality by almost half when $F = f_{1/2}$.

High values for c generate wildly unrealistic rates of growth in herbivore biomass. The maximum rate of growth by a cohort of young growing steers is about a 2–3-fold increase in body mass over the course of a year (see Chapter 7). Accordingly, an upper limit to herbivore growth was set at 1.65% per day, equivalent to 12% per week, or a 2.6-fold increase over one year. The demographic limit for a population of ungulates,

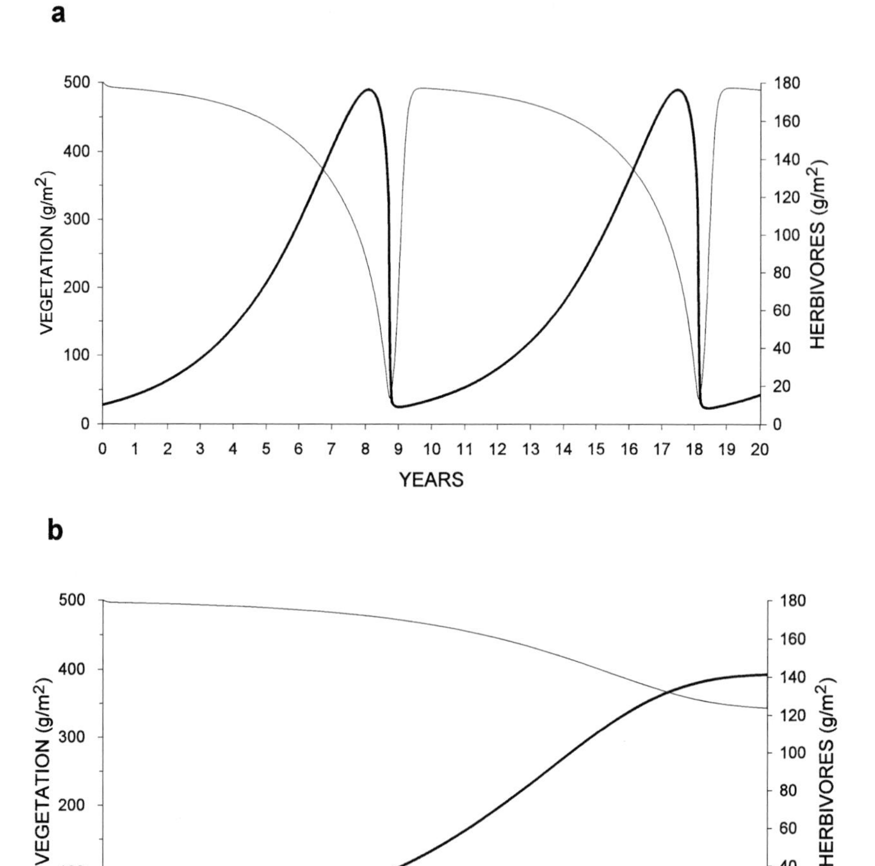

Figure 13.1. Output of the **GMM** model for non-seasonal conditions, showing effects of adjustments in key parameters (see Table 13.1). (a) Output for $g_{1/2} = f_{1/2} = 25$ g/m^2, with forage quality $c = 0.67$. Extinction is prevented by an ungrazable vegetation reserve. (b) Output for conditions as in (a), but with $g_{1/2}$ increased to 37.5 g/m^2. (c) Output for conditions as in (b), but with c increased to 0.7. (d) Output with $c = 0.75$ and $g_{1/2}$ increased to 75 g/m^2.

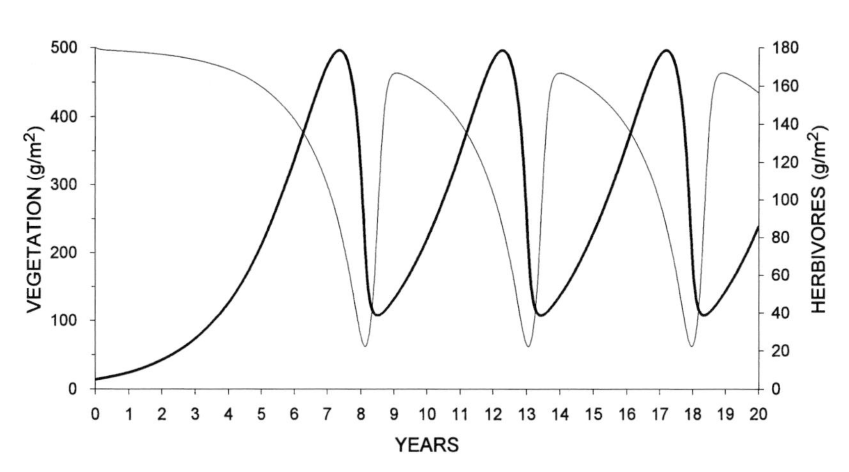

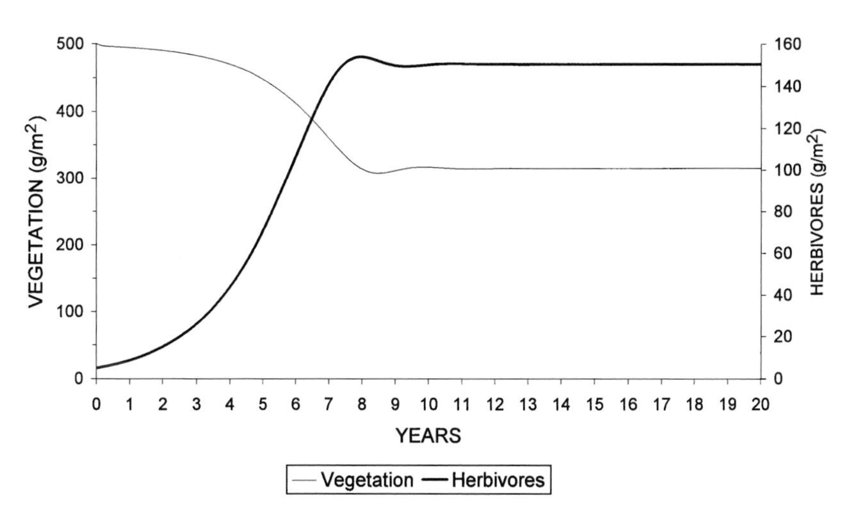

Figure 13.1. (*cont.*)

made up of non-growing adults as well as immature animals, would be a maximum increase rate of about 25% per year, unless more than one offspring is produced annually.

The herbivore biomass generated by the non-seasonal model is also vastly unrealistic, despite seemingly representative parameter values. A biomass amounting to over 140 g/m^2 (140 000 kg/km^2) is more than seven times that recorded in the most productive natural ecosystem, Manyara National Park in Tanzania, and over ten times that supported in the Serengeti region of Tanzania (Owen-Smith 1988; Prins 1996; see also Fritz and Duncan 1994). In livestock terms, it represents 3 animal units per hectare (where 1 animal unit is equal to a 454 kg steer). Assuming a maximum vegetation growth rate of 0.2 per week, this high level of grazing pressure was needed to depress vegetation biomass sufficiently to reduce the net gain in herbivore biomass to zero. Lowering the inherent growth rate of the vegetation to 0.01 per week, to correspond with the annual growth rate of 0.8 assumed by Caughley (1976a), generated a herbivore biomass of the correct magnitude, although the interaction became unstable. However, grass biomass can easily increase at rates of 20% or more per week, as documented in Chapter 8.

Seasonal conditions with a homogeneous food resource

To represent seasonal conditions, a simple seasonal alternation was assumed. Vegetation growth occurred continually during the 26 weeks of the growing season (GS), then ceased during the 26 weeks of the dormant season (DS). Allowance was made additionally for (a) inherent decay in standing vegetation biomass during the course of the DS, and (b) a decline in vegetation quality during the DS (see Table 13.1). The vegetation biomass that remained at the end of the DS generated new growth at the beginning of the next GS. No distinction was made between necromass and biomass (cf. Chapter 8). Hence Equation 13.1 becomes

$$dV/dT = g_V V(1 - V/v_{max}) - v_d V, \qquad (13.6)$$

where v_d = proportional rate of vegetation decay, and g_V is zero for weeks 27–52.

In a seasonal environment, with allowance for a 10% drop in vegetation quality during the DS, a basic value for c of 0.7 generated only a 14% annual growth in herbivore biomass, because growth by 31% during the GS was offset by a DS decline caused by the lowered food quality. An increase in c to 0.72 was necessary to elevate the annual growth in herbivore biomass to 40%. There was a smooth approach towards equilibrium, apart from the seasonal oscillation, and despite the fact that vegetation biomass was reduced below half of its potential maximum before the end of the DS (Fig. 13.2a). With c increased to 0.77, the herbivore population fluctuated

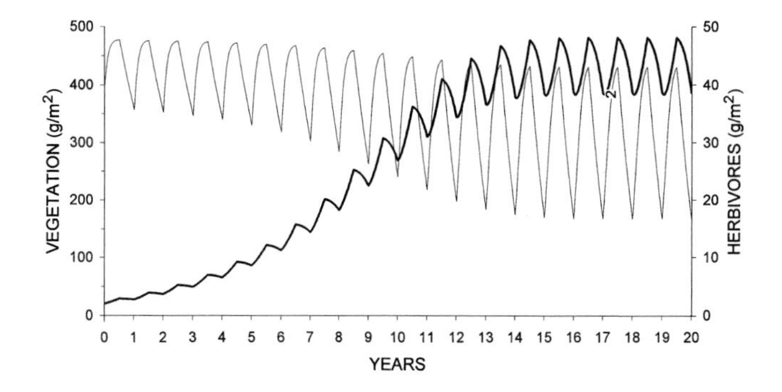

b

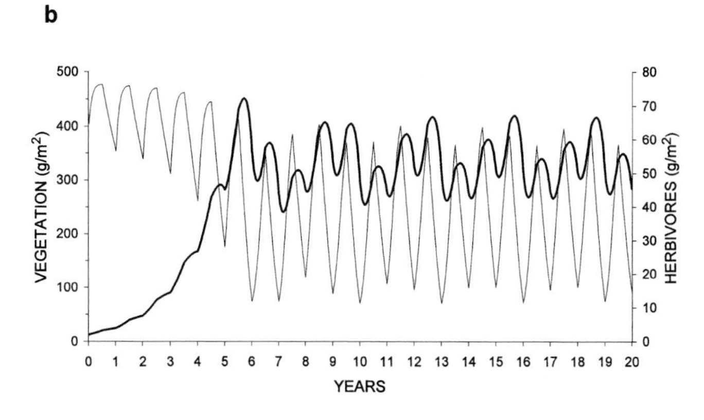

c

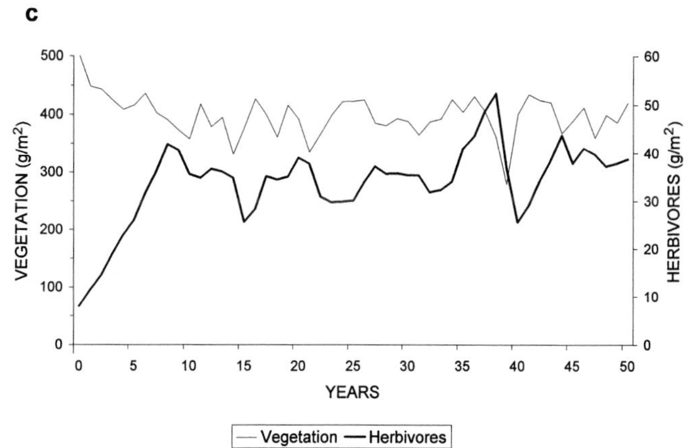

Figure 13.2. Output of the **GMM** model for seasonal vegetation growth, with value of forage quality c varied and other parameters as specified in Table 13.1. (a) Output showing weekly dynamics for $c = 0.72$. (b) Output showing weekly dynamics for $c = 0.77$. (c) Output showing annual dynamics for $c = 0.72$, with stochastically varying rainfall (weekly probability $= 0.67$ during the growing season).

unstably from year to year, but within a limited range (Fig. 13.2b). With food quality set this high, the herbivore population continued growing even through the DS, and more than doubled in biomass over the year. Raising c further to 0.8 created chaotic oscillations in herbivore biomass, with vegetation biomass reduced periodically towards little more than the ungrazable reserve.

In practice rainfall, and other conditions controlling vegetation growth, may vary erratically during the course of the GS. This was represented in the model by making weekly rainfall a stochastic variable, following a one–zero pattern generated by a specified probability level. Vegetation growth was zero if rainfall did not occur during that week. The probability of rain during a DS week was assumed to be one tenth of that for GS weeks. The stability properties of the model with $c = 0.72$ were retained under conditions of stochastically varying rainfall. The mean herbivore biomass of about 35 g/m^2 generated in these circumstances approached realistic values (Fig. 13.2c).

Seasonal conditions with heterogeneous food resources

Resource heterogeneity was represented by partitioning the vegetation among a set of patch types, distinguished by the effective food quality c that they offered and by other features. Each week, herbivores selectively occupied the patch type yielding highest gain, as indexed by G. As initially preferred patches became depleted, herbivores shifted to other patch types, especially during the course of the DS.

With a heterogeneous forage resource, a stable equilibrium was reached even when c exceeded 0.75 in the best patch type. The stability of a system with two patch types, one offering a high c of 0.77 and the other a sub-maintenance c of 0.67, was similar to that for a homogeneous c of 0.72 (cf. Figs. 13.2a and 13.3a). However, at low density the herbivore population grew almost three times as fast in the former situation as in the latter, by selectively concentrating in the best patches (compare Fig. 13.4a and b). There was a corresponding doubling in the maximum growth increment of the herbivore population, affecting the maximum yield that could be obtained by harvesting. In other words, productivity was enhanced, without stability being compromised. Expanding the system to three patches, spanning a wider range in food quality, maintained the stability properties, while increasing the productive potential further (Figs. 13.3b, 13.4c).

As a general rule, plants of higher quality are likely to grow faster, but sustain a lower saturation biomass, than species of lower nutritional value

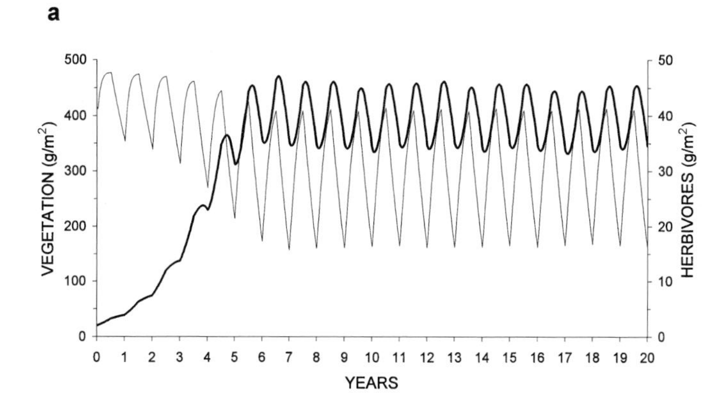

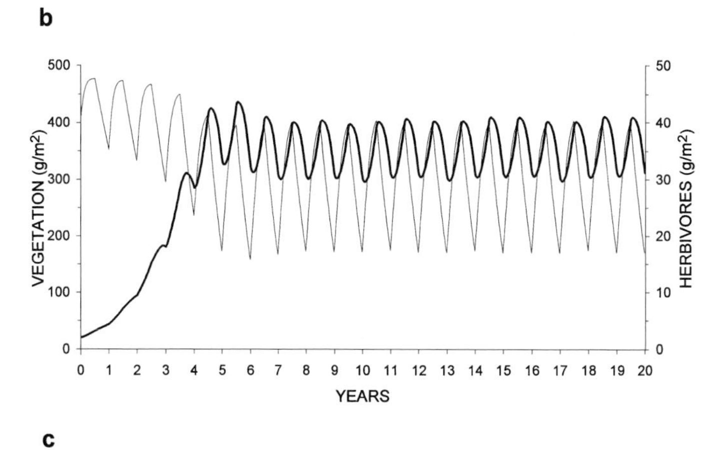

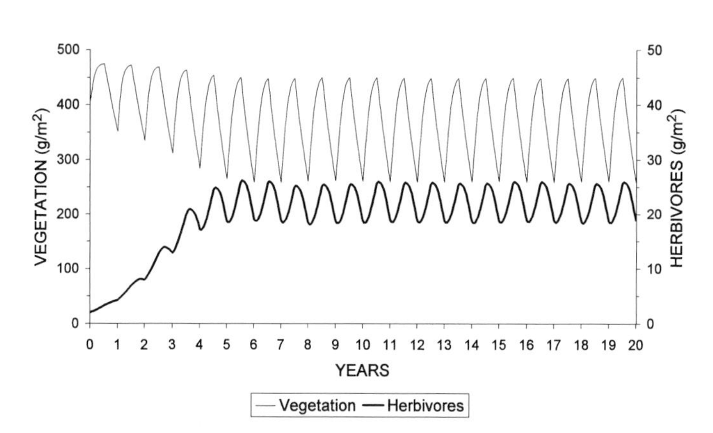

Figure 13.3. Output of the **GMM** model for seasonal vegetation growth, with multiple patch types differing in forage quality c. (a) Two patch types with $c = 0.77$ and 0.67, respectively. (b) Three patch types with $c = 0.8$, 0.72 and 0.64, respectively. (c) Four patch types with $c = 0.8$, 0.72, 0.64 and 0.56, respectively, differing also in $v_{\max}$ and g_V.

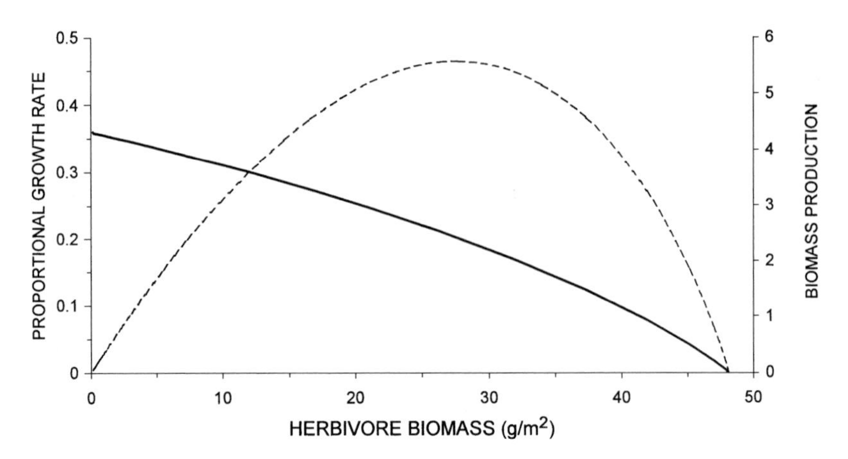

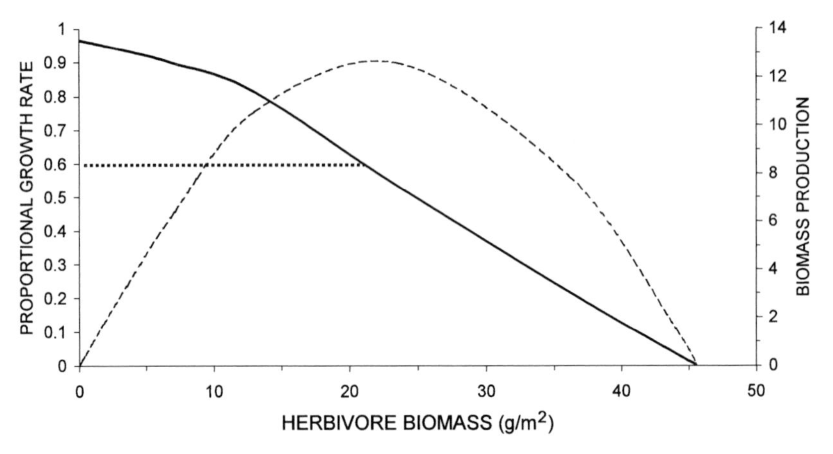

Figure 13.4. Density dependence in relative biomass growth rate and annual biomass production per unit area for modelled herbivore population under different conditions. Dotted line indicates potential upper limit to relative growth rate set by metabolic processing capacity (corresponding effect on growth increment not shown). (a) Single patch type with $c = 0.72$. (b) Two patches with $c = 0.77$ and 0.67, respectively. (c) Three patch types with $c = 0.8$, 0.72 and 0.64, respectively. (d) Four patch types with $c = 0.8$, 0.72, 0.64 and 0.56, respectively, differing also in $v_{\max}$ and g_V.

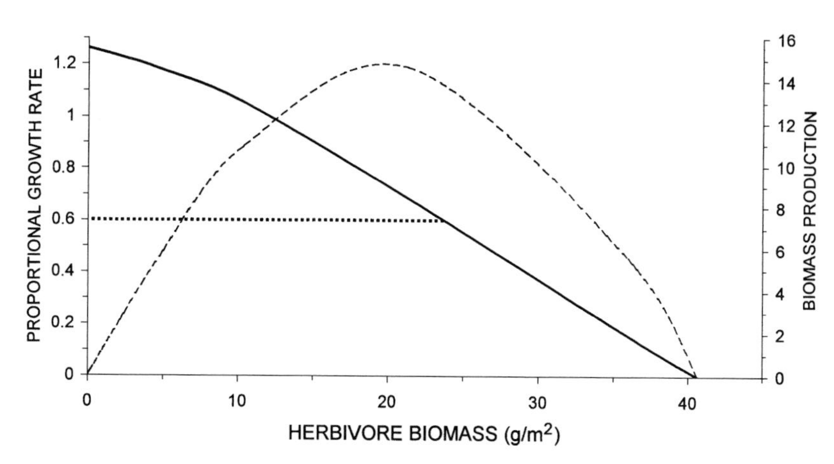

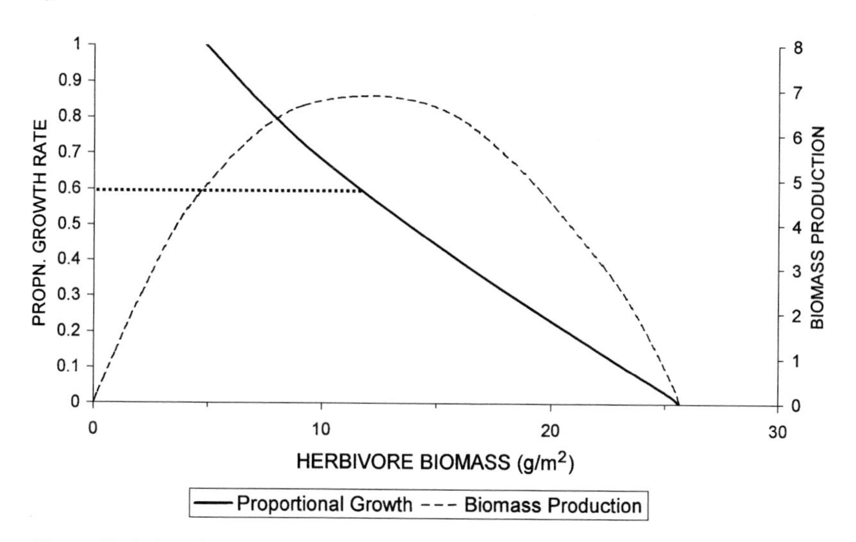

Figure 13.4. (*cont.*)

(see Chapter 8). Accordingly, patch types may differ in v_{max} and g_V as well as in c. The consequences depend on how the half-saturation parameters $f_{1/2}$ and $g_{1/2}$ are related to $v_{\max}$. If $g_{1/2}$ is a constant independent of $v_{\max}$, lowering $v_{\max}$ reduces the effective food quality. To keep food quality independent of the setting of $v_{\max}$, both $f_{1/2}$ and $g_{1/2}$ need to be made proportional to $V_{\max}$ for each patch type, as specified in Table 13.1. The 'paradox of enrichment', i.e. the destabilizing effect of increasing the productive potential of vegetation as represented by v_{max}, arises only when $f_{1/2}$ is a fixed constant.

To represent such functional variability among plant types, v_{max} was raised and g_V lowered as c decreased in each patch type, while keeping the overall mean v_{max} constant at 500 g/m^2. As a consequence, the saturation biomass attained by herbivores was reduced, because a greater proportion of food production took place in patches of poor quality (Fig. 13.3c). Nevertheless, the potential maximum yield from the herbivore population was maintained above that in a uniform environment (see Fig. 13.4a, d). The low-quality patches functioned as forage buffers, exploited temporarily under severe conditions of food restriction.

Effects of food enrichment

As indicated above, the intrinsic growth rate of the herbivore population depended strongly on the nutritional quality of the vegetation consumed, whereas the population sustained depended also on the maximum vegetation biomass potentially attained. For herbivores, food enrichment occurs mainly via increases in the nutritional quality of vegetation resources, rather than in the productive potential of plants. Plants that are high in nutrients tend to be associated with fertile soils, and thus exhibit a high relative growth rate (Grime 1977). Grasses of this type are also commonly short in stature, and thus attain a correspondingly low maximum biomass (e.g. lawn-forming species). They may be maintained by high grazing pressure, but are replaced by taller grasses that are competitively dominant for light in the absence of grazing.

Two patterns of enrichment were modelled: (a) a general shift in patch quality towards higher values of c, and (b) an increase in the

Table 13.2. *Mean and variance in herbivore population density under modelled conditions varying in food quality distribution among four patch types*

	Patch quality conditions			
	High	Medium	Low	Uniform
Conversion coefficients c	0.8,0.75, 0.7,0.65	0.8, 0.72, 0.64, 0.56	0.72, 0.67, 0.62,0.57	0.72
Herbivore biomass				
Mean	17.9	11.5	4.0	30.5
Standard deviation	2.85	1.69	0.46	4.36
Coefficient of variation	0.159	0.147	0.115	0.143

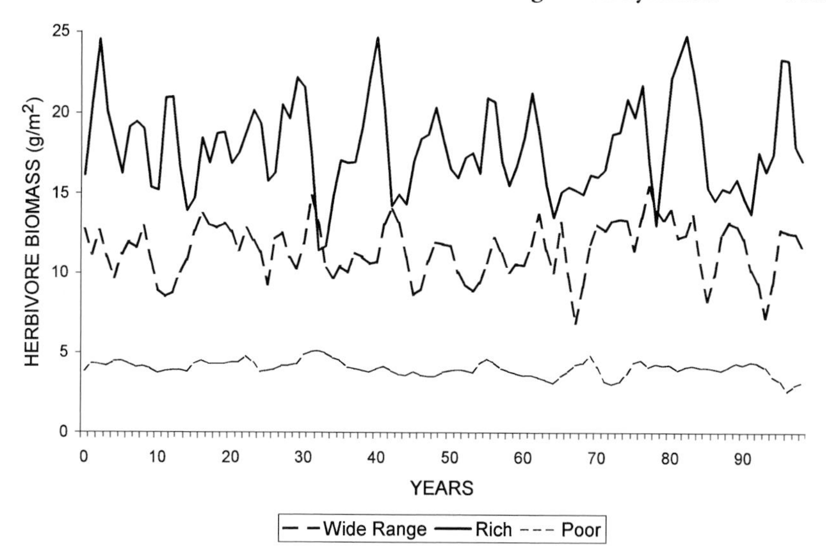

Figure 13.5. Effects of enriching forage quality on variability in herbivore biomass at the end of the growing season in stochastically varying environment (uncorrelated). The environment was constituted by four patch types, with forage quality distributed as follows: (i) rich environment, $c = 0.8, 0.72, 0.7$ and 0.65, respectively; (ii) environment with wide range in food quality, $c = 0.8, 0.72, 0.64$ and 0.56, respectively; (iii) poor environment, with $c = 0.72, 0.67, 0.62$ and 0.57, respectively. The mean and variance in herbivore density for each situation is summarized in Table 13.2.

nutritional quality of the best-quality patches, with that of poor patches left unchanged. Weekly rainfall was varied stochastically according to probability levels as outlined above. A general increase in vegetation quality by about 12% substantially elevated the herbivore biomass attained, but also greatly amplified the annual fluctuations in herbivore biomass (Fig. 13.5; Table 13.2). An increase in nutritional quality applied only to the best patch types had a similar destabilizing effect, despite a reduced influence on herbivore biomass. This was because, with a wide quality differential, vegetation biomass in the best patches was severely depleted before herbivores transferred to alternative patch types. Notably, the herbivore biomass supported now approached the levels exhibited in productive real-world ecosystems.

Supporting observations

Ellis and Swift (1988) noted the wide variability in livestock numbers supported in an arid region of northern Kenya, and suggested that herbivore–vegetation systems are intrinsically disequilibrial where rainfall

a

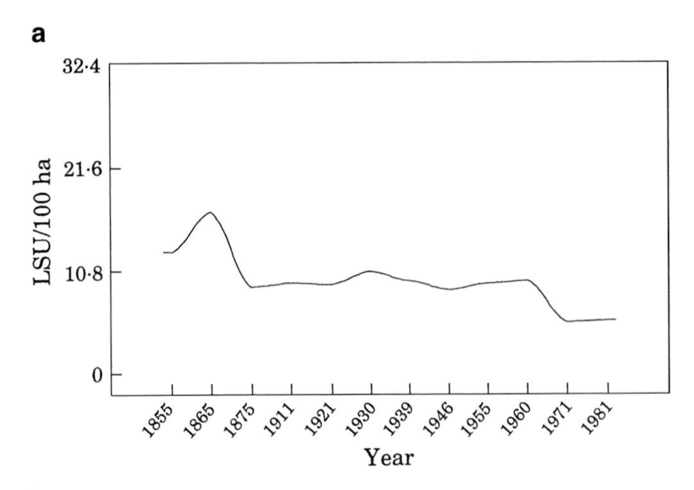

b

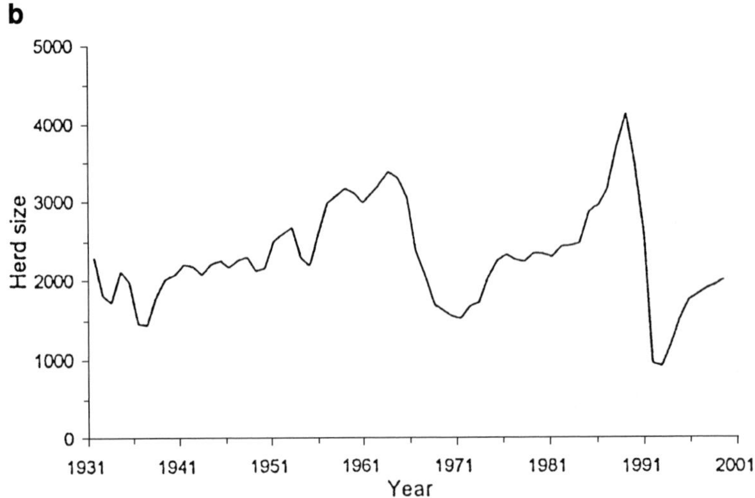

Figure 13.6. Patterns of livestock dynamics documented over extended periods. (a) Historical stocking level of all domestic livestock in semi-arid and arid districts of the Karoo and adjoining regions in the former Cape Province of South Africa (from Dean and Macdonald 1994). (b) Reconstructed changes in the total numbers ('herd size') of cattle in a semi-arid savanna region of KwaZulu–Natal, South Africa (from Hatch and Stafford Smith 1997). (c) Cattle dynamics in three regions of Zimbabwe (from Scoones 1993).

is low and highly variable from year to year. However, wide fluctuations in cattle numbers have also been reported from regions where rainfall is not so low, but soil fertility relatively high (Hatch and Stafford Smith 1997) (Fig. 13.6a). In contrast, sheep numbers in the Karoo region of

Figure 13.6. (cont.)

South Africa have remained fairly stable over nearly a century, despite low and highly variable rainfall (Dean and Macdonald 1994) (Fig. 13.6b). The crucial difference in the latter region may be the potential for sheep to switch from the fluctuating grass component to generally less nutritious, but productively more stable, dwarf shrubs in drought years. Buffering effects of landscape-scale heterogeneity on herbivore population dynamics have been documented for free-ranging cattle in central Australia (Low *et al.* 1981), and for communally managed cattle in southeast Zimbabwe (Scoones 1992, 1995). In Zimbabwe, areas with fertile clay soils showed higher cattle mortality during droughts, but a faster recovery after the drought, than infertile sandy soil regions (Fig. 13.6c).

Soay sheep example

The cause of the persistent instability shown by the Soay sheep population occupying the island of Hirta in the Scottish Hebrides has been the subject of much research. Repeated censuses over two separate periods, spanning 35 years in total, documented cyclic fluctuations in abundance, with a period of 3–4 years and amplitude sometimes exceeding 2.5-fold (Clutton-Brock *et al.* 1991, 1997; Grenfell *et al.* 1992). The animals occupy a wet, lush habitat where rainfall variability is unlikely to be influential. The persistent instability has been ascribed to their rapid growth to reproductive maturity, coupled with overcompensating (accelerating) density dependence in mortality, and relatively little influence of density on fecundity. Specifically, the reproductive segment can increase numerically by over 50% during a single summer at low population density. As a result, the population density soon increases above the level that can be sustained by winter food resources, precipitating an overwinter dieoff. A detailed simulation model developed by Illius and Gordon (1998; see also Clutton-Brock *et al.* 1997), based on these features, closely replicated the observed dynamics of the sheep population.

Other influential factors include heavy parasitic infestations, exacerbated by high population density, which intensify the effects of food shortages (Gulland 1992; Grenfell 1992). Episodes of extreme weather, notably gales during March at the end of winter, synchronize population fluctuations of sheep between neighbouring islands in the St Kilda archipelago (Grenfell *et al.* 1998). Temperature conditions during April, affecting the rate of grass growth during early spring, are an additional influence.

Nevertheless, the basic question remains: why are these sheep populations on certain islands in the Outer Hebrides so susceptible to repeated irruptions and crashes? Bighorn sheep censussed in two mountain ranges in Alberta, over periods of 13 years and 20 years, respectively, have remained relatively stable (Jorgenson *et al.* 1997). Wild goats occupying an island habitat in Greece appear to have remained fairly constant in abundance (Papageorgiou 1979). Red deer inhabiting the Isle of Rum, also in the Scottish Hebrides, have maintained a stable population. Clutton-Brock *et al.* (1997) attributed the demographic contrast between these deer and Soay sheep to the higher inherent growth rate of the sheep population. But if the low population growth rate associated with large body size automatically suppresses population instability, there would be little justification for culling to prevent oscillations in the dynamics of elephants and woodlands (cf. Caughley 1976c).

Demographically structured sheep model

The wide population fluctuations exhibited by the Soay sheep were associated with large differences in the susceptibility of different age and sex classes to mortality, causing marked changes in population structure over time. To represent this, the demographic segments constituting the sheep population need to be differentiated.

From a metaphysiological perspective, population segments are most appropriately distinguished in terms of their growth potential, as was suggested in Chapter 10. The most basic segments comprise (1) adult males, with zero growth potential except via changes in fat reserves, (2) adult females, which may divert surplus gains to support the growth of dependent offspring, (3) still-growing immatures of both sexes, and (4) juveniles, which have their growth subsidized by maternal inputs in the form of milk prior to the time of weaning. For modelling, they were distinguished by parameters defining the metabolic maintenance requirement, the upper limit to weekly growth potential, and sensitivity to mortality. Births were represented by transferring a fraction of maternal biomass to constitute the nucleus of juvenile biomass, at the beginning of the growing season. To ensure that mothers of dead offspring did not subsidize surviving juveniles, the proportion of the adult females contributing to juvenile growth was attenuated over time in accordance with juvenile mortality.

For simplification, all sheep aged over 1 yr were regarded as adults. The twinning exhibited by a small proportion of adults, and reproduction by some lambs less than one year old, was ignored. Allowance was made for the lesser fertility of yearlings compared with females aged 2 yrs or more, defined by the proportion contributing to juvenile production and growth. Juveniles were generated in the first week of the growing season, and received maternal supplements through the six months of the GS. Observations show that lambs can increase seven-fold in body mass before the end of their first summer, and yearlings can nearly double in body mass over a year, under ideal conditions. Corresponding maximum limits in relative growth rate were specified in the model. Adult females were allowed to regain the 10% of their biomass lost to constitute newborn juveniles over the course of the year, and replace lost fat reserves. With the parameter values shown in Table 13.3, the model replicated the convex density dependence, and high maximum growth rate, shown by the real Soay sheep population (Fig. 13.7a).

The standing crop of herbaceous vegetation on Hirta fluctuates seasonally from a peak of about 100–250 g/m^2 in late summer, to a low of 5–10 g/m^2 at the end of winter (with the range representing variability among sampling stations (Milner and Gwynne 1974)). In meadows favoured by the sheep, the peak herbaceous biomass at the end of summer was 476 g/m^2 in a year when the sheep population was low, compared with 310 g/m^2 when the sheep population was at its peak (Clutton-Brock *et al.* 1991). Organic matter digestibility of green leaves

Table 13.3. *Parameter values used for the Soay sheep model*

Parameter	Value	Justification
v_{max}	100–200 g/m^2	Variation in peak standing crop biomass among sites
g_V	0.2–0.22 per week	Approximates observed growth rate
v_u	10 g/m^2	Minimum vegetation biomass
d_g	0.01 per week	Decay of leaf biomass
i_m	0.28–0.32 per week	Range for adult to juvenile sheep, body mass 30–10 kg
$f_{1/2}$	$0.04 \times K_v$	Steeply saturating intake response for small herbivore
$g_{1/2}$	equal to $f_{1/2}$	Quality gradient determined solely by patch choice
p	0.2–0.25 per week	Relative to basal metabolism for 30–15 kg sheep; active metabolism $= 1.77 \times$ basal metabolism
c	range 0.5–1.2	represents outcome of digestion, assimilation and growth efficiency plus transformation from plant dry mass to animal live mass; dormant season $c = 0.9 \times$ growing season c
q_0	0.1–0.15 per year	Potential lifespan = 10 yrs for adult females, 7 yrs for adult males
z_0	-1 to -1.2	nutritionally related mortality when $p = G$ is 30% per yr for adult females, 35% for yearlings, 40% for adult males and 50% for juveniles
z	1.4–1.6	—
Maximum weekly gain	0.29 (juv), 0.235 (yrl), 0.202 (AdF), 0.2 (AdM)	Adjusted relative to p so that juveniles can increase 7 times in body mass over 1 year, yearlings can double in body mass, adult females can replace reproductive losses and adult males show zero growth

of the favoured *Agrostis–Festuca* grassland was 0.77, compared with 0.68 for the little-used *Molinia* grassland (Illius and Gordon 1998). Seasonally, the digestibility of 'pinch' samples of amalgamated vegetation varied from about 0.7 over the summer months to 0.55 during winter (Milner and Gwynne 1974). *Calluna* heathland was grazed to some extent by the sheep during winter, but exhibits a low digestibility (about 0.35, (I. J. Gordon, personal communication)). Bryophytes may also be consumed when forage is sparse (Milner *et al.* 1999b). The quality of the forage

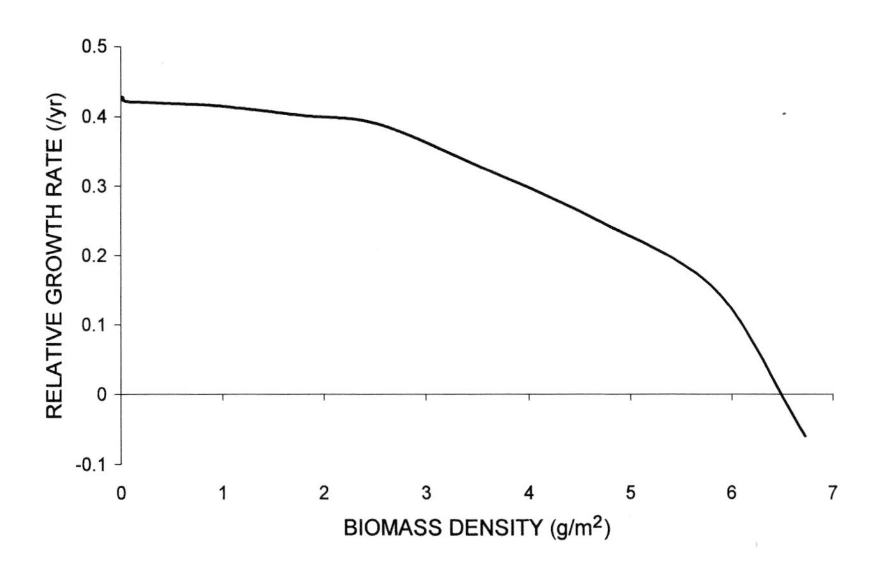

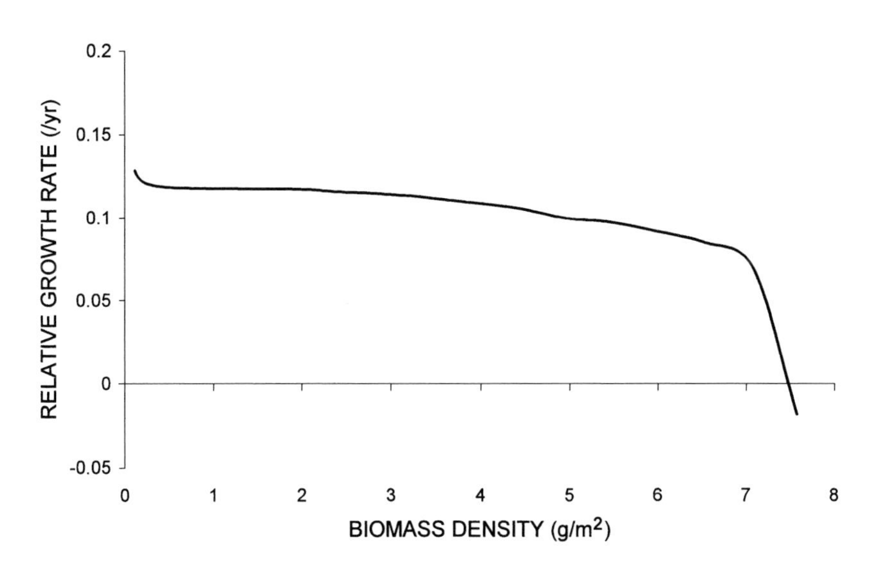

Figure 13.7. Form of density dependence generated by the demographically structured sheep model. (a) Basic sheep model. (b) Modified model representing a larger, slower-growing herbivore.

ingested by sheep may vary somewhat less than indicated by the vegetation samples.

For modelling purposes, the vegetation was constituted by four patch types differing in nutritional quality, as indexed by c. Values for c were adjusted relative to the mass-specific metabolic requirements of the 20–30 kg (adult mass) sheep, in order to allow the high growth potential of the immature and juvenile segments to be achieved. Three of these patch types represented preferred, staple and reserve resources, respectively; the fourth represented a low-quality buffer (perhaps heather). Vegetation growth took place continually through the six summer months, then ceased completely during the six winter months. Iteration was weekly. The model was deterministic, i.e. environmental conditions influencing plant growth remained constant from year to year apart from the seasonal cycle.

Model output

To generate the irregular period of the fluctuations shown by the Soay sheep population, fine-tuning of model parameters was needed such that the herbivore dynamics became gently chaotic (Fig. 13.8a). Oscillations developed when consumers depleted forage biomass in the selected patch types to critically low levels, such that food intake and hence nutritional gain declined precipitously. This situation arose following the attainment of a population high during summer, with low vegetation biomass being reached towards the end of the following winter. The result was a sharp rise in mortality, varying in severity among population segments. Appropriate settings of the mortality parameters replicated the observed pattern, with the survival rates of juveniles, yearlings and adult males being most sensitive to resource deficiencies, and adult females somewhat more resistant (Fig. 13.8b; cf. Grenfell *et al.* 1992; Clutton-Brock *et al.* 1997). The severe mortality among juveniles resulted in a paucity of yearlings the following year. The consequent 2-year lag in recruitment to the adult segment helped produce the 3–4-year period of the oscillations.

Having replicated the oscillatory pattern shown by the Soay sheep population, we next consider the changes in environmental conditions needed to suppress the instability. Reducing vegetation quality uniformly lowered the inherent growth rate of the population below its potential, dampening the oscillations. However, oscillations could also be suppressed, while retaining the same potential growth rate, by lowering the quality of just the reserve food type by about 10% (Fig. 13.9c). In contrast, if the quality of the buffer resource was reduced, oscillations were amplified. Raising the quality of the buffer resource reduced the population instability, without lowering the population density attained.

To reveal the functional response patterns underlying population instability, weekly food intake and nutritional gains were extracted from the model output and plotted relative to changing food availability over the winter period (Fig. 13.9). The effective food availability was assumed to be the mean standing biomass averaged over the three best patch types, i.e. excluding the buffer habitat, which was hardly utilized. Seasonal responses are depicted specifically for the adult female segment. The intake response, determined directly by the value for $f_{1/2}$ specified in the model, was sharply threshold in form (Fig. 13.9a). For an ungrazable biomass amounting to 10 g/m^2, daily food intake would not attain half of its maximum until standing grass biomass was only 13–14 g/m^2. The corresponding functional response for rate of nutritional gain emerged through the effective seasonal decline in diet quality as a result of changing patch selection. However, identifying the half-saturation biomass of vegetation is somewhat problematic. The weekly gain in herbivore biomass, expressed as relative growth potential, is only meaningful relative to the zero growth rate set by metabolic maintenance requirements.

Instability resulted when low food availability caused gains to fall markedly below requirements towards the end of winter (Fig. 13.9b), generating precipitous mortality and hence a population crash. Reducing the quality of the reserve resource caused the biomass growth potential of the sheep to become just marginally positive early in winter. As a result, food availability in the utilized patch types did not reach critical starvation levels (Fig. 13.9b). Raising the quality of the buffer resource had no influence on the gains achieved by the sheep through most of the winter period, while the latter exploited other patch types. However, at the critical stage when the sheep turned to the buffer resource, its quality was sufficient to hold weekly nutritional gain above the critical level generating severe mortality (Fig. 13.9c).

Clutton-Brock *et al.* (1997) suggested that the Soay sheep population was pre-disposed to high instability because of its high potential growth rate. To evaluate how the rate of population increase influences stability, the sheep model was adjusted to represent a herbivore with a much slower intrinsic growth. To achieve this, the time step in the model was simply lengthened from one to two weeks, effectively reducing all rates by half.

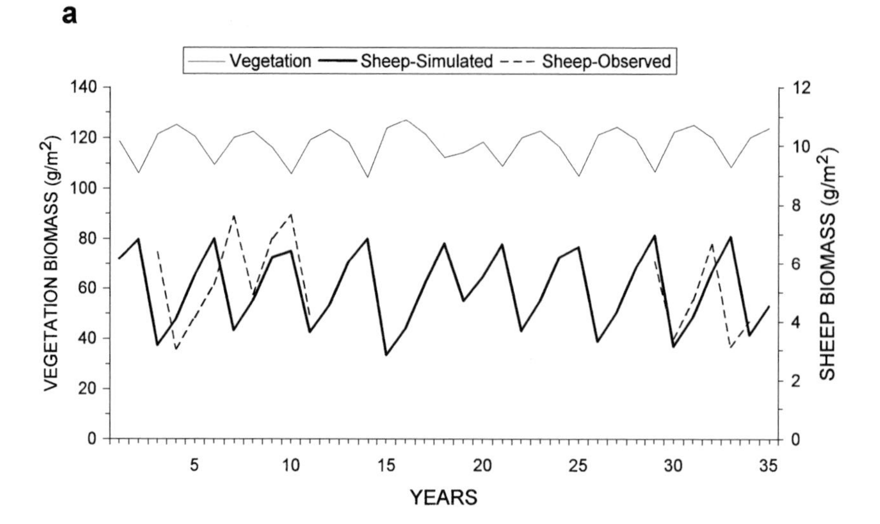

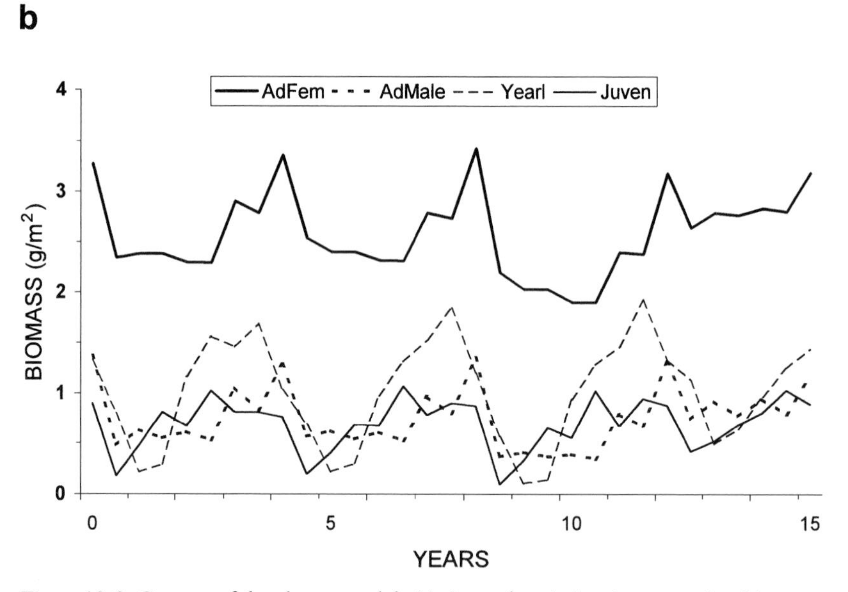

Figure 13.8. Output of the sheep model. (a) Annual variation in vegetation biomass and sheep biomass at the end of summer, for simulation with patch quality values c equal to 1.2, 1.1, 1 and 0.6, respectively. Also indicated is the observed sheep biomass calculated from census totals for the periods 1959–67 and 1985–90, assuming that one third of the island area does not support sheep (from Morton Boyd 1974 and Clutton-Brock *et al.* 1991). (b) Biomass of sheep population segments generated by the model (end of summer, end of winter). (c) Changes in sheep biomass dynamics produced by (i) lowering c for the reserve patch type from 1.0 to 0.9, (ii) elevating c for the buffer patch type from 0.6 to 0.8, compared with output from (a) above. (d) Output from the modified model with greatly slowed population growth (cf. Fig. 13.7b).

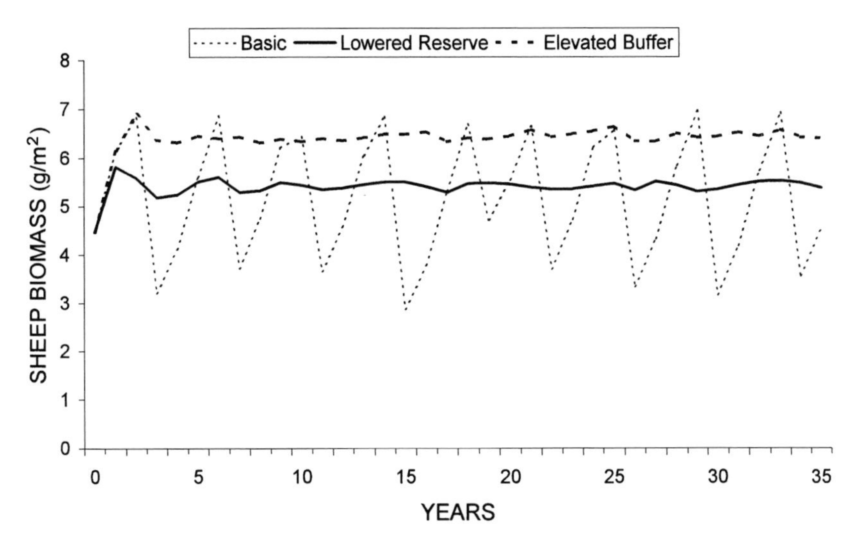

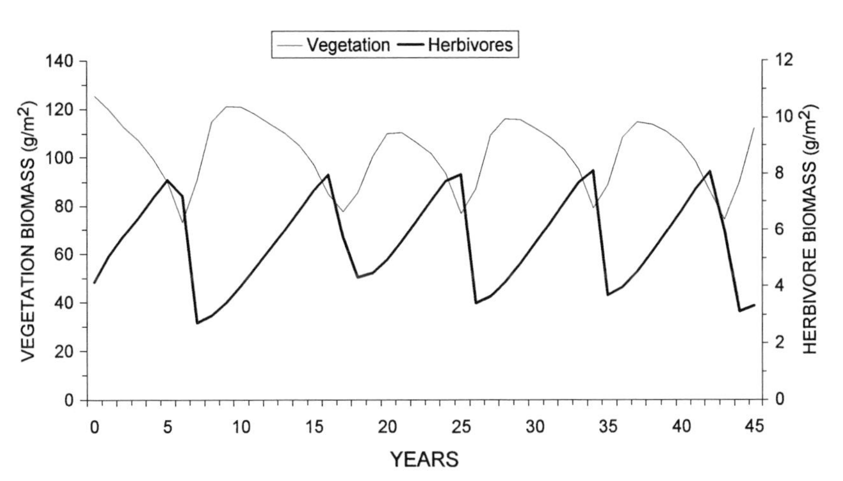

Figure 13.8. (cont.)

In addition, the number of population segments was increased from four to five, to extend the period between birth and full reproductive maturity from two to three years. These changes reduced maximum population growth rate from 42% to 12% per year (Fig. 13.7b). The effective density dependence remained sharply convex, arising through the

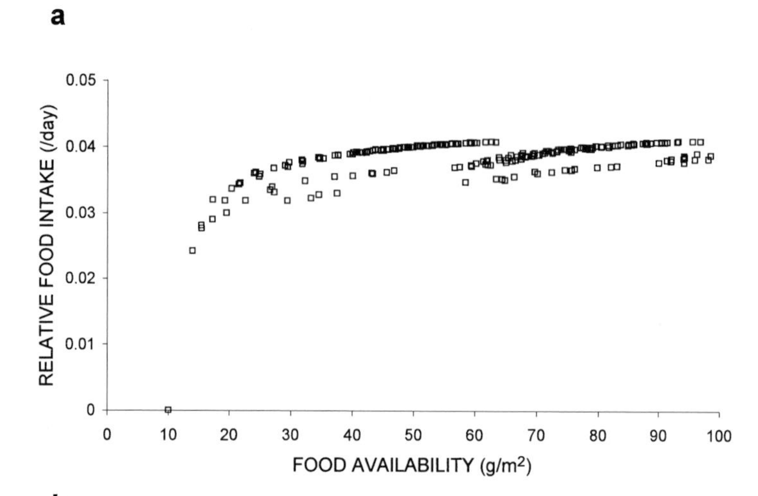
a
RELATIVE FOOD INTAKE (/day)
FOOD AVAILABILITY (g/m2)
0.05
0.04
0.03
0.02
0.01
0
0
10
20
30
40
50
60
70
80
90
100

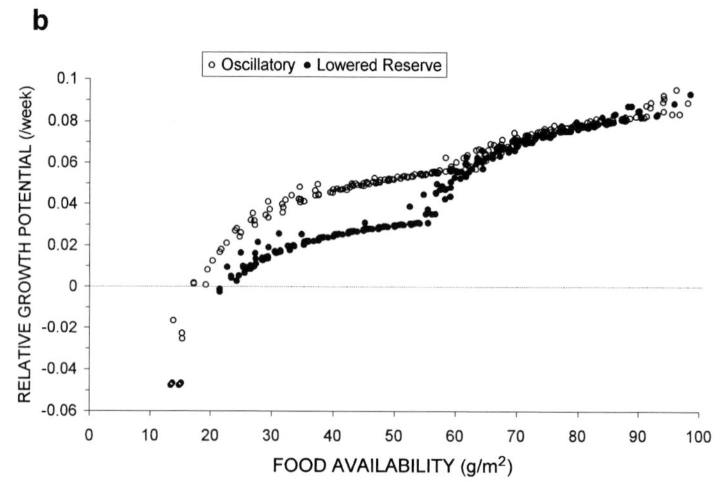
b
Oscillatory
Lowered Reserve
RELATIVE GROWTH POTENTIAL (/week)
FOOD AVAILABILITY (g/m2)
0.1
0.08
0.06
0.04
0.02
0
-0.02
-0.04
-0.06
0
10
20
30
40
50
60
70
80
90
100

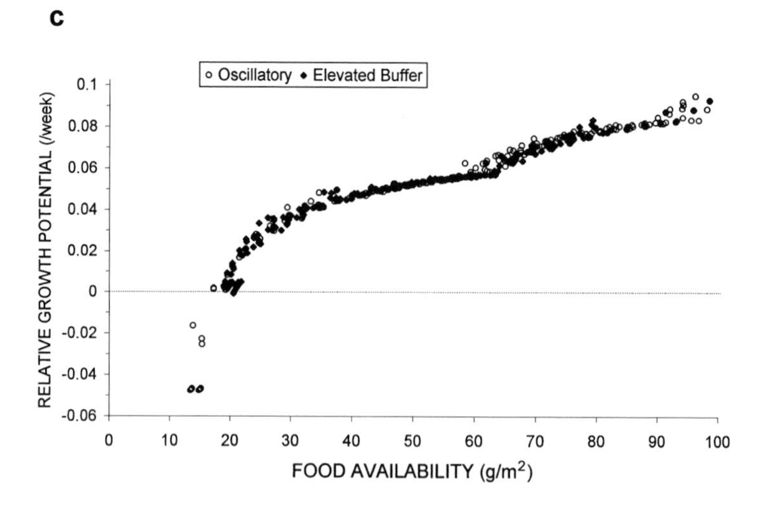
c
Oscillatory
Elevated Buffer
RELATIVE GROWTH POTENTIAL (/week)
FOOD AVAILABILITY (g/m2)
0.1
0.08
0.06
0.04
0.02
0
-0.02
-0.04
-0.06
0
10
20
30
40
50
60
70
80
90
100

consumer–resource interaction. Despite the slowed inherent growth rate, oscillations were generated similar in amplitude to those shown by the sheep population, but with an 8–9-year period (Fig. 13.8d). It merely took longer for the population to reach the level where consumption caused the resource base to be reduced below the critical level before the end of winter.

Overall assessment

The modelling exercise suggests that population instability depends fundamentally upon whether resource gains fall below the critical starvation threshold during the course of the dormant season when resources are non-renewing. This is dependent partly on the quality of the preferred resources, which enables the population to grow to high density levels, increasing the consumption offtake. However, also critically important is the nutritional quality of the reserve and buffer resources supporting animals through the adverse season. Specifically, if the reserve resources upon which the animals depend during the dormant season are too high in quality, they become depleted, and oscillations result through overall lack of food. If the buffer resources that the animals turn to finally are too low in quality, they are not exploited until better resources have been reduced to critically low levels, and then are inadequate to prevent rapid starvation. The magnitude of the nutritional deficit is crucial in determining whether animals survive through the DS, or die of starvation. Hence, it is not merely the presence of reserve and buffer resources that is important, but rather their functional effectiveness in averting mortality.

The measurements of Milner and Gwynne (1974) indicate that vegetation biomass in the grassy meadows favoured by the Soay sheep is

Figure 13.9. Effective intake and biomass gain responses to changing food availability during the dormant season extracted from the sheep model output, for the adult female segment. Food availability is represented by the mean vegetation biomass averaged over the three better patch types. (a) Daily food intake response for oscillatory population. (b) Resource gains expressed as weekly growth potential for the oscillatory population, compared with circumstances where the population was stabilized by lowering the effective quality c of the reserve patch type. (c) Resource gains for oscillatory population, compared with gains for the situation with the circumstances where the population was stabilized by elevating the effective quality c of the buffer patch type.

lowered by the end of winter to levels sufficient to substantially reduce daily food intake. Whether the nutritional value of alternative resources, i.e. coarser grasses and heather, is indeed inadequate to avert oscillations is unclear.

Modelling focussed only on the effects of resource components on population instability. No allowance was made for year-to-year changes in vegetation composition, and hence in food production, as a result of grazing pressure by the sheep, which could also affect the adequacy of the resource base. There is mention of an increase in bryophytes, potentially at the expense of grass, following years when the sheep population had been high (Milner *et al.* 1999b). Nematode parasitism and thermal stress from extreme weather have been identified as additional influences on sheep dynamics. High levels of parasitic infestation exacerbate food shortages (Gulland 1992; Grenfell 1992). Both factors could be represented in the model in terms of their effects on physiological costs (M_P). However, high sheep densities also promote a buildup in parasite numbers and hence in transmission. Cold stress from extreme winds in late winter heighten physiological costs, perhaps causing them to pass critical thresholds at a stage when body reserves have been depleted (Grenfell *et al.* 1998).

Physiologically, starvation-induced mortality occurs when body reserves have been depleted below critical levels. Illius and Gordon (1998) modelled this mechanism specifically in their simulation of Soay sheep dynamics, with due allowance for statistical variability in fat levels within the population. This approach does not change the outcome, merely the timing. Severe nutritional deficits draw body reserves down to critical levels, eventually precipitating mortality.

Clutton-Brock *et al.* (1997) suggested that the stability of the red deer population on Rum, despite a similar island environment to that of the Soay sheep, was related to differences in fecundity and maturation rates, governing the maximum population growth rate. The alternative possibility raised by the sheep model is that nutritional limitations intensify more gradually over late winter for the red deer than for the sheep. This could occur despite similar vegetation resources, if the larger red deer are better able to digest and hence extract energy from the buffer resource (heather or other vegetation components) than the sheep. Indeed, red deer do digest heather better than domestic sheep (Milne *et al.* 1976). The characteristics of the red deer population described by Clutton-Brock *et al.* (1997) indicate chronic limitations in food quality: slow population growth, late attainment of maturity, failure by most females to reproduce annually, and small body size relative to continental red deer.

This is a possible testament to their partial dependence on low-quality buffer resources.

Recently, the sheep population on Hirta has persisted at a high level for several years without a crash. Notably, this situation was associated with unresponsiveness in adult female mortality to high population density (Milner *et al.* 1999a). Whether this is due to a change in vegetation, in parasites or in weather, or is just a chance interlude, is unclear.

Other ungulate populations

The crash of the reindeer population on St Matthew Island in the Bering Sea followed the elimination of lichens, which had constituted the food reserve during winter (Klein 1968). The sedges and grasses that replaced the lichens apparently functioned as a food reserve, but became inaccessible when deep snow covered them during a severe winter. The high nutritional value of the forage available to these reindeer during summer probably contributed to the population buildup to high density. Similar conditions led to a precipitous decline by reindeer inhabiting the nearby St Paul Island (Scheffer 1951).

The moose population on Isle Royale in Lake Superior has fluctuated five-fold in density over the 35-year period spanned by censuses (Peterson *et al.* 1984; Peterson 1999; see Fig. 1.2b). The oscillations were originally ascribed to a predator–prey interaction with wolves. However, wolf predation primarily affects calves, and for a population crash adult moose must also incur heavy mortality. Recent evidence suggests that the instability originates from the interaction between moose and woody vegetation. The moose crash of 1995–97 was associated with clear evidence of starvation, ascribed to depletion of balsam fir, which had formerly served as the principal food source during winter. Density-dependent effects on the reproduction and mortality of the moose were too slow-acting to avert the crash. Severe winter weather, as well as tick loads, exacerbated the effects of food shortages. Reduced predation, owing to a disease-induced collapse by the wolf population, could have been a factor allowing the moose to reach population densities where they severely impacted the balsam fir resource. Nevertheless, the directly pre-disposing factor seems to be the depletion of this winter food reserve.

Calamitous drought-related mortality among wild ungulates occupying the Klaserie Private Nature Reserve in South Africa, amounting to 90% reductions in the populations of some species, apparently originated from over-provision of waterpoints (Walker *et al.* 1987; Owen-Smith 1996a). Populations of the same species in the adjoining Kruger National Park, where waterpoints were more widely spaced, suffered relatively little mortality during the same drought. The severe mortality was precipitated by elimination of the forage buffer that would have remained in places remote from water had waterpoints been more widely spaced. In the Kruger Park, animals were able to access the grass persisting in areas more than 3 km from waterpoints as soon as the first rainshowers broke the drought, alleviating starvation at this crucial stage. Interventions that restrict the spatial scale of animal movements, such as fencing, can hamper access to buffer habitats.

Instability in populations of white-tailed and mule deer in North America appears to result from the severe impact that high deer abundance has on the woody plant species serving as the primary food resource (McCullough 1997). The deer then become dependent largely on herbaceous plants, including annuals that do not provide forage through the winter period.

Fertile ecosystems supporting largely palatable plant species seem to be more susceptible to herbivore population crashes than regions where unpalatable plants provide a forage buffer, at least for domestic herbivores (Dean and Macdonald 1994; Shackleton 1993; Owen-Smith 1998b). Actions aimed at enhancing productivity, through providing supplementary forage and additional waterpoints, eliminating predators and controlling parasites, tend to promote instability, unless ungulate populations are tightly controlled (e.g. rapid de-stocking when drought conditions take hold). Especially in communal-tenure rangelands, the long-term result is reduced food security. Similar consequences follow from interventions that restrict the spatial scale of herbivore movements, for wild as well as domestic ungulates.

Alternative modelling approaches

Through incorporating population structure, the sheep model converged with demographic population models. However, whereas the latter estimate population changes from numerical births and deaths, the metaphysiological approach focuses on resource fluxes through each population segment. Variability in resource supplies can also be incorporated into demographic models, by making birth and death rates a function of food availability. I used such an approach to model the dynamics of populations of kudus, based on the rainfall-dependent mortality functions documented in Chapter 10 (Owen-Smith 2000). This model closely replicated the observed dynamics of the kudu population over the 10 years spanned by the study, provided the initial age structure was adjusted appropriately (Owen-Smith 2000). However, over a longer period it failed to project the low population density of kudus observed at the start of the study. This failure was ascribed to its neglect of the amplifying effect that episodic cold spells could have on mortality during the late dry season. This effect was subsequently added arbitrarily, improving the fit. A more fundamental shortcoming is that such models fail to allow for the impact that herbivore populations exert on their food resources, not only over long time periods but even seasonally.

Population crashes occur when animals become vulnerable to direct, starvation-induced mortality, even among prime-aged adults. Such mortality occurs during the dry or winter season, perhaps precipitated directly by physiological stress due to extreme cold weather or other influences. Demographic models may represent such processes phenomenologically via density-dependent functions that are strongly 'overcompensating' in form (see, for example, Grenfell *et al.* 1992). However, they do not reveal the origins of such overcompensation shown by some populations, but not others.

Coupled consumer–resource models tend to generate overcompensating density dependence, if interference competition is lacking and the functional response patterns governing resource intake and gains rise steeply towards asymptotic levels. This underlies the unstable dynamics generated when they are applied to large herbivore populations interacting strongly with vegetation. Stability can result when the gain response deviates from the intake response because food quality declines with diminishing food abundance, as has been shown in this chapter. The intake response could also become more gradually saturating in circumstances where diminishing grass height increasingly restricts bite size, as outlined in Chapter 3. A rising physiological cost, owing to greater foraging effort as resources diminish in availability, could also help stabilize consumer biomass dynamics, an effect noted but not treated in this book. Predation could be stabilizing, by restricting population growth so that biomass levels impacting severely on vegetation resources are not reached. These processes are readily incorporated into metaphysiological models, but not into standard demographic models.

The **GMM** approach could be elaborated further into a detailed physiological model, by sub-dividing population biomass more finely into appropriate age, size or state classes, even down to the individual organism level. Whether such an approach is helpful depends on how reliably all the parameter adjustments could be estimated. The mechanisms governing population dynamics tend to become obscured by the wealth of detail.

For strategic purposes a simple model tracking resource fluxes through aggregated population biomass, and estimating the consequences for mortality losses, could be quite adequate. Even the seasonal cycle could perhaps be simplified, to a two-phase oscillation between a productive and a dormant season. However, ignoring the seasonality that is a basic feature of vegetation growth in most real-world environments is likely to miss some of the fundamental mechanisms governing the dynamics of large herbivore populations.

Of course, much of the instability manifested by herbivore populations may be a consequence of their long-term impacts on vegetation structure and composition. The latter processes have not been treated in this book, because an equivalent metaphysiological model of plant population dynamics would be needed to do so reliably. The **GMM** model developed in this book demonstrates the consequences for herbivore population dynamics of particular features of their vegetation resource base, whatever the mechanisms producing such vegetation patterns.

Overview

This chapter was concerned with how features of resource heterogeneity might influence the stability or instability of herbivore populations. It emphasized the critical contribution of the alternative food types functioning as reserve and buffer resources during crucial periods of the seasonal cycle. Fertile ecosystems presenting generally high-quality resources are intrinsically prone to instability, unless spatial or other buffering effects operate. Actions to promote higher productivity by augmenting water, supplementing forage, eliminating predators and treating parasites can be at the cost of stability, and hence the long-term sustainability of the enterprise. Instability can be counteracted by managerial intervention, through culling, de-stocking and forage cultivation, but only where managers have the resources to do so. For the vast areas of rangeland with communal tenure systems, as well as for areas managed for wildlife, maintaining appropriate heterogeneity promotes stability in the sense of a more even flow of benefits. Notably, the form of dynamic stability in mind is an emergent feature of an interactive system that is intrinsically disequilibrial in its basic operation.

The **GMM** modelling approach accommodates the interactive influences of resources, physiological costs and mortality on population biomass dynamics. In particular, it focusses attention on critical thresholds generating instability if surpassed. It can be elaborated down to a detailed physiological or individual level, if warranted tactically. It could also be applied more strategically, to capture the basic resource dependency of growth, reproduction and mortality. Demographic models can describe population changes phenomenologically, but fail to reveal the mechanisms governing population stability by ignoring the consumer–resource interaction that is fundamental in herbivore–vegetation systems, as well as seasonal variation. To represent the long-term processes of change in vegetation composition and structure that may occur in response to herbivore impacts, an equivalent metaphysiological model would be needed for plant population dynamics.

Further reading

A comprehensive evaluation of herbivore–vegetation models and their stability properties is given by Crawley (1983). Fryxell and Lundberg (1994) show how only a narrow range of parameter combinations leads to stable equilibria for generalized predator–prey systems with adaptively foraging predators, but do not take seasonal variation into account. Implications of wide temporal variability in vegetation growth for sustainable management of herbivores are outlined by Stafford Smith (1996). Illius and O'Connor (2000) analyse how annual variability in resource production governs herbivore impacts on vegetation, and consequences for herbivore dynamics, in seasonal environments where animals alternate their grazing between a wet-season range and a dry-season range.

Appendix 13A. Formulation of the sheep model

The sheep model is a TrueBASIC program governed by functional relations defined by Equations 13.1, 13.2, 13.4 and 13.5 in the text. The model incorporated a weekly iteration with vegetation growing during the 26 weeks of the growing season, then ceasing during the 26 weeks of the winter dormant season. The vegetation comprised four patch types, each covering an equal area, differing in forage quality c, inherent growth rate g_V and saturation vegetation biomass v_{max}, with the half-saturation level for intake set as a constant fraction of v_{max}. Vegetation biomass also decayed at a constant rate year-round independent of consumption.

Forage quality showed a stepwise decline by a constant proportion at the start of the dormant season. At each iteration, the herbivores selected the patch type yielding the highest gain for foraging, i.e. consumption was restricted to this patch type. This was repeated for each herbivore segment, representing juveniles, yearlings, adult females and adult males, respectively. These segments were distinguished by their proportional metabolic maintenance costs, maximum weekly food intake and coefficients governing their susceptibility to mortality. Upper limits were also set for weekly growth in biomass. Adjustments were made to the mortality function to limit how rapidly weekly mortality accelerated for conditions where the annual mortality was projected to exceed 1.0 per year. Births were represented by transferring 10% of adult female biomass plus 4.5% of yearling biomass (allowing for sex ratio and lower fertility) to constitute the nucleus of juvenile biomass during the first week of the growing season. Juveniles and yearlings also advanced one segment during this week. Biomass gains achieved by adult females during the 26 weeks of the growing season contributed to juvenile biomass gains, in addition to food consumed by these juveniles. However, this maternal supplementation was adjusted for the survival proportion of juveniles from one week to the next. Total consumption by all sheep segments was subtracted from the patch type where this consumption occurred before calculating vegetation re-growth. Output representing the weekly biomass of sheep and vegetation for each year was optionally graphed on the computer screen, then saved to a file for later display in a spreadsheet.

14 · *An adaptive resource ecology: foundation and prospects*

In this concluding chapter, I review retrospectively the conceptual foundation established by preceding sections, and outline prospects for further developments of the resource-centred metaphysiological approach. I also look beyond the specific models developed for large mammalian herbivores in this book, to raise some general theoretical issues concerning the dynamics of consumer–resource systems. Thereby I introduce the vision of a broader *Adaptive Resource Ecology*.

Retrospective review

In starting this book, I set out to establish a conceptual framework for modelling herbivore–vegetation systems that would accommodate spatiotemporal variability in the resource base. The approach entailed integrating models of adaptive resource use by individual herbivores into a metaphysiological formulation of population dynamics. The basic principle was not to develop an elaborate, multi-level simulation, but rather to incorporate the functional outcome of lower-level processes, such as adaptive behaviour, into higher-level dynamics.

The starting foundation was Caughley's (1976a) modification of the classical Lotka–Volterra equations for herbivore–vegetation systems. I extended Caughley's analysis by recognizing how variation in food quality could cause the consumer gain function to deviate from the intake ('functional') response to changing resource abundance. Furthermore, I specified mortality losses as being non-linearly dependent on nutritional gains (following Getz 1991, 1993). In addition, I allowed for physiological expenditures as an independent biomass loss. The **GMM** label captured mnemonically the basic processes governing consumer biomass dynamics, at all levels from individuals to populations: **G**rowth, **M**etabolism and **M**ortality. These processes also relate to different components of the environment: biomass growth controlled directly by resources gained; metabolic costs influenced by ambient thermal conditions; and mortality losses dependent on predation as well as nutritional balance. More fundamentally, I took seasonal variation in vegetation growth as well as in forage quality into account. Three features oriented the model specification towards large mammalian herbivores, as opposed to other kinds of consumers: the wide seasonal fluctuation in food availability; the central importance of food quality; and the quantitative superabundance of potential food, albeit of variable quality.

More radically, the approach recognized the reality that no equilibrium persists in seasonally variable environments. Herbivore biomass expands in response to vegetation renewal during the benign season, then shrinks as a consequence of resource depression, and also lowered food quality, during the adverse season. Because analytic treatments of the dynamics of systems governed by non-linear equations in variable environments with changing parameter values are intractable, computational solutions were adopted. Despite the underlying disequilibrial nature of the dynamics, higher-level stability can emerge through the adaptive responses of consumers to changing resource availability, and their consequences for the form of functional relations. I outlined a dynamic optimization approach to resource allocation via state dependence in body condition that accommodated seasonal variability, but did not develop it further. Nevertheless, this intellectual excursion emphasized the importance of timing in life histories.

The modelling currency took the form of aggregated population biomass, as opposed to the numerical currency that is conventional in demographic models. Attention was thereby focussed primarily on the environmental components determining population biomass change, rather than on the population components contributing to it. Nevertheless, I showed further how population biomass could be disaggregated into segments differing in their growth potential, metabolic costs and susceptibility to mortality, as a secondary elaboration.

Models were applied strategically to three specific problems, having both theoretical and applied aspects. I illustrated how habitat suitability, and hence population performance, could depend not just on resource production, but on the seasonal persistence of key, reserve and buffer resources with particular quality attributes. I demonstrated the importance of habitat heterogeneity in conjunction with seasonal variation for promoting coexistence among potentially competing species of herbivore, and showed how seasonal facilitation might ameliorate, but not over-ride, competition. I outlined further how population instability could be associated with features of the food resource base, depending in particular on how the distribution of nutritional quality among functionally distinct vegetation components influenced the form of the biomass gain response.

Prospects

This book provides merely an outline of the **GMM** metaphysiological approach, supported by some specific examples and applications. Much more work remains to be done to consolidate this approach and evaluate what further insights it might give. Below I recognize some of the limitations of what has been covered in preceding pages, and make some suggestions for further modelling and assessment.

Adaptation and optimality

By invoking the term 'adaptive', I tried to avoid some of the problematic connotations of what is 'optimal'. Foraging optimization is commonly assessed in the short term, often based on rate-maximizing principles, leaving uncertain whether longer-term consequences for biological fitness follow (Stephens and Krebs 1986). Hypothesized optima may not be observed, because of the opportunity costs of obtaining sufficient information to guide alternative choices, inherent environmental uncertainty, and multiple considerations that have to be reconciled (Ward 1992). I have in mind that adaptive responses are merely better than certain alternatives, e.g. not reacting to environmental change. My concern was how such responses influence the form of functional relations, rather than whether the outcome maximizes or minimizes an appropriate measure of fitness. Accordingly, adaptive responses in habitat or food choice were usually based simply on maximizing the immediate rate of biomass gain achieved by consumers. The underlying principle is that more resources confer greater power to grow, survive and reproduce.

A more fundamental issue is whether state-dependent measures would be more appropriate for assessing what is adaptive. In Chapter 7, I recognized that it might not be optimal to maximize resource gains when body condition is optimal, in the sense that greater fat stores would diminish fitness. A novel specification of the dynamic fitness function was introduced, in terms of the total biomass produced through both individual growth and reproductive output. The appropriateness of this formulation needs to be evaluated. The state-dependent perspective was not followed up in remaining chapters, because of the challenging complexity of dynamic optimization procedures.

A potential resolution would be to assess the optimal body state at a particular stage in the annual cycle, as was done in Chapter 7, and then adjust immediate food gains to maintain this state. This could introduce some saving in the physiological costs involved in foraging, and in the mortality risks incurred in the process, via reduced foraging time. Such an approach would require partitioning aggregated population biomass between lean body mass and stored reserves, as outlined by Getz and Owen-Smith (1999). With allowance also for differing optimal states among different population segments, this leads into adaptive life history modelling, linking with the evolutionary realm of analysis.

For the ecological issues addressed in this book, I believe that this elaboration is unnecessary. This assumption remains to be tested.

Flux vs. state-dependent specification of mortality

Allied to the above is whether estimates of mortality risk should be based on body condition, rather than on short-term biomass fluxes (see, for example, Illius and Gordon 1998; Illius and O'Connor 2000). Obviously, stored reserves provide an important buffer against short-term fluctuations in resources gained. However, when resource availability becomes inadequate, animals may be expected to work harder to try to maintain their optimal body condition, and incur heightened risk of predation and increased exposure to adverse conditions as a result. Hence, mortality risk may rise before body condition deteriorates, and the extent to which this occurs will depend on the degree to which animals are exposed to predation. The models developed in this book tried to reach a compromise by estimating mortality risk from resource gains averaged over a weekly period, but this may have been inadequate to avoid spuriously accelerating mortality when resources were temporarily deficient.

This equivocation could be overcome by partitioning population biomass between a basic constitutive component and labile reserves, following Getz and Owen-Smith (1999). Mortality risk could then be based directly on this relative ratio for the population, as it changes through the seasonal cycle. However, a starting requirement would be to assess the optimal extent of the stored reserves carried at some cost through the benign season, in anticipation of resource deficits likely to be incurred during the adverse season. Moreover, how mortality risk depends on body state at different stages in the annual cycle can only be guessed. Overall, this points to a need for more empirical research on how and when mortality responds to changing resource supplies in relation to body condition.

Vegetation dynamics

In contrast to the **GMM** formulation of herbivore population dynamics, vegetation dynamics was modelled more crudely through a modified logistic function. This was intended to represent largely the annual growth dynamics in vegetation biomass. Nevertheless, plants that can produce more biomass over the annual cycle have the potential to increase more between years in their abundance, whether vegetatively or reproductively, than species that are more negatively impacted by herbivore grazing (see Chapter 8). However, whether this potential is realised depends crucially on whether there is space to accommodate such expansion. Much of plant dynamics occurs through gap colonization, rather than through direct competition for resources besides light (Tilman 1997).

The logistic production function appeared to be over-compensatory in its response to vegetation depression as a result of grazing, as well as to interruptions in growth due to rainfall deficits during the growing season. Omitted is the important limitation that soil nutrients can impose on vegetation growth. In wet–dry savannas, a pulse of nutrient release is associated with the early rains, which enable microbial decomposition of the plant litter that has accumulated through the dry season (Scholes and Walker 1993). Thereafter, continuing plant growth becomes strongly dependent on further nutrient inputs, especially the nitrogen that may be recycled from the urine and faeces of mammalian herbivores (Hamilton *et al.* 1998).

Furthermore, long-term changes in vegetation composition may be mediated through the effects of herbivores on soil nutrient availability and spatial distribution (Pastor *et al.* 1997). In addition, seed dispersal may be facilitated or inhibited by herbivores (O'Connor 1991; Miller 1996). A full systems model of herbivore–plant dynamics would need to accommodate these feedbacks. Because of the limitations of the logistic production function, I restricted applications of the **GMM** model largely to within-year dynamics, or to periods spanning only a few years.

Spatial heterogeneity

In developing the **GMM** model from the herbivore perspective, spatial heterogeneity was considered only implicitly, in terms of functional distinctions among resource components and their relative extent in the environment. This can be justified because large herbivores are highly mobile, and can move readily over substantial distances. Nevertheless, there are metabolic costs associated with movement, and the spatial proximity of alternative resources could be an important consideration in their use. Principles of landscape ecology have been used to develop spatially explicit models of resource use by large herbivores, notably by Turner *et al.* (1993).

Because plants are sessile, except as propagules, an explicit spatial context is demanded if vegetation dynamics is to be modelled effectively. Interactions take place mainly with neighbours, and population changes are manifested largely as changes in the area covered by particular plant species. Grid-based models offer promising approaches for handling such spatial dynamics (Weber *et al.* 1998; Wiegand *et al.* 1999). Nevertheless, a metaphysiological approach would remain valid for evaluating the growth dynamics of the plants occupying each cell.

Large versus small herbivores

Models in this book were developed with large mammalian herbivores in mind. Such species are generally accepted to be food-limited, and also to have considerable impacts on vegetation (Coe *et al.* 1976; Owen-Smith and Danckwerts 1997; Peterson 1999). Although it has been claimed that the vast majority of smaller herbivores are limited more commonly by predation and disease (Hairston *et al.* 1960), even populations of foliage-feeding insects are not divorced from resource limitations (Hunter *et al.* 1992).

However, rodent and insect herbivores may be more strongly dependent on high-quality vegetation components, and hence on temporal variability in the availability of these plant parts and growth stages. The effects of strong seasonality and non-linear density dependence on the dynamics of a small rodent in an African savanna environment were well captured by Leirs *et al.* (1997). Many foliage-feeding Lepidoptera schedule the emergence of their larval feeding stages to coincide with the new leaf flush on their food plants (Coley and Barone 1996). During the dormant season for plant growth, many insects likewise enter dormant stages as eggs or pupae (Price 1984). The extent to which the cyclic or non-cyclic dynamics of rodent and lagomorph populations in the far north depends on resource interactions remains unclear (Krebs *et al.* 1998). I suggest that timing in life-history processes may be a more important consideration in the dynamics of insect and small mammal populations than short-term rates of resource gain, in comparison with large mammals.

Broader issues

Resource dependence

The central thesis of this book is that population processes, including predation, disease transmission and life-history schedules, can only be understood in the context of resource availability. This contrasts with the pervasive demographic emphasis on density dependence in the literature without clarification of the supposed mechanisms. Obviously, density dependent feedbacks are important regulators of population dynamics. However, the density level about which regulation might occur is strongly influenced by resource supplies, which vary spatially as well as over time. Density-dependent signals become clear only when resource variability is controlled (Owen-Smith 1990; Mduma *et al.* 1999).

I suggest that attempts to detect density dependence in long-term records of the population dynamics of various insects and other species, without corresponding information on fluctuations in resource supplies as well as in conditions, may be somewhat hopeless (Woiwod and Hanski 1992; Sale and Tolimieri 2000). Although some part of predation may be directly dependent on the density of prey, an additional component is likely to be interactive with resources, as was outlined in Chapter 10.

Disequilibrial dynamics

Equilibrium thinking – the 'balance of nature' paradigm – still permeates too much of ecological theory. I see too many formal models occupying the pages of journals like *The American Naturalist* that seek analytic solutions around near-equilibrium states. As a former physical scientist (in my early student days!), I perceive such modelling attempts as little more than physics masquerading as biology. Even textbooks depict the physical analogy of a ball perched in a basin to represent community stability (Begon *et al.* 1996, p. 839). As summarized in Chapter 2, the challenging departure of biology from most of physics encompasses two basic elements: (a) biological systems are open, capturing energy to maintain order against the forces of disintegration, always far from equilibrium (Nicolis and Prigogine 1977; Weber *et al.* 1988); (b) biological entities respond adaptively to counteract or accommodate change, at multiple levels of integration (Kaufmann 1993).

Much of what we perceive as equilibria are merely the outcome of inertia; another part is due to the active steering of organisms to maintain some form of homeostasis. Nevertheless, the underlying disequilibrium in the dynamic processes does not preclude some form of stability from emerging at higher levels of integration. However, a more appropriate depiction of this stability would be an object jiggling to stay in place in a flowing stream, rather than a ball resting in a static notch.

Seasonal oscillators

Also affecting apparent stability are functional relations that are inherently, and often quite strongly, non-linear. Hence, over a wide range in conditions little changes, then over a crucial region change can become drastic. This may arise partly from limits of adaptive compensation being reached. Thus system dynamics becomes strongly sensitive to threshold transitions, as captured theoretically by 'catastrophe theory' (Thom 1975), and practically by concepts such as event-driven management (Westoby *et al.* 1989).

Rather than seeking some equilibrium state, such as ecological carrying capacity, the system dynamics would be better represented as seasonally alternating between an upper saturation density that is never reached because of inadequate time during the benign season, and a lower density that would ultimately be zero if dormant season conditions persisted long enough (Fig. 14.1). Furthermore, inside the system is a 'demon' that slackens its control on the upswings and pulls back against the downswings, thereby counteracting the environmental perturbations that may act to disrupt the dynamic persistence of the oscillations.

This concept stands in radical contrast to the kind of instability generated by discrete-time formulations of the logistic model, whereby populations are tossed way above equilibrium in one time step, then finding themselves materially unsupported crash way below equilibrium in the next step, leading, not surprisingly, to chaos. This kind of dynamics emerges when resource limitations on material biomass are not specified, and when there is no adaptive control by the organisms constituting the population. Chaotic oscillations may well emerge when functional relations are strongly non-linear and there is an appropriate lag between cause and response, but expectations of their occurrence in real-world biota are rather exaggerated, I believe. Instead, more attention needs to be directed towards identifying variation in environmental drivers that may disrupt stability, relative to the spatial or functional heterogeneity that can enable counteractive responses.

I suspect that an appropriate mathematical framework could be developed to handle the dynamics of systems that function as seasonal oscillators

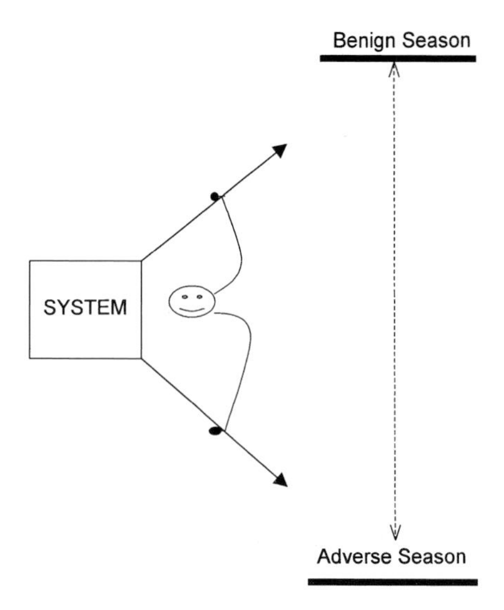

Figure 14.1. Representation of a seasonal oscillator alternating between but never reaching two attracting states: the high abundance potentially supported by benign season resources, and the zero abundance that would eventually be reached if vegetation dormancy persisted long enough. Within the system is a 'demon' counteracting the downswings and slackening off on the upswings.

between alternative attracting states, with neither being closely attained because of limitations of inertia and time. The seasonality could itself be dynamically stabilizing or destabilizing, depending on what it influences and how. An analysis of the factors governing the distinction between such alternative outcomes could be illuminating, and relevant to more fundamental understanding of herbivore–vegetation dynamics.

Dynamic niche

The Hutchinsonian concept of the ecological niche as some location of a species within a multi-dimensional hyperspace in terms of resources and conditions (Hutchinson 1958) is misleadingly static. I believe that the earlier Eltonian view of niche as the functional role ('profession') of a species within its community (Elton 1927) is more appropriate, although it is less readily quantified. The latter captures the dynamic reality that organisms do not remain constant in the environmental space they occupy, because conditions and circumstances continually change, seasonally and between years. The food eaten, temperatures experienced, and hazards encountered are correspondingly variable.

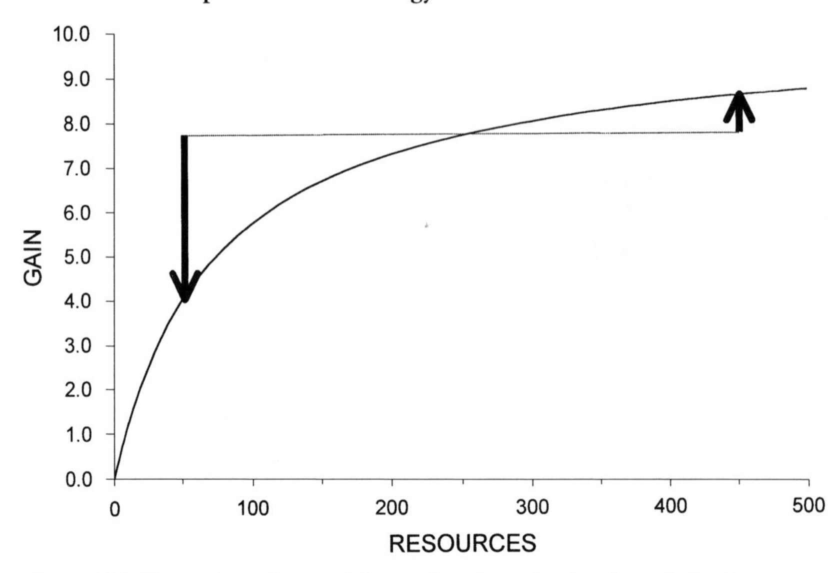

Figure 14.2. Illustration of Jensen's inequality. For a decelerating relation between nutritional gains and resource availability in a variable environment, the gains under favourable conditions are more than counterbalanced by the losses when conditions are adverse, as indicated by the arrows.

Moreover, Jensen's inequality applies (see Ruel and Ayres 1999), whereby the gains when conditions are favourable are more than counterbalanced by the losses when adverse conditions ensue (Fig. 14.2). Rather than being perched on some adaptive peak (Wright 1932; cf. Futuyma 1979, p.329), within the ecological realm organisms traverse a rugose plateau landscape, disrupted by diverse pitfalls and regular seasonal ravines. This landscape shifts past in time like a continual escalator, impelling organisms to steer adaptively to maintain their energizing elevation and find appropriate stepping stones through the seasonal sumps (Fig. 14.3). More specifically, while drawing energizing and material resources from the trophic firmament below, organisms must also evade becoming resources for the trophic level above, and avoid excessive physiological drains on their energy reserves. Furthermore, the resource firmament subsides whenever contacted. Populations may sink into oblivion when key stepping stones, bridges or buffers are missing, or offer inadequate escape from the suction of predators and parasites. Add to this herbivore model counterpart constructs for adaptively responding vegetation and predators, and we have a dynamic resource-dependent depiction of the 'game of life'.

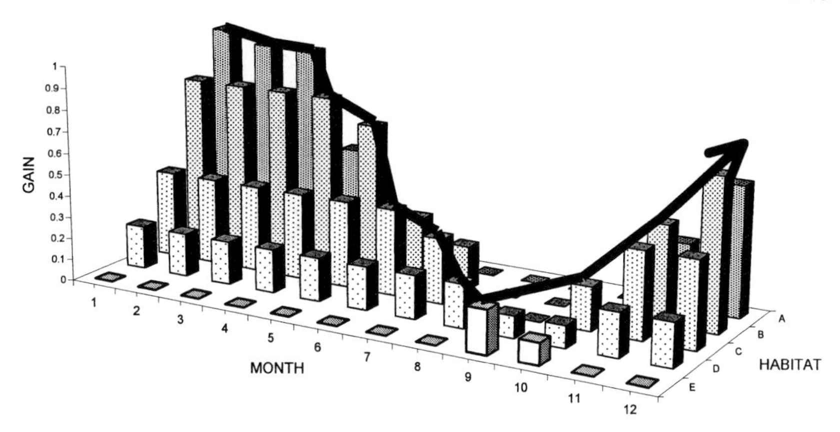

Figure 14.3. Visual representation of idealized trajectory of consumer biomass through time within a heterogeneous and interacting resource hyperspace. The optimal trajectory via habitat 'stepping stones' across a seasonal ravine is indicated by arrows. Losses to predators and parasites are not depicted.

Overview

This final chapter summarizes the paradigm of disequilibrium resource dynamics in a temporally changing and spatially variable world that I attempted to capture through the **GMM** metaphysiological model. Persistence under such conditions depends crucially on adaptive steering to dodge the pitfalls and traverse seasonal ravines. My final statement is a call for models assuming homogeneous, constant conditions, near-equilibrium situations, or linear response functions, to be abandoned as credible representations of biological systems.

References

Abrams, P. A. 1994. The fallacies of "ratio dependent" predation. *Ecology* **75**: 1842–50.

Abrams, P. 1997. Anomalous predictions of ratio-dependent models of predation. *Oikos* **80**: 163–71.

Abrams, P. A. & L. R. Ginzburg. 2000. The nature of predation: prey dependent, ratio dependent or neither. *Trends Ecol. Evol.* **15**: 337–41.

Adamczewski, J. Z., C. C. Gates, R. J. Hudson & M. A. Price. 1987. Seasonal changes in body composition of mature female caribou and their calves on an Arctic island with limited winter resources. *Can. J. Zool.* **65**: 1149–57.

Adamczewski, J. Z., R. J. Hudson & C. C. Gates. 1993. Winter energy balance and activity of female caribou on Coats Island, Northwest Territories: the relative importance of foraging and body reserves. *Can. J. Zool.* **71**: 1221–9.

Akcakaya, H. R. 1992. Population cycles of mammals: evidence for ratio-dependent predation hypothesis. *Ecol. Monogr.* **62**: 119–42.

Akcakaya, H. R., R. Arditi & L. R. Ginzburg. 1995. Ratio-dependent predation: an abstraction that works. *Ecology* **76**: 995–1004.

Alexander, R. McN. 1991. Optimization of gut structure and diet for higher vertebrate herbivores. *Phil. Trans. R. Soc. Lond.* B **333**: 249–55.

Allden, W. G. & I. A. D. McD. Whittaker. 1970. Determinants of herbage intake by grazing sheep: the interrelationships of factors influencing herbage intake and availability. *Aust. J. Agric. Res.* **21**: 755–66.

Allen-Rowlandson, T. S. 1980. The social and spatial organisation of the greater kudu in the Andries Vosloo Kudu Reserve, Eastern Cape. MSc thesis, Rhodes University.

Anderson, E. W. & R. J. Scherzinger. 1975. Improving quality of winter forage for elk by cattle grazing. *J. Range Mgmt* **28**: 120–5.

Andrew, M. H. 1988. Grazing impact in relation to livestock watering points. *Trends Ecol. Evol.* **3**: 336–9.

Arditi, R & B. Dacorogna. 1988. Optimal foraging on arbitrary patch distributions and the definition of habitat patches. *Am. Nat.* **131**: 837–46.

Arditi, R. & L. R. Ginzburg. 1989. Coupling in predator-prey dynamics: ratio dependence. *J. Theor. Biol.* **139**: 311–26.

Ash, A. J., J. G. McIvor, J. P. Corfield & W. H. Winter. 1996. How land condition alters plant–animal relationships in Australia's tropical rangelands. *Agric. Ecosyst. Environ.* **56**: 77–92.

Astrom, A., P. Lundberg & K. Danell. 1990. Partial prey consumption by browsers: trees as patches. *J. Anim. Ecol.* **59**: 287–300.

Baile, C. A. & J. M. Forbes. 1974. Control of feed intake and regulation of energy balance in ruminants. *Physiol. Rev.* **54**: 160–216.

Bailey, D. W., J. E. Gross, E. A. Laca, L. R. Rittenhouse, M. B. Coughenour, D. M. Swift & P. L. Sims. 1996. Mechanisms that result in large herbivore grazing distribution patterns. *J. Range Mgmt* **49**: 386–400.

Bailey, D. W., J. W. Walker & L. R. Rittenhouse. 1990. Sequential analysis of cattle location: day to day movement patterns. *Appl. Anim. Behav. Sci.* **25**: 137–48.

Baker, D. L. & N. T. Hobbs. 1987. Strategies of digestion: digestive efficiency and retention time of forage diets in montane ungulates. *Can. J. Zool.* **65**: 1978–84.

Bartmann, R. M., G. C. White & L. H. Carpenter. 1992. Compensatory mortality in a Colorado mule deer population. *Wildl. Monogr.* No. 121, 39pp.

Bednekoff, P. A. & A. I. Houston. 1994. Optimizing fat reserves over the entire winter: a dynamic model. *Oikos* **71**: 408–15.

Begon, M., J. Harper & C. Townsend. 1996. *Ecology: Individuals, Populations and Communities*, 3rd edition. Blackwell, Oxford.

Behnke, R. H., I. Scoones & C. Kerven (eds). 1993. *Range Ecology at Disequilibrium*. ODI Publications, London.

Bell, R. H. V. 1970. The use of the herb layer by grazing ungulates in the Serengeti. In *Animal Populations in Relation to their Food Resources*, ed. A. Watson (*Symp. Brit. Ecol. Soc.*), pp. 111–24. Blackwell, Oxford.

Bell, R. H. V. 1971. A grazing ecosystem in the Serengeti. *Scient. Am.* **225**: 86–93.

Bellows, T. S. 1981. The descriptive properties of some models for density dependence. *J. Anim. Ecol.* **50**: 139–56.

Belovsky, G. E. 1978. Diet optimization in a generalist herbivore: the moose. *Theor. Popul. Biol.* **14**: 105–34.

Belovsky, G. E. 1984. Herbivore optimal foraging: a comparative test of three models. *Am. Nat.* **124**: 97–115.

Belovsky, G. E. 1986. Optimal foraging and community structure: implications for a guild of generalist grassland herbivores. *Oecologia* **70**: 35–52.

Belovsky, G. E. 1990. How important are nutrient constraints in optimal foraging models or are spatial/temporal factors more important? In *Behavioral Mechanisms of Food Selection*, ed. R. N. Hughes, pp. 255–80. Springer-Verlag, Berlin and Heidelberg.

Belovsky, G. E. 1994. How good must models and data be in ecology? *Oecologia* **100**: 475–80.

Belovsky, G. E. & O. J. Schmitz. 1993. Owen-Smith's evaluation of herbivore foraging models: what is constraining? *Evol. Ecol.* **7**: 525–9.

Belovsky, G. E. & J. B. Slade. 1986. Time budgets of grassland herbivores: body size similarities. *Oecologia* **70**: 53–62.

Bergman, C. M., J. M. Fryxell & C. C. Gates. 2000. The effect of tissue complexity and sward height on the functional response of Wood Bison. *Funct. Ecol.* **14**: 61–9.

Berryman, A. A. 1992. The origins and evolution of predator-prey theory. *Ecology* **73**: 1530–5.

Berryman, A. A. 1999. Alternative perspectives on consumer-resource dynamics: a reply to Ginzburg. *J. Anim. Ecol.* **68**: 1263–6.

Berryman, A. A., A. P. Gutierrez & R. Arditi. 1995. Credible, parsimonious and useful predator-prey models – a reply to Abrams, Gleeson and Sarnelle. *Ecology* **76**: 1980–5.

Boggs, C. L. 1992. Resource allocation: exploring connections between foraging and life history. *Funct. Ecol.* **6**: 508–18.

Boomker, E. A. 1987. Fermentation and digestion in the kudu. PhD thesis, University of Pretoria.

Borner, M., C. D. FitzGibbon, M. Borner, T. M. Caro, W. K. Lindsay. D. A. Collins & M. E. Holt. 1987. The decline of the Serengeti Thomson's gazelle population. *Oecologia* **73**: 32–40.

Botkin, D. B. 1990. *Discordant Harmonies. A New Ecology for the Twenty First Century.* Oxford University Press, Oxford.

Bradbury, J. W., S. L. Vehrencamp, K. E. Clifton & L. M. Clifton. 1996. The relationship between intake rate and local forage abundance in wild Thomson's gazelles. *Ecology* **77**: 2237–55.

Brooks, P. M. & J. A. W. Macdonald. 1983. The Hluhluwe-Umfolozi Reserve: An ecological case history. In *Management of Large Mammals in African Conservation Areas*, ed. R. N. Owen-Smith, pp. 51–77. Haum, Pretoria.

Brown, J. H., P. A. Marquet & M. L. Taper. 1993. Evolution of body size: consequences of an energetic definition of fitness. *Am. Nat.* **142**: 573–84.

Burgman, M., S. Ferson & H. R. Akcakaya. 1992. *Risk Assessment in Conservation Biology*. Chapman & Hall, London.

Burlison, A. J., J. Hodgson & A. W. Illius. 1991. Sward canopy structure and the bite dimensions and bite weight of grazing sheep. *Grass Forage Sci.* **46**: 29–38.

Cahill, G. F. 1970. Starvation in man. *New Engl. J. Med.* **282**: 668–75.

Calder, W. A. III. 1984. *Size, Function and Life History*. Harvard University Press, Cambridge, MA.

Campling, R. C. 1970. Physical regulation of voluntary intake. In *Physiology of Digestion and Metabolism in the Ruminant*, ed. A. J. Phillipson, pp. 226–34. Oriel Press, Newcastle.

Case, T. J. 2000. *An Illustrated Guide to Theoretical Ecology*. Oxford University Press, Oxford.

Caughley, G. 1970. Eruption of ungulate populations with emphasis on Himalayan Thar in New Zealand. *Ecology* **51**: 53–72.

Caughley, G. 1976a. Plant-herbivore systems. In *Theoretical Ecology*, ed. R. M. May, pp. 94–113. Blackwell, Oxford.

Caughley, G. 1976b. Wildlife management and the dynamics of ungulate populations. In *Advances in Applied Biology*, Vol. 1, ed. T. H. Coaker, pp. 183–245. Academic Press, London.

Caughley, G. 1976c. The elephant problem – an alternative hypothesis. *E. Afr. Wildl. J.* **14**: 265–83.

Caughley, G. 1982. Vegetation complexity and the dynamics of modelled grazing systems. *Oecologia* **54**: 309–12.

Caughley, G. 1994. Directions in conservation biology. *J. Anim. Ecol.* **63**: 215–44.

Caughley, G. & J. H. Lawton. 1981. Plant-herbivore systems. In *Theoretical Ecology*, 2nd edition, ed. R. M. May, pp. 132–66. Blackwell, Oxford.

Caughley, G. & A. R. E. Sinclair. 1994. *Wildlife Ecology and Management*. Blackwell, Oxford.

Chan-McLeod, A. C. A., R. G. White & D. F. Holleman. 1994. Effects of protein and energy intake, body condition, and season on nutrient partitioning and milk production in caribou and reindeer. *Can. J. Zool.* **72**: 938–47.

Chapin, F. S. III. 1980. The nutrition of wild plants. *A. Rev. Ecol. Syst.* **11**: 233–60.

Charnov, E. L. 1976. Optimal foraging: attack strategy of a mantid. *Am. Nat.* **110**: 141–51.

Clark, C. W. 1990. *Mathematical Bioeconomics. The Optimal Management of Renewable Resources*. 2nd edition. John Wiley, New York.

Clark, C. W. 1993. Dynamic models of behavior: an extension of life history theory. *Trends Ecol. Evol.* **8**: 205–9.

Clark, C. W. & M. Mangel. 1999. *Dynamic State Variable Models in Ecology*. Oxford University Press, Oxford.

Clutton-Brock, T. H., S. D. Albon & F. E. Guinness. 1989. Fitness costs of gestation and lactation in wild mammals. *Nature* **337**: 260–2.

Clutton-Brock, T. H., A. W. Illius, K. Wilson, B. T. Grenfell, A. MacColl & S. D. Albon. 1997. Stability and instability in ungulate populations: an empirical analysis. *Am. Nat.* **149**: 196–219.

Clutton-Brock, T. H., M. Major & F. E. Guinness. 1985. Population regulation in male and female red deer. *J. Anim. Ecol.* **54**: 831–46.

Clutton-Brock, T. H., O. F. Price, S. D. Albon & P. A. Jewell. 1991. Persistent instability and population regulation in Soay sheep. *J. Anim. Ecol.* **60**: 593–608.

Coe, M. J., D. H. M. Cumming & J. Phillipson. 1976. Biomass and production of large herbivores in relation to rainfall and primary production. *Oecologia* **22**: 341–54.

Coley, P. D. & J. A. Barone. 1996. Herbivory and plant defenses in tropical forests. *A. Rev. Ecol. Syst.* **27**: 305–35.

Coley, P. D., J. P. Bryant & F. S. Chapin III. 1985. Resource availability and plant anti-herbivore defense. *Science* **230**: 895–9.

Cooper, S. M. 1985. Factors influencing the utilization of woody plants and forbs by ungulates. PhD thesis, University of the Witwatersrand.

Cooper, S. M. 1990. The hunting behaviour of spotted hyenas in a region containing both sedentary and migratory prey. *Afr. J. Ecol.* **28**: 131–41.

Cooper, S. M. & N. Owen-Smith. 1986. Effects of plant spinescence on large mammalian herbivores. *Oecologia* **68**: 446–55.

Cooper, S. M., N. Owen-Smith & J. P. Bryant. 1988. Foliage acceptability to browsing ruminants in relation to seasonal changes in the leaf chemistry of woody plants in a South Africa savanna. *Oecologia* **75**: 336–42.

Coppock, D. L., D. M. Swift, J. E. Ellis & K. Galvin. 1986. Seasonal patterns of energy allocation to basal metabolism, activity and production for livestock in a nomadic pastoral ecosystem. *J. Agric. Sci., Camb.* **107**: 357–65.

Coughenour, M. B. 1991. Spatial components of plant herbivore interactions in pastoral, ranching and natural ungulate ecosystems. *J. Range Mgmt* **44**: 530–42.

Coughenour, M. B. & F. J. Singer. 1996. Elk population processes in Yellowstone National Park under the policy of natural regulation. *Ecol. Appl.* **6**: 573–93.

Crawley, M. J. 1983. *Herbivory: The Dynamics of Animal-Plant Interactions.* (Studies in Ecology Series No. 10.) Blackwell, Oxford.

Danckwerts, J. E., A. J. Aucamp & H. J. Barnard. 1983. Herbaceous species preference by cattle in the false thornveld of eastern Cape. *Proc. Grass. Soc. Sth. Afr.* **18**: 89–94.

Dean, W. R. J. & I. A. W. Macdonald. 1994. Historical changes in stocking rates of domestic livestock as a measure of semi-arid and arid rangeland degradation in the Cape Province, South Africa. *J. Arid Envir.* **26**: 281–98.

DeAngelis, D. L. 1992. *Dynamics of Nutrient Cycling and Food Webs.* Chapman & Hall, London.

DeAngelis, D. L., R. A. Goldstein & R. V. O'Neil. 1975. A model for trophic interaction. *Ecology* **56**: 881–92.

DeAngelis, D. L. & L. J. Gross (eds). 1992. *Individual-Based Models and Approaches in Ecology.* Chapman & Hall, London.

Detling, J. K., W. J. Parton & H. W. Hunt. 1979. A simulation model of *Bouteloua gracilis* biomass dynamics on the North American shortgrass prairie. *Oecologia* **38**: 167–91.

de Vos, V., R. C. Bengis & H. J. Coetzee. 1983. Population control of large mammals in the Kruger National Park. In *Management of Large Mammals in African Conservation Areas*, ed. R. N. Owen-Smith, pp. 213–31. Haum, Pretoria.

Dinius, D. A. & B. R. Baumgardt. 1970. Regulation of food intake in ruminants: 6. Influence of caloric density of pelleted rations. *J. Dairy Sci.* **53**: 311–16.

Distel, R. A., E. A. Laca, T. C. Griggs & M. W. Demment. 1995. Patch selection by cattle: maximization of intake rate in horizontally heterogeneous pastures. *Appl. Anim. Behav. Sci.* **45**: 11–21.

Duncan, P. 1975. Topi and their food supply. PhD thesis, University of Nairobi.

Duncan, P., T. J. Foose, I. J. Gordon, C. G. Gakahu & M. Lloyd. 1990. Comparative nutrient extraction from forages by bovids and equids: a test of the nutritional model of equid/bovid competition and coexistence. *Oecologia* **84**: 411–18.

Dunham, K. M. 1982. The foraging behaviour of impala. *S. Afr. J. Wildl. Res.* **12**: 36–40.

Dunham, K. M. & M. G. Murray. 1982. The fat reserves of impala. *Afr. J. Ecol.* **20**: 81–7.

du Toit, J. T. 1988. Patterns of resource use within the browsing ruminant guild in the central Kruger National Park. PhD thesis, University of the Witwatersrand.

du Toit, J. T. & N. Owen-Smith. 1989. Body size, population metabolism and habitat specialization among African large herbivores. *Am. Nat.* **133**: 736–40.

Dyer, M. I., C. L. Turner & T. R. Seastedt. 1993. Herbivory and its consequences. *Ecol. Applic.* **3**: 10–16.

Edwards, G. R., A. J. Parsons, P. D. Penning & J. A. Newman. 1995. Relationship between vegetation state and bite dimensions of sheep grazing contrasting plant species and its implications for intake rate and diet selection. *Grass Forage Sci.* **50**: 378–88.

Ellis, J. E., & D. M. Swift. 1988. Stability of African pastoral ecosystems: alternative paradigms and implications for development. *J. Range Mgmt* **41**: 450–9.

Elser, J. J., R. W. Sterner, E. Gorokhova, W. F. Fagan, T. A. Markow, J. B. Cotner, J. F. Harrison, S. E. Hobbie, G. M. Odell & L. J. Weides. 2000. Biological stoichiometry from genes to ecosystems. *Ecol. Lett.* **3**: 540–50.

Elton, C. 1927. *Animal Ecology*. Sidgwick & Jackson, London.

Eltringham, S. K. 1974. Changes in the large mammal community of Mweya Peninsula, Rwenzori National Park, following removal of hippopotamus. *J. Appl. Ecol.* **11**: 855–65.

Eltringham, S. K. 1980. A quantitative assessment of range usage by large African mammals with particular reference to the effects of elephants on trees. *Afr. J. Ecol.* **18**: 53–71.

Farnsworth, K. D. & A. W. Illius. 1996. Large grazers back in the fold: generalizing the prey model to incorporate mammalian herbivores. *Funct. Ecol.* **10**: 678–80.

Farnsworth, K. D. & A. W. Illius. 1998. Optimal diet choice for large herbivores: an extended contingency model. *Funct. Ecol.* **12**: 74–81.

Ferrar, A. A. & B. H. Walker. 1974. An analysis of herbivore/habitat relationships in Kyle National Park, Rhodesia. *J. S. Afr. Wildl. Mgmt Ass.* **4**: 137–47.

Finch, V. A. 1972. Thermoregulation and heat balance of the East African eland and hartebeest. *Am. J. Physiol.* **222**: 1374–9.

FitzGibbon, C. D. & J. H. Fanshawe. 1988. Stotting in Thomson's gazelles: an honest signal of condition. *Behav. Ecol. Sociobiol.* **23**: 69–74.

Forbes, T. D. A. 1988. Researching the plant-animal interface: the investigation of ingestive behaviour in grazing animals. *J. Anim. Sci.* **66**: 2369–79.

Fowler, C. W. 1981. Density dependence as related to life history strategy. *Ecology* **62**: 602–10.

Fowler, C. W. 1987. A review of density dependence in populations of large mammals. In *Current Mammalogy*, Vol. 1, ed. H. H. Genoways, pp. 401–41. Plenum Press, New York.

Freeland, W. J. & D. H. Janzen. 1974. Strategies in herbivory by mammals: the role of plant secondary compounds. *Am. Nat.* **108**: 269–89.

Fretwell, S. D. 1972. *Populations in a Seasonal Environment*. Princeton University Press, Princeton, NJ.

Fritz, H. & M. de Garine-Witchatitsky. 1996. Foraging in a social antelope: effects of group size on foraging choices and resource perception in impala. *J. Anim. Ecol.* **65**: 736–42.

Fritz, H., M. de Garine-Witchatitsky & G. Letessier. 1996. Habitat use by sympatric wild and domestic herbivores in an African savanna woodland: the influence of cattle spatial behaviour. *J. Appl. Ecol.* **33**: 589–98.

Fritz, H. & P. Duncan. 1994. On the carrying capacity for large ungulates of African savanna ecosystems. *Proc. R. Soc. Lond.* B **256**: 77–82.

Fryxell, J. M. 1991. Forage quality and aggregation by large herbivores. *Am. Nat.* **138**: 478–98.

Fryxell, J. M., J. Greever & A. R. E. Sinclair. 1988. Why are migratory ungulates so abundant? *Am. Nat.* **131**: 781–98.

Fryxell, J. M. & P. Lundberg. 1994. Diet choice and predator prey dynamics. *Evol. Ecol.* **8**: 407–21.

Fryxell, J. M. & P. Lundberg. 1998. *Individual Behavior and Community Dynamics*. Chapman & Hall, London.

Futuyma, D. J. 1979. *Evolutionary Biology*. Sinauer Associates, Sunderland, MA.

Gaillard, J. M., M. Festa-Bianchet, N. G. Yoccoz & A. Loison. 2000. Temporal variation in vital rates and population dynamics of large herbivores. *A. Rev. Ecol. Syst.* **31**: 367–93.

Gallivan, G. I., J. Culverwell & R. Girdwood. 1995. Body condition indices of impala: effect of age class, sex, season and management. *S. Afr. J. Wildl. Res.* **25**: 23–31.

Garten, C. T. Jr., 1978. Multivariate perspectives on the ecology of plant mineral element composition. *Am. Nat.* **112**: 533–44.

Geist, V. 1974. On the relationship of social evolution and ecology in ungulates. *Am. Zool.* **14**: 205–20.

Gerhart, K. L., R. G. White, R. D. Cameron & D. E. Russell. 1996. Body composition and nutrient reserves of arctic caribou. *Can. J. Zool.* **74**: 136–46.

Getz, W. M. 1984. Population dynamics: a per-capita resource approach. *J. Theor. Biol.* **108**: 623–43.

Getz, W. M. 1991. A unified approach to multispecies modelling. *Natur. Res. Model.* **5**: 393–421.

Getz, W. M. 1993. Metaphysiological and evolutionary dynamics of populations exploiting constant and interactive resources: r-K selection revisited. *Evol. Ecol.* **7**: 287–305.

Getz, W. M. 1994. A metaphysiological approach to modelling ecological populations and communities. In *Frontiers in Mathematical Biology*, ed. S. A. Levin, pp. 411–42. Springer-Verlag, Berlin.

Getz, W. M. & N. Owen-Smith. 1999. A metaphysiological population model of storage in variable environments. *Natur. Res. Model.* **12**: 197–230.

Giesecke, D. & N. O. van Gylswyk. 1975. A study of feeding types and certain rumen functions in six species of South African wild ruminants. *J. Agric. Sci., Camb.* **85**: 75–83.

Gillard, P. & R. Monypenny. 1990. A decision support model to evaluate the effects of drought and stocking rate on beef cattle properties in Northern Australia. *Agric. Syst.* **34**: 37–52.

Gillingham, M. P., K. L. Parker & T. A. Hanley. 1997. Forage intake by black tailed deer in a natural environment: bout dynamics. *Can. J. Zool.* **75**: 1118–28.

Ginnett, T. F., J. A. Dankosky, G. Deo & M. W. Demment. 1999. Patch depression in grazers: the roles of biomass distribution and residual stems. *Funct. Ecol.* **13**: 37–44.

Ginnett, T. F. & M. W. Demment. 1995. The functional response of herbivores: analysis and test of a simple mechanistic model. *Funct. Ecol.* **9**: 376–84.

Ginnett, T. F. & M. W. Demment. 1997. Sex differences in giraffe foraging behaviour at two scales. *Oecologia* **110**: 291–300.

Ginzburg, L. R. 1998. Assuming reproduction to be a function of consumption raises doubts about some popular predator-prey models. *J. Anim. Ecol.* **67**: 325–7.

Gleeson, S. K. 1995. Density dependence is better than ratio dependence. *Ecology* **75**: 1834–5.

Goddard, J. 1968. Food preferences of two black rhinoceros populations. *E. Afr. Wildl. J.* **6**: 1–18.

Gordon, I. J. 1988. Facilitation of red deer grazing by cattle and its impact on red deer performance. *J. Appl. Ecol.* **25**: 1–10.

Gordon, I. J. & A. W. Illius. 1988. Incisor arcade structure and diet selection in ruminants. *Funct. Ecol.* **2**: 15–22.

Gordon, I. J. & A. W. Illius. 1994. The functional significance of the browser-grazer dichotomy in African ruminants. *Oecologia* **98**: 167–75.

Gotelli, N. J. 1995. *A Primer of Ecology*. Sinauer, Sunderland, MA.

Greenwood, G. B. & M. W. Demment. 1988. The effect of fasting on short-term cattle grazing behaviour. *Grass Forage Sci.* **43**: 377–86.

Grenfell, B. T. 1992. Parasitism and the dynamics of grazing systems. *Am. Nat.* **139**: 907–29.

Grenfell, B. T., O. F. Price, S. D. Albon & T. H. Clutton-Brock. 1992. Overcompensation and population cycles in an ungulate. *Nature* **355**: 823–6.

Grenfell, B. T., K. Wilson, B. F. Finkenstadt, T. N. Coulson, S. Murray, S. D. Albon, J. M. Pemberton, T. H. Clutton-Brock & M. J. Crawley. 1998. Noise and determinism in synchronized sheep dynamics. *Nature* **394**: 674–7.

Grime, J. P. 1977. Evidence for the existence of three primary strategies in plants and its relevance to ecological and evolutionary theory. *Am. Nat.* **111**: 1169–94.

Gross, J. E., O. V. Alkon & M. W. Demment. 1996. Nutritional ecology of dimorphic herbivores: digestion and passage rates in Nubian ibex. *Oecologia* **102**: 170–8.

Grossman, D., P. L. Holden & R. F. H. Collinson. 1999. Veld management on the game ranch. In *Veld Management in South Africa*, ed. N. M. Tainton, pp. 261–79. University of Natal Press, Pietermaritzburg.

Grunow, J. O., H. T. Groeneveld & H. C. du Toit. 1980. Above ground dry matter dynamics of the grass layer in the Nylsvley tree savanna. *J. Ecol.* **68**: 877–89.

Gulland, F. M. D. 1992. Role of nematode infections in mortality of Soay sheep during a population crash on St. Kilda. *Parasitology* **105**: 493–503.

Gurney, W. S. C. & R. M. Nisbet. 1998. *Ecological Dynamics*. Oxford University Press, Oxford.

Gustafson, L. L., W. L. Franklin, R. J. Sarno, R. L. Hunter, K. M. Young, W. E. Johnson & M. J. Behl. 1998. Predicting early mortality of newborn guanaco by birth mass and haematological parameters: a provisional model. *J. Wildl. Mgmt* **62**: 24–35.

Hagerman, A. E., C. T. Robbins, Y. Weerasuriya, T. C. Wilson & C. McArthur. 1992. Tannin chemistry in relation to digestion. *J. Range Mgmt* **45**: 57–62.

Hairston, N. G., F. E. Smith & L. B. Slobodkin. 1960. Community structure, population control and competition. *Am. Nat.* **94**: 421–5.

Hamilton, E. W. III, M. S. Giovannini, S. A. Moses, J. S. Coleman & S. J. McNaughton. 1998. Biomass and mineral element responses of a Serengeti short-grass species to nitrogen supply and defoliation: compensation requires a critical [N]. *Oecologia* **116**: 407–18.

Hammond, K. A. & J. Diamond. 1997. Maximal sustained energy budgets in humans and animals. *Nature* **386**: 457–62.

Hanley, T. A. 1997. A nutritional view of understanding and complexity in the problem of diet selection by deer (Cervidae). *Oikos* **79**: 209–18.

Hanski, J. 1989. Metapopulation dynamics: does it help to have more of the same? *Trends Ecol. Evol.* **4**: 113–14.

Hanski, J. 1999. *Metapopulation Ecology*. Oxford University Press.

Hanson, J. D., W. J. Parton & G. S. Innis. 1985. Plant growth and production of grassland ecosystems: a comparison of modelling approaches. *Ecol. Model.* **29**: 131–44.

Hansson, L., L. Fahrig & G. Merriam (eds). 1995. *Mosaic Landscapes and Ecological Processes*. Chapman & Hall, London.

Hardy, M. B., C. R. Hurt & O. J. H. Bosch. 1999. Veld condition assessment. Grassveld. In *Veld Management in South Africa*, ed. N. M. Tainton, pp. 194–206. University of Natal Press, Pietermaritzburg.

Harper, J. L. 1977. *Population Biology of Plants*. Academic Press, London and New York.

Hassell, M. P., J. H. Lawton & R. M. May. 1976. Patterns of dynamical behaviour in single-species populations. *J. Anim. Ecol.* **45**: 471–86.

Hastings, A. 1997. *Population Biology. Concepts and Models*. Springer-Verlag, New York.

Hatch, G. P. & D. M. Stafford Smith. 1997. The bioeconomic implications of various drought management strategies for a communal cattle herd in a semi-arid savanna of KwaZulu-Natal. *Afr. J. Range Forage Sci.* **14**: 17–25.

Heady, H. F. & R. D. Child. 1994. *Rangeland Ecology and Management*. Westview Press, Boulder, CO.

Heitschmidt, R. K. & J. W. Stuth (eds). 1991. *Grazing Management: An Ecological Perspective*. Timber Press, Portland, OR.

Herbers, J. M. 1981. Time resources and laziness in animals. *Oecologia* **49**: 252–62.

Hewison, A. J. M., J. M. Angibault, J. Boutin, E. Bideau, J. P. Vincent & A. Sempere. 1996. Annual variation in body composition of roe deer in moderate environmental conditions. *Can. J. Zool.* **74**: 245–53.

Hik, D. S. 1995. Does risk of predation influence population dynamics? Evidence from the cyclic decline of snowshoe hares. *Wildl. Res.* **22**: 115–29.

Hirakawa, H. 1997. Digestion – constrained optimal foraging in generalist mammalian herbivores. *Oikos* **78**: 37–47.

Hobbs, N. T. 1990. Diet selection by generalist herbivores: a test of the linear programming model. In *Behavioural Mechanisms of Food Selection*, ed. R. N. Hughes, pp. 395–413. Springer-Verlag, Heidelberg.

Hobbs, N. T., D. L. Baker, G. D. Bear & D. C. Bowden. 1996a. Ungulate grazing in sagebrush grassland: mechanisms of resource competition. *Ecol. Applic.* **6**: 200–17.

Hobbs, N. T., D. L. Baker, G. D. Bear & D. C. Bowden. 1996b. Ungulate grazing in sagebrush grassland: effects of resource competition on secondary production. *Ecol. Applic.* **6**: 218–27.

Hodgson, J. & A. W. Illius. 1996. *The Ecology and Management of Grazing Systems*. CAB International, Wallingford.

Hofmann, R. R. 1973. *The Ruminant Stomach*. (*East Afr. Monogr. Biol.* Vol. 2.) East African Literature Bureau, Nairobi.

Hofmann, R. R. 1989. Evolutionary steps of ecophysiological adaptation and diversification of ruminants: A comparative review of their digestive system. *Oecologia* **78**: 443–57.

Holland, O. 1992. Seasonal variation in body composition of European roe deer. *Can. J. Zool.* **70**: 502–4.

Holling, C. S. 1959. Some characteristics of simple types of predation and parasitism. *Can. Entomol.* **91**: 385–98.

Holling, C. S. 1965. The functional response of predators to prey density and its role in mimicry and population regulation. *Mem. Entomol. Soc. Can.* **45**: 1–60.

Houston, A. J. & J. M. McNamara. 1999. *Models of Adaptive Behaviour. An Approach Based on State.* Cambridge University Press, Cambridge.

Houston, A. J., N. J. Welton & J. M. McNamara. 1997. Acquisition and maintenance costs in the long-term regulation of avian fat reserves. *Oikos* **78**: 331–40.

Hudson, R. J. & S. Frank. 1987. Foraging ecology of bison in aspen boreal habitats. *J. Range Mgmt* **40**: 71–5.

Hudson, R. J. & W. G . Watkins. 1986. Foraging rates of wapiti on green and cured pastures. *Can. J. Zool.* **64**: 1705–8.

Huff, D. E. & J. D. Varley. 1999. Natural regulation in Yellowstone National Park's northern range. *Ecol. Applic.* **9**: 17–29.

Hungate, R. E. 1975. The rumen microbial ecosystem. *A. Rev. Ecol. Syst.* **6**: 39–66.

Hunter, M. D., T. Ohgushi & P. W. Price. 1992. *Effects of Resource Distribution on Animal-Plant Interactions.* Academic Press, London and New York.

Hunter, M. D. & P. W. Price. 1992. Playing chutes and ladders: heterogeneity and the relative roles of bottom-up and top-down forces in natural communities. *Ecology* **73**: 724–32.

Hutchinson, G. E. 1958. Concluding remarks. *Cold Spring Harbor Symp. Quant. Biol.* **22**: 415–27.

Hutchinson, G. E. 1959. Homage to Santa Rosalia, or why are there so many kinds of animals? *Am. Nat.* **93**: 145–59.

Illius, A. W. & I. J. Gordon. 1987. The allometry of food intake in grazing ruminants. *J. Anim. Ecol.* **56**: 989–99.

Illius, A. W. & I. J. Gordon. 1991. Prediction of intake and digestion in ruminants by a model of rumen kinetics integrating animal size and plant characteristics. *J. Agric. Sci., Cambr.* **116**: 145–57.

Illius, A. W. & I. J. Gordon. 1992. Modelling the nutritional ecology of ungulate herbivores: evolution of body size and competitive ability. *Oecologia* **89**: 428–34.

Illius, A. W. & I. J. Gordon. 1998. Scaling up from functional response to numerical response in vertebrate herbivores. In *Herbivores, Plants and Predators*, ed. H. Olff, V. K. Brown & R. T. Drent, pp. 397–427. Blackwell Science, Oxford.

Illius, A. W., I. J. Gordon, D. A. Elston & J. D. Milne. 1999. Diet selection in goats: a test of intake-rate maximization. *Ecology* **80**: 1008–18.

Illius, A. W. & N. S. Jessop. 1995. Modelling metabolic costs of allelochemical ingestion by foraging ungulates. *J. Chem. Ecol.* **21**: 693–719.

Illius, A. W. & N. S. Jessop. 1996. Metabolic constraints on voluntary intake in ruminants. *J. Anim. Sci.* **74**: 3052–62.

Illius, A. W. & T. G. O'Connor. 1999. On the relevance of nonequilibrium concepts to arid and semi-arid grazing systems. *Ecol. Applic.* **9**: 798–813.

Illius, A. W. & T. G. O'Connor. 2000. Resource heterogeneity and ungulate population dynamics. *Oikos* **89**: 283–94.

Irby, L. R. 1977. Food habits of Chandler's mountain reedbuck in a Rift Valley ranch. *E. Afr. Wildl. J.* **15**: 289–94.

Ivlev, V. S. 1961. *Experimental Ecology of the Feeding of Fishes*. Yale University Press, New Haven, MA.

James, C. D., J. Landsberg & S. R. Morton. 1999. Provision of watering points in the Australian Arid zone: a review of effects on biota. *J. Arid Envir.* **41**: 87–121.

Janis, C. 1976. The evolutionary strategy of the Equidae and the origins of rumen and cecal digestion. *Evolution* **30**: 757–76.

Jarman, P. J. 1971. Diets of large mammals in the woodlands around Lake Kariba. *Oecologia* **8**: 157–78.

Jarman, P. J. 1974. The social organization of antelope in relation to their ecology. *Behaviour* **48**: 215–67.

Jeschke, J. M., M. Kopp & R. Tollrian. 2001. Predator functional responses: discriminating between handling and digesting prey. *Ecology*. (In press.)

Jiang, Z. & R. J. Hudson. 1992. Estimating forage intake and energy requirements of free-ranging wapiti. *Can. J. Zool.* **70**: 675–79.

Jiang, Z. & R. J. Hudson. 1993. Optimal grazing of wapiti on grassland: patch and feeding station departure rules. *Evol. Ecol.* **7**: 488–98.

Jiang, Z. & R. J. Hudson. 1994. Bite characteristics of wapiti in seasonal *Bromus-Poa* swards. *J. Range Mgmt* **47**: 127–32.

Jones, R. J. & R. L. Sandland. 1974. The relation between animal gain and stocking rate: derivation of the relation from the results of grazing trials. *J. Agric. Sci., Cambr.* **83**: 335–41.

Joos-Vandewalle, M. E. 2000. Movement of migratory zebra and wildebeest in northern Botswana. PhD thesis, University of the Witwatersrand.

Jorgenson, J. T., M. Festa-Bianchet, J.-M. Gaillard & W. D. Wishart. 1997. Effects of age, sex, disease and density on survival of bighorn sheep. *Ecology* **78**: 1019–32.

Joubert, D. M. 1954. The influence of winter nutritional depressions on the growth, reproduction and production of cattle. *J. Agric. Sci., Cambr.* **44**: 5–66.

Judson, O. P. 1994. The rise of the individual based model in ecology. *Trends Ecol. Evol.* **9**: 9–14.

Kaufmann, S. A. 1993. *The Origins of Order: Self-Organization and Selection in Evolution*. Oxford University Press, Oxford.

Kaufmann, S. A. 1995. *At Home in the Universe. The Search for the Laws of Complexity*. Oxford University Press, New York.

Kemeny, J. G. & T. E. Kurtz. 1990. *True BASIC. A Powerful Structured Language System by the Original Authors of BASIC*. True BASIC Inc., West Lebanon, NH.

Klein, D. R. 1968. The introduction, increase and crash of reindeer on St. Matthew Island. *J. Wildl. Mgmt* **32**: 350–67.

Kooijman, S. A. L. M. 1990. Biomass conversion at the population level. In *Individual-Based Models and Approaches in Ecology*, ed. D. L. DeAngelis & L. J. Gross, pp. 338–58. Chapman & Hall, New York and London.

Kooijman, S. A. L. M. 1993. *Dynamic Energy Budgets in Biological Systems*. Cambridge University Press, Cambridge.

Krebs, J. R. & R. H. McLeery. 1984. Optimization in behavioural ecology. In *Behavioural Ecology. An Evolutionary Approach*, ed. J. R. Krebs & N. B. Davies, 2nd edn., pp. 91–21. Blackwell, Oxford.

Krebs, C. J., A. R. E. Sinclair, R. Boonstra, S. Boutin, K. Martin & J. N. M. Smith. 1998. Community dynamics of vertebrate herbivores: how can we untangle the web? In *Herbivores: Between Plants and Predators*, ed. H. Olff, V. K. Brown & R. H. Drent, pp. 447–73. Blackwell Science, Oxford.

Kruuk, H. 1972. *The Spotted Hyena*. University of Chicago Press, Chicago.

Laca, E. A. & M. W. Demment. 1991. Herbivory: the dilemma of foraging in a spatially heterogeneous food environment. In *Plant Defences Against Mammalian Herbivory*, ed. R. T. Palo & C. T. Robbins. pp. 29–44. CRC Press, Boca Raton, FL.

Laca, E. A. & M. W. Demment. 1993. Modelling intake of a grazing ruminant in a heterogeneous environment. In *Proceedings of the International Symposium on Vegetation–Herbivore Relationships*, ed. T. Okubo, B. Hubest & G. Arnold, pp. 57–76. Academic Press, London.

Laca, E. A. & M. W. Demment. 1996. Foraging strategies of grazing animals. In *The Ecology and Management of Grazing Systems*, ed. J. Hodgson & A. W. Illius, pp. 137–58. CAB International, Wallingford, UK.

Laca, E. A., R. A. Distel, T. C. Griggs & M. W. Demment. 1994a. Effects of canopy structure on patch depression by grazers. *Ecology* **75**: 706–16.

Laca, E. A., R. A. Distel, T. C. Griggs, G. Deo & M. W. Demment. 1993. Field test of optimal foraging with cattle: the marginal value theorem successfully predicts patch selection and utilisation. *Proc. XVII Int. Grassl. Congr.*, pp. 709–10.

Laca, E .A., E. D. Ungar & M. W. Demment. 1994b. Mechanism of handling time and intake rate of a large mammalian grazer. *Appl. Anim. Behav. Sci.* **39**: 3–19.

Laca, E. A., E. D. Ungar, N. Seligman & M. W. Demment. 1992. Effects of sward height and bulk density on bite dimensions of cattle grazing homogeneous swards. *Grass Forage Sci.* **47**: 91–102.

Lange, R. T. 1969. The piosphere: sheep track and dung patterns. *J. Range Mgmt* **22**: 396–400.

Langvatn, R. & T. A. Hanley. 1993. Feeding patch choice by red deer in relation to foraging efficiency. *Oecologia* **95**: 164–70.

Laws, R. M. & I. S. C. Parker. 1968. Recent studies on elephant populations in East Africa. *Symp. Zool. Soc., Lond.* No. 21, pp. 319–59.

Lechner-Doll, M., T. Rutagwenda, H. J. Schwartz, W. Schultka & W. v. Engelhardt. 1990. Seasonal changes of ingestion mean retention time and forestomach fluid volume in indigenous camels, cattle, sheep and goats grazing a thornbush savannah pasture in Kenya. *J. Agric. Sci., Cambr.* **115**: 409–20.

Ledger, H. P. 1968. Body composition as a basis for a comparative study of some East African mammals. In *Comparative Nutrition of Wild Animals*, ed. M. A. Crawford (*Symp. Zool. Soc., Lond.* No. 21), pp. 289–310.

Leirs, H., N. C. Stenseth, J. D. Nicholl, J. E. Hines, R. Verhagen & W. Verhagen. 1997. Stochastic seasonality and nonlinear density-dependent factors regulate population size in an African rodent. *Nature* **389**: 176–80.

Lemaire, G. & D. Chapman. 1996. Tissue flows in grazed plant communities. In *The Ecology and Management of Grazing Systems*, ed. J. Hodgson & A. W. Illius, pp. 3–36. CAB International, Wallingford, UK.

Loison, A., M. Festa-Bianchet, J.-M. Gaillard, J. T. Jorgenson & J.-M. Jullien. 1999. Age-specific survival in five populations of ungulates: evidence of senescence. *Ecology* **80**: 2539–54.

Lotka, A. J. 1925. *Elements of Mathematical Biology*. Dover, New York.

Lott, D. F. 1991. *Intraspecific Variation in the Social Systems of Wild Vertebrates*. Cambridge University Press, Cambridge.

Low, W. A., M. L. Dudzinski & W. J. Miller. 1981. The influence of forage and climatic conditions on range community preferences of shorthorn cattle in central Australia. *J. Appl. Ecol.* **18**: 11–26.

Ludwig, D., R. Hilborn & C. Walters. 1993. Uncertainty, resource exploitation, and conservation: lessons from history. *Science* **260**: 17–36.

Mangel, M. & C. W. Clark. 1988. *Dynamic Modelling in Behavioral Ecology*. Princeton University Press, Princeton, NJ.

Manser, M. B. & P. N. M. Brotherton. 1995. Environmental constraints on the foraging behaviour of a dwarf antelope. *Oecologia* **102**: 404–12.

Martin, R. B., G. C. Craig & V. R. Booth. 1989. *Elephant Management in Zimbabwe*. Department of National Parks & Wild Life Management, Harare, Zimbabwe.

May, R. M. 1973. *Stability and Complexity in Model Ecosystems*. Princeton University Press, Princeton, NJ.

May, R. M. 1981. Models for single populations. In *Theoretical Ecology*, ed. R. M. May, pp. 5–29. Blackwell, Oxford.

May, R. & G. Oster. 1976. Bifurcations and dynamic complexity in simple ecological models. *Am. Nat.* **110**: 573–99.

Maynard Smith, J. & M. Slatkin. 1973. The stability of predator-prey systems. *Ecology* **54**: 384–91.

McCabe, J. T. 1987. Drought and recovery: Livestock dynamics among the Ngisonyoka Turkana of Kenya. *Human Ecol.* **15**: 317–29.

McCullough, D. R. 1992. Concepts of large herbivore population dynamics. In *Wildlife 2001. Populations*, ed. D. R. McCullough & R. H. Barrett, pp. 967–84. Elsevier, London.

McCullough, D. R. 1997. Irruptive behavior in ungulates. In *The Science of Overabundance, Deer Ecology and Population Management*, ed. W. J. McShea, H. B. Underwood & J. H. Rappole, pp. 69–99. Smithsonian Institution Press, Washington, DC.

McLeod, S. R. 1997. Is the concept of carrying capacity useful in variable environments? *Oikos* **79**: 529–42.

McNamara, J. M. & A. I. Houston. 1987a. Partial preferences and foraging. *Anim. Behav.* **35**: 1084–99.

McNamara, J. M. & A. I. Houston. 1987b. Starvation and predation as factors limiting population size. *Ecology* **68**: 1515–19.

McNamara, J. M. & A. I. Houston. 1990. The value of fat reserves and the trade-off between starvation and predation. *Acta Biotheor.* **38**: 37–61.

McNamara, J. M. & A. I. Houston. 1996. State-dependent life histories. *Nature* **380**: 215–21.

McNaughton, S. J. 1976. Serengeti migratory wildebeest : facilitation of energy flow by grazing. *Science* **191**: 92–3.

McNaughton, S. J. 1979a. Grazing as an optimization process: grass-ungulate relationships in the Serengeti. *Am. Nat.* **113**: 691–703.

McNaughton, S. J. 1979b. Grassland herbivore dynamics. In *Serengeti: Dynamics of an Ecosystem*, ed. A. R. E. Sinclair & M. Norton-Griffiths, pp. 46–81. University of Chicago Press, Chicago, IL.

McNaughton, S. J. 1983. Compensatory plant growth as a response to herbivory. *Oikos* **40**: 329–36.

McNaughton, S. J. 1984. Grazing lawns: animals in herds, plant form and coevolution. *Am. Nat.* **124**: 863–86.

McNaughton, S. J., F. F. Banyikwa & M. M. McNaughton. 1997. Promotion of the cycling of diet-enhancing nutrients by African grazers. *Science* **278**: 1798–800.

McNaughton, S. J., F. F. Banyikwa & M. M. McNaughton. 1998. Root biomass and productivity in a grazing ecosystem: the Serengeti. *Ecology* **79**: 587–92.

Mduma, S. A. R., A. R. E. Sinclair & R. Hilborn. 1999. Food regulates the Serengeti wildebeest: a 40-year record. *J. Anim. Ecol.* **68**: 1101–22.

Meissner, H. H. & D. V. Paulsmeier. 1995. Plant composition affecting between-plant and animal species prediction of forage intake. *J. Anim. Sci.* **73**: 2447–57.

Meissner, H. H., P. J. K. Zacharias & P. J. O'Reagain. 1999. Forage quality (feed value). In *Veld Management in South Africa*, ed. N. M.. Tainton, pp. 139–68. University of Natal Press, Pietermaritzburg.

Mentis, M. T. 1981. Acceptability and palatability. In *Veld and Pasture Management in South Africa*, ed. N. M. Tainton, pp. 186–91. Shuter & Shooter, Pietermaritzburg.

Mertens, D. R. 1977. Dietary fiber components: relationships to the rate and extent of ruminal digestion. *Fedn Am. Socs. Exp. Biol.* **36**: 187–92.

Mertens, D. R. 1987. Predicting intake and digestibility using mathematical models of rumen function. *J. Anim. Sci.* **64**: 1548–58.

Messier, F. 1994. Ungulate population models with predation: a case study with North American moose. *Ecology* **75**: 478–88.

Messier, F. 1995. Trophic interaction in two northern wolf-ungulate systems. *Wildl. Res.* **22**: 131–46.

Metz, J. A. J. & O. Dieckmann. 1986. *The Dynamics of Physiologically Structured Populations*. Springer-Verlag, New York.

Milinski, M. & G. Parker. 1991. Competition for resources. In *Behavioural Ecology. An Evolutionary Approach*, 3rd edn, ed. J. R. Krebs & N. B. Davies, pp. 137–68. Blackwell, Oxford.

Miller, M. F. 1996. Dispersal of *Acacia* seeds by ungulates and ostriches in an African savanna. *J. Trop. Ecol.* **12**: 345–56.

Mills, M. G. L., H. C. Biggs & I. J. Whyte. 1995. The relationship between rainfall, lion predation and population trends in African herbivores. *Wildl. Res.* **22**: 75–88.

Milne, J. A., J. C. MacRae, A. M. Spence & S. Wilson. 1976. Intake and digestion of hill-land vegetation by the red deer and the sheep. *Nature* **263**: 763–4.

Milner, C. & D. Gwynne. 1974. The Soay sheep and their food supply. In *Island Survivors. The Ecology of the Soay Sheep of St. Kilda*, ed. P. A. Jewell, C. Milner & J. Morton Boyd, pp. 273–325. Athlone Press, University of London.

Milner, J. M., S. D. Albon, A. W. Illius, J. M. Pemberton & T. H. Clutton-Brock. 1999a. Repeated selection of morphometric traits in the Soay sheep on St. Kilda. *J. Anim. Ecol.* **68**: 472–88.

Milner, J. M., D. A. Elston & S. D. Albon. 1999b. Estimating the contributions of population density and climatic fluctuations to interannual variation in survival of Soay sheep. *J. Anim. Ecol.* **68**: 1235–47.

Milner-Gulland, E. J. & R. Mace. 1998. *Conservation of Biological Resources*. Blackwell Science, Oxford.

Moen, A. N. 1973. *Wildlife Ecology*. Freeman, San Francisco.

Molvar, E. M., R. T. Bowyer & V. Van Ballenbergh. 1993. Moose herbivory, browse quality and nutrient cycling in an Alaskan treeline community. *Oecologia* **94**: 472–9.

Morton Boyd, J. 1974. Introduction. In *Island Survivors. The Ecology of the Soay Sheep of St Kilda*, ed. P. A. Jewell, C. Milner & J. Morton Boyd, pp. 1–7. Athlone Press, University of London.

Mrosovsky, N. & T. L. Prowley. 1977. Set points for body weight and fat. *Behav. Biol.* **20**: 205–23.

Murray, M. G. 1993. Comparative nutrition of wildebeest, hartebeest and topi in the Serengeti. *Afr. J. Ecol.* **31**: 172–7.

Murray, M. G. & D. Brown. 1993. Niche separation of grazing ungulates in the Serengeti: an experimental test. *J. Anim. Ecol.* **62**: 380–9.

Murray, M. G. & A. W. Illius. 1996. Multispecies grazing in the Serengeti. In *The Ecology and Management of Grazing Systems*, ed. J. Hodgson & A. W. Illius, pp. 247–72. CAB International, Wallingford, UK.

Murray, M. G. & A. W. Illius. 2000. Vegetation modification and resource competition in grazing ungulates. *Oikos* **89**: 501–8.

Newman, J. A., A. J. Parsons & A. Harvey. 1993. Not all sheep prefer clover: diet selection revisited. *J. Agric. Sci., Cambr.* **119**: 275–83.

Nicolis G. & I. Prigogine. 1977. *Self-organization in Non-equilibrium Systems*. Wiley, New York.

Novellie, P. A. 1978. Comparison of the foraging strategies of blesbok and springbok on the Transvaal highveld. *S. Afr. J. Wildl. Res.* **8**: 137–44.

Novellie, P. 1990. Habitat use by indigenous grazing ungulates in relation to sward structure and veld condition. *J. Grassl. Soc. Sth. Afr.* **7**: 16–23.

Noy-Meir, I. 1975. Stability of grazing systems: an application of predator-prey graphs. *J. Ecol.* **63**: 459–81.

Noy-Meir, I. 1978. Grazing and production in seasonal pastures: analysis of a simple model. *J. Appl. Ecol.* **15**: 809–35.

O'Connor, T. G. 1991. Local extinction in perennial grasslands: a life history approach. *Am. Nat.* **137**: 753–73.

Oftedal, O. T. 1984. Milk composition, milk yield and energy output at peak lactation. In *Lactation Strategies*. (*Symp. Zool. Soc. Lond.* No. 51), pp. 33–85.

Oksanen, L., J. Moen & P. A. Lundberg. 1992. The time-scale problem in exploiter-victim models: does the solution lie in ratio-dependent exploitation? *Am. Nat.* **140**: 938–60.

Oliver, M. D. N., N. R. M. Short & J. Hanks. 1978. Population ecology of oribi, grey rhebuck and mountain reedbuck in Highmoor State Forest Land, Natal. *S. Afr. J. Wildl. Res.* **8**: 95–105.

O'Reagain, P. J. 1993. Plant structure and the acceptability of different grasses to sheep. *J. Range Mgmt* **46**: 232–6.

O'Reagain, P. J. 1994. The effect of sward structure and species composition on dietary quality and intake in cattle and sheep grazing the Dohne sourveld. PhD thesis, University of the Witwatersrand.

O'Reagain, P. J. 2001. Foraging strategies on rangelands: Effects of intake and animal performance. In *Proceedings of the XIX International Grassland Congress, Sao Pedro,*

Brazil, Feb. 2001, ed. J. A. Gomide, W. R. S. Mattos and S. C. da Silva, pp. 277–84. Brazilian Society of Animal Production, Sao Paulo, Brazil.

O'Reagain, P. J., B. C. Goetsch & R. N. Owen-Smith. 1995. Ruminal degradation characteristics of some African rangeland grasses. *J. Agric. Sci., Cambr.* **125**: 189–97.

O'Reagain, P. J., B. C. Goetsch & R. N. Owen-Smith. 1996. Effect of species composition and sward structure on the ingestive behaviour of cattle and sheep grazing South African sourveld. *J. Agric. Sci., Cambr.* **127**: 271–80.

O'Reagain, P. J. & E. R. Grau. 1995. Sequence of species selection by cattle and sheep on South African sourveld. *J. Range Mgmt* **48**: 314–21.

O'Reagain, P. J. & M. T. Mentis. 1989a. Sequence and process of species selection by cattle in relation to optimal foraging theory in an old land in the Natal Sour Sandveld. *J. Grassl. Soc. Sth. Afr.* **6**: 71–6.

O'Reagain, P. J. & M. T. Mentis. 1989b. The effect of plant structure on the acceptability of different grass species to cattle. *J. Grassl. Soc. Sth. Afr.* **6**: 163–70.

O'Reagain, P. J. & M. J. Mentis. 1990. The effects of veld condition on the quality of diet selected by cattle grazing the Natal Sour Sandveld. *J. Grassl. Soc. Sth. Afr.* **7**: 190–5.

O'Reagain, P. J. & R. N. Owen-Smith. 1996. Effect of species composition and sward structure on dietary quality in cattle and sheep grazing South Africa sourveld. *J. Agric. Sci., Cambr.* **127**: 261–70.

O'Reagain, P. J. & J. Schwartz. 1995. Dietary selection and foraging strategies of animals on rangeland. Coping with spatial and temporal variability. In *Recent Developments in the Nutrition of Herbivores (Proceedings of IV International Symposium on the Nutrition of Herbivores)*, ed. M. Journet, E. Grenet, M.-H. Farce, M. Thierez & C. Demarquilly, pp. 407–23. INRA Editions, Paris.

Orians, G. H. & N. E. Pearson. 1979. On the theory of central place foraging. In *Analysis of Ecological Systems*, ed. D. J. Horn, R. D. Mitchell & G. R. Stairs, pp. 154–77. Ohio State University Press, Columbus.

Owen-Smith, N. 1973. The behavioural ecology of the white rhinoceros. PhD thesis, University of Wisconsin.

Owen-Smith, N. 1974. The social system of the white rhinoceros. In *The Behaviour of Ungulates and Its Relation to Management* (*IUCN Public.*, NS No. 24) ed. V. Geist & F. Walther, pp. 341–51. IUCN, Morges.

Owen-Smith, N. 1975. The social ethology of the white rhinoceros. *Z. Tierpsychol.* **38**: 337–84.

Owen-Smith, N. 1979. Assessing the forage efficiency of a large herbivore, the kudu. *S. Afr. J. Wildl. Res.* **9**: 102–10.

Owen-Smith, N. 1981. The white rhinoceros overpopulation problem, and a proposed solution. In *Problems in Management of Locally Abundant Wild Mammals*, ed. P. A. Jewell, S. Holt & D. Hart, pp. 129–50. Academic Press, New York.

Owen-Smith, N. 1982. Factors influencing the consumption of plant products by large herbivores. In *The Ecology of Tropical Savannas*, ed. B. J. Huntley & B. H. Walker, pp. 359–404. Springer-Verlag, Berlin and Hamburg.

Owen-Smith, N. 1983. Dispersal and the dynamics of large herbivore populations in enclosed areas. In *Management of Large Mammals in African Conservation Areas*, ed. R. N. Owen-Smith, pp. 127–43. Haum, Pretoria.

Owen-Smith, N. 1985. Niche separation among African ungulates. In *Species and Speciation*, ed. E. S. Vrba (*Transvaal Museum Monograph* No. 4), pp. 167–71.

Owen-Smith, R. N. 1988. *Megaherbivores. The Influence of Very Large Body Size on Ecology*. (Cambridge Studies in Ecology). Cambridge University Press, Cambridge.

Owen-Smith, N. 1989. Megafaunal extinctions: the conservation message from 11 000 years BP. *Cons. Biol.* **3**: 405–12.

Owen-Smith, N. 1990. Demography of a large herbivore, the greater kudu, in relation to rainfall. *J. Anim. Ecol.* **59**: 893–913.

Owen-Smith, N. 1991. Veld condition and animal performance: application of an optimal foraging model. *J. Grassl. Soc. Sth. Afr.* **8**: 77–81.

Owen-Smith, N. 1993a. Evaluating optimal diet models for an African browsing ruminant, the kudu: how constraining are the assumed constraints? *Evol. Ecol.* **7**: 499–524.

Owen-Smith, N. 1993b. Comparative mortality rates of male and female kudus: the costs of sexual size dimorphism. *J. Anim. Ecol.* **62**: 428–40.

Owen-Smith, N. 1993c. Age, size, dominance and reproduction among male kudus: mating enhancement by attrition of rivals. *Behav. Ecol. Sociobiol.* **32**: 177–84.

Owen-Smith, N. 1994. Foraging responses of kudus to seasonal changes in food resources: elasticity in constraints. *Ecology* **75**: 1050–62.

Owen-Smith, N. 1996a. Ecological guidelines for waterpoints in extensive protected areas. *S. Afr. J. Wildl. Res.* **26**: 107–12.

Owen-Smith, N. 1996b. Circularity in linear programming models of optimal diet. *Oecologia* **108**: 259–61.

Owen-Smith, N. 1997. Control of energy balance by a wild ungulate, the kudu, through adaptive foraging behaviour. *Proc. Nutr. Soc.* **56**: 15–24.

Owen-Smith, N. 1998a. How high ambient temperature affects the daily activity and foraging time of a subtropical ungulate, the greater kudu. *J. Zool., Lond.* **246**: 183–92.

Owen-Smith, N. 1998b. Dynamics of herbivore-vegetation systems and the overgrazing issue. In *Communal Rangelands in Southern Africa: A Synthesis of Knowledge. Proceedings of a Symposium on Policy-making for the Sustainable Use of Southern African Communal Rangelands*, ed. T. D. de Bruyn & P. F. Scogings, pp. 124–34. Dept. of Livestock & Pasture Science, University of Fort Hare, Alice, South Africa.

Owen-Smith, N. 2000. Modelling the population dynamics of a subtropical ungulate in a variable environment: rain, cold and predators. *Natur. Res. Model.* **13**: 57–87.

Owen-Smith, N. & S. M. Cooper. 1985. Comparative consumption of vegetation components by kudus, impalas and goats in relation to their commercial potential as browsers in savanna regions. *S. Afr. J. Sci.* **81**: 72–6.

Owen-Smith, N. & S. M. Cooper. 1987a. Palatability of woody plants to browsing ungulates in a South African savanna. *Ecology* **68**: 319–31.

Owen-Smith, N. & S. M. Cooper. 1987b. Acceptability indices for assessing the food preferences of ungulates. *J. Wildl. Mgmt* **51**: 372–8.

Owen-Smith, N. & S. M. Cooper. 1989. Nutritional ecology of a browsing ruminant, the kudu, through the seasonal cycle. *J. Zool., Lond.* **219**: 29–43.

Owen-Smith, N. & J. E. Danckwerts. 1997. Herbivory. In *Vegetation of Southern Africa*, ed. R. M. Cowling, D. M. Richardson & S. M. Pierce, pp. 397–420. Cambridge University Press.

Owen-Smith, N. & P. Novellie. 1982. What should a clever ungulate eat? *Am. Nat.* **119**: 151–78.

Panaretto, B. A. 1968. Estimation of body composition by the dilution of hydrogen isotopes. In *Body Composition in Animals and Man*, pp. 200–17. National Academy of Sciences, Washington, DC.

Papageorgiou, N. 1979. Population energy relationships of the Agrimi on Theodorou Island, Greece. *Mammalia Depicta*. Verlag Paul Parey, Berlin.

Parker, K. L. 1988. Effects of heat, cold and rain on coastal black-tailed deer. *Can. J. Zool.* **66**: 2475–83.

Parker, K. L., M. P. Gillingham, T. A. Hanley & C. T. Robbins. 1993. Seasonal patterns in body weight, body composition, and water transfer rates of free ranging and captive black tailed deer in Alaska. *Can. J. Zool.* **71**: 1397–404.

Parker, K. L., M. P. Gillingham, T. A. Hanley & C. T. Robbins. 1996. Foraging efficiency: energy expenditure versus energy gain in free ranging black tailed deer. *Can. J. Zool.* **74**: 442–50.

Parker, K. L., M. P. Gillingham, T. A. Hanley & C. T. Robbins. 1999. Energy and protein balance of free-ranging black-tailed deer in a natural forest environment. *Wildl. Monogr.* No. 143, 48pp.

Parsons, A. J., J. A. Newman, P. D. Penning, A. Harvey & R. J. Orr. 1994. Diet preference of sheep: effects of recent diet, physiological state and species abundance. *J. Anim. Ecol.* **63**: 465–78.

Pastor, J., R. A. Moen & Y. Cohen. 1997. Spatial heterogeneity, carrying capacity and feedbacks in animal-landscape interactions. *J. Mamm.* **78**: 1040–52.

Peel, M. J. S., H. Biggs & P. J. K. Zacharias. 1999. The evolving use of stocking rate indices currently based on animal number and type in semi-arid heterogeneous landscapes and complex land-use systems. *Afr. J. Range Forage Sci.* **15**: 117–27.

Pellew, R. A. 1984. The feeding ecology of a selective browser, the giraffe. *J. Zool., Lond.* **202**: 57–81.

Penry, D. L. 1993. Digestive constraints on diet choice. In *Diet Selection*, ed. R. N. Hughes, pp. 32–55. Blackwell, Oxford.

Penry, D. L. & P. A. Jumars. 1986. Chemical reactor analysis and optimal digestion. *BioScience* **36**: 310–15.

Penry, D. L. & P. A. Jumars. 1987. Modelling animal guts as chemical reactors. *Am. Nat.* **129**: 69–96.

Perez-Barberia, F. J. & I. J. Gordon. 2001. Relationships between oral morphology and feeding style in the Ungulata: a phylogenetically controlled evaluation. *Proc. R. Soc. Lond.* B **268**: 1021–30.

Peters, R. H. 1983. *The Ecological Implications of Body Size*. Cambridge University Press, Cambridge.

Peterson, C. C., K. A. Nagy & J. Diamond. 1990. Sustained metabolic scope. *Proc. Nat. Acad. Sci. USA* **87**: 2324–8.

Peterson, R. O. 1999. Wolf-moose interaction on Isle Royale: the end of natural regulation? *Ecol. Applic.* **9**: 10–16.

Peterson, R. O., R. E. Page & K. M. Dodge. 1984. Wolves, moose and the allometry of population cycles. *Science* **224**: 1350–2.

Pielou, E. 1969. *An Introduction to Mathematical Ecology*. Wiley-Interscience, New York.

Prigogine, I. & J. M. Wiame. 1946. Biologie et thermodynamique des phenomenes irreversibles. *Experientia* **2**: 451–3.

Price, P. W. 1984. *Insect Ecology*. 2nd edn. John Wiley & Sons, New York.

Prins, H. H. T. 1996. *Ecology and Behaviour of the African Buffalo*. Chapman & Hall, London.

Prins, H. H. T. & H. Olff. 1998. Species richness of African grazer assemblages: towards a functional explanation. In *Dynamics of Tropical Ecosystems*, ed. D. N. Newberry, H. H. T. Prins & N. Brown, pp. 449–90. Blackwell, Oxford.

Provenza, F. D. 1995. Postingestive feedback as an elementary determinant of food preference and intake in ruminants. *J. Range Mgmt* **48**: 2–17.

Pulliam, H. R. 1988. Sources, sinks and population regulation. *Am. Nat.* **132**: 652–61.

Rasmussen, D. I. 1941. Biotic communities of the Kaibab Plateau. *Ecol. Monogr.* **3**: 229–75.

Real, L. A. 1977. The kinetics of functional response. *Am. Nat.* **111**: 289–300.

Reimers, E., J. Ringberg & R. Sorumgard. 1982. Body composition of Svalbard reindeer. *Can. J. Zool.* **60**: 1812–21.

Renecker, L. A. & R. J. Hudson. 1986. Seasonal foraging rates of free-ranging moose. *J. Wildl. Mgmt* **50**: 143–7.

Ricklefs, R. E., M. Konarzewski & S. Daan. 1996. The relationship between basal metabolic rate and daily energy expenditure in birds and mammals. *Am. Nat.* **147**: 1047–71.

Ringberg, T. M., R. G. White, D. F. Holleman & J. R. Luick. 1981. Body growth and carcase composition of lean reindeer from birth to sexual maturity. *Can. J. Zool.* **59**: 1040–61.

Risenhoover, K. L. 1987. Winter foraging strategies of moose in subarctic and boreal forest habitats. PhD thesis, Michigan Technological University.

Risenhoover, K. L. & R. O. Peterson. 1986. Mineral licks as a sodium source for Isle Royal moose. *Oecologia* **71**: 121–6.

Robbins, C. T. 1993. *Wildlife Feeding and Nutrition*. Academic Press, New York.

Robbins, C. T., R. L. Prior, A. N. Moen & W. J. Vesek. 1974. Nitrogen metabolism of white-tailed deer. *J. Anim. Sci.* **38**: 186–91.

Robbins, C. T., R. S. Podbielancik-Norman, D. L. Wilson & E. P. Mould. 1981. Growth and nutrient consumption of elk calves compared to other ungulate species. *J. Wildl. Mgmt* **45**: 172–86.

Robbins, C. T., D. E. Spalinger & W. Van Hoven. 1995. Adaptions of ruminants to browse and grass diets: are anatomical-based browser-grazer interpretations valid? *Oecologia* **103**: 208–13.

Rosenzweig, M. L. 1971. Paradox of enrichment: destabilization of exploitation systems. *Science* **171**: 385–7.

Rosenzweig, M. L. 1981. A theory of habitat selection. *Ecology* **62**: 327–35.

Rosenzweig, M. L. 1991. Habitat selection and population interactions: the search for mechanism. *Am. Nat.* **137** (Suppl.): 5–28.

Rosenzweig, M. L. & Z. Abramsky. 1997. Two gerbils of the Negev: a long-term investigation of optimal habitat selection and its consequences. *Evol. Ecol.* **11**: 733–56.

Roughgarden, J. 1997. Production functions for ecological populations: a survey with emphasis on spatially explicit models. In *Spatial Ecology*, ed. D. Tilman & P. Kareiva, pp. 296–317. Princeton University Press, Princeton, NJ.

Ruel, J. J. & M. P. Ayres. 1999. Jensen's inequality predicts effects of environmental variation. *Trends Ecol. Evol.* **14**: 361–6.

Rutherford, M. C. 1980. Annual plant production–precipitation relations in arid and semi-arid regions. *S. Afr. J. Sci.* **76**: 53–6.

Rutherford, M. C. 1984. Relative allocation and seasonal phasing of growth of woody plant components in a South African savanna. *Progr. Biometeorol.* **3**: 200–21.

Rutherford, M. C. & M. D. Panagos. 1982. Seasonal woody plant shoot growth in *Burkea africana – Ochna pulchra* savanna. *S. Afr. J. Bot.* **1**: 104–16.

Sale, P. P. & N. Tolimieri. 2000. Density dependence at some time and place. *Oecologia* **124**: 166–71.

Scheffer, V. B. 1951. The rise and fall of a reindeer herd. *Scient. Monthly* **73**: 356–62.

Schmidt, K. A., J. M. Eambardt, J. S. Brown & R. D. Holt. 2000. Habitat selection under temporal heterogeneity: exorcizing the ghost of competition past. *Ecology* **81**: 2622–30.

Schmidt-Nielsen, K. 1975. *Animal Physiology. Adaptation and Environment.* Cambridge University Press, Cambridge.

Schmitz, O. J., A. P. Beckerman & S. Litman. 1997. Functional responses of adaptive consumers and community stability with emphasis on the dynamics of plant–herbivore systems. *Evol. Ecol.* **11**: 773–84.

Scholes, R. J. & B. H. Walker. 1993. *An African Savanna. Synthesis of the Nylsvley Study.* Cambridge University Press, Cambridge.

Scoones, I. 1992. Coping with drought: responses of herders and livestock in contrasting savanna environments in Southern Zimbabwe. *Human Ecol.* **20**: 293–314.

Scoones, I. 1993. Why are there so many animals? Cattle population dynamics in the communal areas of Zimbabwe. In *Range Ecology at Disequilibrium*, ed. R. H. Behnke, I. Scoones & C. Kerven, pp. 62–76. ODI Publications, London.

Scoones, I. 1995. Exploiting heterogeneity: habitat use by cattle in dryland Zimbabwe. *J. Arid Envir.* **29**: 221–37.

Senft, R. L., M. A. Stilwell & L. R. Rittenhouse. 1987. Nitrogen and energy budgets of free-roaming cattle. *J. Range Mgmt* **40**: 421–4.

Shackleton, C. M. 1993. Are the communal grazing lands in need of saving? *Development Southern Africa* **10**: 65–78.

Shipley, L. A., J. G. Gross, D. E. Spalinger, N. T. Hobbs & B. A. Wunder. 1994. The scaling of the functional response of mammalian herbivores. *Am. Nat.* **143**: 1055–82.

Shipley, L. A. & D. E. Spalinger. 1995. Influence of size and density of browse patches on intake rates and foraging decisions of young moose and white-tailed deer. *Oecologia* **104**: 112–21.

Short, J. 1985. The functional response of kangaroos, sheep and rabbits in an arid grazing system. *J. Appl. Ecol.* **22**: 435–47.

Shugart, H. H. 1998. *Terrestrial Ecosystems in Changing Environments.* Cambridge University Press, Cambridge.

Sibly, R. M. 1981. Strategies of digestion and defecation. In *Physiological Ecology*, ed. C. R. Townsend & P. Calow, pp. 109–39. Blackwell, Oxford.

Sinclair, A. R. E. 1974. The natural regulation of buffalo populations in East Africa. IV. The food supply as a regulating factor and competition. *E. Afr. Wildl. J.* **12**: 291–321.

Sinclair, A. R. E. 1977. *The African Buffalo. A Study of Resource Limitation in Populations*. University of Chicago Press, Chicago.

Sinclair, A. R. E. 1985. Does interspecific competition or predation shape the African ungulate community? *J. Anim. Ecol.* **54**: 899–918.

Sinclair, A. R. E. 1989. Population regulation in animals. In *Ecological Concepts*, ed. J. M. Cherrett, pp. 197–241. Blackwell, Oxford.

Sinclair, A. R. E. & P. Arcese. 1995. Population consequences of predation-sensitive foraging: the Serengeti wildebeest. *Ecology* **76**: 882–91.

Sinclair, A. R. E., H. Dublin & M. Borner. 1985. Population regulation of Serengeti wildebeest: a test of the food hypothesis. *Oecologia* **65**: 266–8.

Sinclair, A. R. E. & P. Duncan. 1972. Indices of condition in tropical ruminants. *E. Afr. Wildl. J.* **10**: 143–9.

Sinclair, A. R. E. & M. Norton-Griffiths. 1983. Does competition or facilitation regulate migrant ungulate populations in the Serengeti? A test of hypotheses. *Oecologia* **53**: 364–9.

Singer, F. J., A. Harting, K. K. Symonds & M. B. Coughenour. 1997. Density dependence, compensation and environmental effects on elk calf mortality in Yellowstone National Park. *J. Wildl. Mgmt* **61**: 12–25.

Skogland, T. 1985. The effects of density-dependent resource limitations on the demography of wild reindeer. *J. Anim. Ecol.* **54**: 359–74.

Skogland, T. 1988. Tooth wear by food limitation and its life history consequences in wild reindeer. *Oikos* **51**: 238–42.

Skogland, T. 1990. Density dependence in a fluctuating wild reindeer herd: maternal vs offspring effects. *Oecologia* **84**: 442–50.

Smith, N. S. 1970. Appraisal of condition estimation models for East African ungulates. *E. Afr. Wildl. J.* **8**: 123–9.

Sohal, R. S. & R. Weindruch. 1996. Oxidative stress, caloric restriction and aging. *Science* **273**: 59–63.

Solomon, M. E. 1949. The natural control of animal populations. *J. Anim. Ecol.* **18**: 1–35.

Spalinger, D. E. & N. T. Hobbs. 1992. Mechanisms of foraging in mammalian herbivores: new models of functional response. *Am. Nat.* **140**: 325–48.

Spalinger, D. E., C. T. Robbins & T. A. Hanley. 1993. Adaptive rumen function in elk and mule deer. *Can. J. Zool.* **71**: 601–10.

Staddon, J. E. R. 1983. *Adaptive Behavior and Learning*. Cambridge University Press, Cambridge.

Stafford Smith, D. M. 1996. Management of rangelands : paradigms at their limits. In *The Ecology and Management of Grazing Systems*, ed. J. Hodgson & A. Illius, pp. 325–58. CAB International, Wallingford, UK.

Starfield, A. M., K. A. Smith & A. L. Bleloch. 1990. *How to Model It. Problem Solving for the Computer Age*. McGraw-Hill, New York.

Stephens, D. W. & J. R. Krebs. 1986. *Foraging Theory*. Princeton University Press, Princeton, NJ.

Sutherland, W. J. 1996. *From Individual Behaviour to Population Ecology*. Oxford University Press, Oxford.

Swift, M. J., O. W. Heal & J. M. Anderson. 1979. *Decomposition in Terrestrial Ecosystems*. Blackwell Scientific Publications, Oxford.

Tainton, N. M. 1981. The ecology of the main grazing lands of South Africa. *In Veld and Pasture Management*, ed. N. M. Tainton, pp. 25–56. Shuter & Shooter, Pietermaritzburg.

Tainton, N. M. 1988. A consideration of veld condition assessment techniques for commercial livestock production in South Africa. *J. Grassl. Soc. Sth. Afr.* **5**: 76–9.

Tainton, N. M. 1999. *Veld Management in South Africa*. University of Natal Press, Pietermaritzburg.

Thom, R. 1975. *Structural Stability and Morphogenesis*. W. A. Benjamin Co., Reading, MA.

Thornley, J. H. M. & J. R. Johnson. 1990. *Plant and Crop Modelling*. Oxford University Press, Oxford.

Thrash, I. & J. F. Derry. 1999. The nature and modelling of piospheres: a review. *Koedoe* **42**: 73–94.

Tilman, D. 1980. Resources: a graphical–mechanistic approach to competition and predation. *Am. Nat.* **116**: 362–93.

Tilman, D. 1982. *Resource Competition and Community Structure*. Princeton University Press, Princeton, NJ.

Tilman, D. 1986. A consumer-resource approach to community structure. *Am. Zool.* **26**: 3–22.

Tilman, D. 1988. *Plant Strategies and the Dynamics and Structure of Plant Communities*. (*Monographs in Population Biology* No. 26.) Princeton University Press, Princeton, NJ.

Tilman, D. 1996. Biodiversity: population versus ecosystem stability. *Ecology* **77**: 350–63.

Tilman, D. 1997. Community invasability, recruitment limitation and grassland biodiversity. *Ecology* **78**: 81–92.

Tilman, D. & P. Kareiva. 1997. *Spatial Ecology. The Role of Space in Population Dynamics and Interspecific Interactions*. Princeton University Press, Princeton, NJ.

Tolkamp, B. J. & J. J. M. H. Ketelaars. 1992. Towards a new theory of feed intake regulation in ruminants 2. Costs and benefits of feed consumption: An optimization approach. *Livestock Production Science* **30**: 97–317.

Torbit, S. C., L. H. Carpenter, A. W. Allridge & D. M. Swift. 1985. Mule deer body composition – a comparison of methods. *J. Wildl. Mgmt* **49**: 86–91.

Torbit, S. C., L. H. Carpenter, D. M. Swift & A. W. Alldredge. 1982. Differential loss of fat and protein by mule deer during winter. *J. Wildl. Mgmt* **49**: 80–5.

Trudell, J. & R. G. White. 1981. The effect of forage structure and availability on food intake, biting rate, bite size and daily eating time of reindeer. *J. Appl. Ecol.* **18**: 63–81.

Turner, M. G., Y. Wu, W. H. Romme & L. L. Wallace. 1993. A landscape simulation model of winter foraging by large ungulates. *Ecol. Model.* **69**: 163–84.

Tyler, N. J. C. 1987. Body composition and energy balance of pregnant and non-pregnant Svalbard reindeer during winter. *Symp. Zool. Soc. Lond.* **57**, 203–29.

Underwood, R. 1982. Vigilance behaviour in grazing African ungulates. *Behaviour* **79**: 81–107.

Underwood, R. 1983. The feeding behaviour of grazing African ungulates. *Behaviour* **84**: 195–243.

Ungar, E. D., N. G. Seligman & M. W. Demment. 1992. Graphical analysis of sward depletion by grazing. *J. Appl. Ecol.* **29**: 427–35.

Van der Meer, J. & B. J. Ens. 1997. Models of interference and their consequences for the spatial distribution of ideal and free predators. *J. Anim. Ecol.* **66**: 846–58.

Van Soest, P. J. 1982. *Nutritional Ecology of the Ruminant.* O & B Books Inc., Corvallis, OR.

van Wieren, S. E. 1992. Factors limiting food intake in ruminants and non-ruminants in the temperate zone. In *Ongules/Ungulates 91*, ed. F. Spitz, G. Janeau, G. Gonzalez & S. Aulagnier, pp. 139–145. SFEPM-IRGM, Paris, Tolouse.

van Wieren, S. E. 1997. Do large herbivores select a diet that maximizes short-term energy-intake rate? *For. Ecol. Mgmt* **88**: 149–56.

Verlinden, C. & R. H. Wiley. 1989. The constraints of digestive rate: an alternative model of diet selection. *Evol. Ecol.* **3**: 264–73.

Vesey-Fitzgerald, D. T. 1960. Grazing succession among East African game animals. *J. Mamm.* **41**: 161–72.

Vivas, H. J., B.-E. Saether & R. Anderson. 1991. Optimal twig-size selection of a generalist herbivore, the moose: implications for plant-herbivore interactions. *J. Anim. Ecol.* **60**: 395–408.

Volterra, V. 1926. Fluctuations in the abundance of species considered mathematically. *Nature* **118**: 558–60.

Walker, B. H., R. H. Emslie, N. Owen-Smith & R. J. Scholes. 1987. To cull or not to cull: lessons from a Southern African drought. *J. Appl. Ecol.* **24**: 381–402.

WallisDeVries, M. & C. Daleboudt. 1994. Foraging strategy of cattle in patchy grassland. *Oecologia* **100**: 98–106.

WallisDeVries, M. F., E. O. Laca & M. W. Demment. 1998. From feeding station to patch: scaling up food intake measurements in grazing cattle. *Appl. Anim. Behav. Sci.* **60**: 301–15.

Walters, C. 1986. *Adaptive Management of Renewable Natural Resources.* Macmillan, New York.

Ward, D. 1992. The role of satisficing in foraging theory. *Oikos* **63**: 312–17.

Watson, L. H. 1997. The feeding ecology of eland in Mountain Zebra National Park. PhD thesis, University of the Witwatersrand.

Weber, B. H., D. J. Depew & J. D. Smith (eds). 1988. *Entropy, Energy and Evolution. New Perspectives on Physical and Biological Evolution.* MIT Press, Cambridge, MA.

Weber, G. E., F. Jeltsch, N. van Rooyen & S. J. Milton. 1998. Simulated long-term vegetation response to grazing heterogeneity in semi-arid rangelands. *J. Appl. Ecol.* **35**: 687–99.

Western, D. 1975. Water availability and its influence on the structure and dynamics of a savannah large mammal community. *E. Afr. Wildl. J.* **13**: 265–86.

Westoby, M. 1978. What are the biological bases of varied diets? *Am. Nat.* **112**: 627–31.

Westoby, M., B. Walker & I. Noy-Meir. 1989. Opportunistic management for rangelands not at equilibrium. *J. Range Mgmt* **42**: 266–74.

White, R. G. 1983. Foraging patterns and their multiplier effects on productivity of northern ungulates. *Oikos* **40**: 377–84.

Whyte, J., R. van Aarde & S. L. Pimm. 1998. Managing the elephants of Kruger National Park. *Anim. Conserv.* **1**: 77–83.

Wickstrom, M. L., C. T. Robbins, T. A. Hanley, D. E. Spalinger & S. M. Parish. 1984. Food intake and foraging energetics of elk and mule deer. *J. Wildl. Mgmt* **48**: 1285–301.

Wiegand, T., F. Jeltsch, S. Bauer & K. Kellner. 1999. Simulation models for semi-arid rangelands of Southern Africa. *Afr. J. Range Forage Sci.* **15**: 48–60.

Wilmshurst, J. F., J. M. Fryxell & C. M. Bergman. 2000. The allometry of patch selection in ruminants. *Proc. R. Soc. Lond.* B **267**: 345–9.

Wilmshurst, J. F., J. M. Fryxell & P. E. Colucci. 1999a. What constrains daily intake in Thomson's gazelles? *Ecology* **80**: 2338–47.

Wilmshurst, J. F., J. M. Fryxell & R. J. Hudson. 1995. Forage quality and patch choice by wapiti. *Behav. Ecol.* **6**: 209–17.

Wilmshurst, J. F., J. M. Fryxell, B. P. Farm, A. R. E. Sinclair & C. P. Henschel. 1999b. Spatial distribution of Serengeti wildebeest in relation to resources. *Can. J. Zool.* **77**: 1223–32.

Wilson, E. O. & W. H. Bossert. 1971. *A Primer of Population Biology*. Sinauer, Sunderland, MA.

Witter, M. S. & I. C. Cuthill. 1993. The ecological costs of avian fat storage. *Phil. Trans. Roy. Soc. Lond.* B **340**: 73–92.

Woods, S. C., R. J. Leely, D. Porte & M. W. Schwartz. 1998. Signals that regulate food intake and energy homeostasis. *Science* **280**: 1378–83.

Woiwod, I. P. & I. Hanski. 1992. Patterns of density dependence in moths and aphids. *J. Anim. Ecol.* **61**: 619–29.

Wright, S. 1932. The roles of mutation, inbreeding, crossbreeding and selection in evolution. *Proc. XI Int. Congr. Genet.* **1**: 356–66.

Wu, J. & O. L. Loucks. 1995. From balance of nature to hierarchical patch dynamics: A paradigm shift in ecology. *Q. Rev. Biol.* **70**: 439–66.

Yodzis, P. 1989. *Introduction to Theoretical Ecology*. Harper & Row, New York.

Ziv, Y. 2000. On the scaling of habitat specificity with body size. *Ecology* **81**: 2932–38.

Index

NORTHERN MICHIGAN UNIVERSITY
3 1854 007 439 287